Differential Calculus in Several Variables

The aim of this book is to lead the reader out from the ordinary routine of computing and calculating by engaging in a more dynamic process of learning. This Learning-by-Doing Approach can be traced back to Aristotle, who wrote in his *Nicomachean Ethics* that "For the things we have to learn before we can do them, we learn by doing them".

The theory is illustrated through many relevant examples, followed by a large number of exercises whose requirements are rendered by action verbs: find, show, verify, check and construct. Readers are compelled to analyze and organize analytical skills.

Rather than placing the exercises in bulk at the end of each chapter, sets of practice questions after each theoretical concept are included. The reader has the possibility to check their understanding, work on the new topics and gain confidence during the learning activity. As the theory unfolds, the exercises become more complex – sometimes they span over several topics. Hints have been added in order to guide the reader in the process.

This book stems from the Differential Calculus course which the author taught for many years. The goal of this book is to immerse the reader in the subtleties of Differential Calculus through an active perspective. Particular attention was paid to continuity and differentiability topics, presented in a new course of action.

Marius Ghergu is an Associate Professor at University College Dublin. He holds a Ph.D. from Universite de Savoie, France. His interests lie at the interface of Calculus and Partial Differential Equations. He is the author and co-author of four research monographs. He has published over 60 research articles in major journals in the field and has been invited to give talks at various international meetings such as conferences and summer schools for graduate students.

Textbooks in Mathematics

Series editors:
Al Boggess, Kenneth H. Rosen

Multiplicative Differential Calculus
Svetlin Georgiev, Khaled Zennir

Applied Differential Equations
The Primary Course
Vladimir A. Dobrushkin

Introduction to Computational Mathematics: An Outline
William C. Bauldry

Mathematical Modeling the Life Sciences
Numerical Recipes in Python and MATLAB™
N. G. Cogan

Classical Analysis
An Approach through Problems
Hongwei Chen

Classical Vector Algebra
Vladimir Lepetic

Introduction to Number Theory
Mark Hunacek

Probability and Statistics for Engineering and the Sciences with Modeling using R
William P. Fox and Rodney X. Sturdivant

Computational Optimization: Success in Practice
Vladislav Bukshtynov

Computational Linear Algebra: with Applications and MATLAB® Computations
Robert E. White

Linear Algebra With Machine Learning and Data
Crista Arangala

Discrete Mathematics with Coding
Hugo D. Junghenn

Applied Mathematics for Scientists and Engineers
Youssef N. Raffoul

Graphs and Digraphs, Seventh Edition
Gary Chartrand, Heather Jordon, Vincent Vatter and Ping Zhang

An Introduction to Optimization with Applications in Data Analytics and Machine Learning
Jeffrey Paul Wheeler

Encounters with Chaos and Fractals, Third Edition
Denny Gulick and Jeff Ford

Differential Calculus in Several Variables: A Learning-by-Doing Approach
Marius Ghergu

https://www.routledge.com/Textbooks-in-Mathematics/book-series/CANDHTEXBOOMTH

Differential Calculus in Several Variables
A Learning-by-Doing Approach

Marius Ghergu

CRC Press
Taylor & Francis Group
Boca Raton London New York

CRC Press is an imprint of the
Taylor & Francis Group, an **informa** business

A CHAPMAN & HALL BOOK

First edition published 2024
by CRC Press
2385 Executive Center Drive, Suite 320, Boca Raton, FL 33431

and by CRC Press
4 Park Square, Milton Park, Abingdon, Oxon, OX14 4RN

CRC Press is an imprint of Taylor & Francis Group, LLC

ISBN: 978-1-032-58339-6 (hbk)
ISBN: 978-1-032-58254-2 (pbk)
ISBN: 978-1-003-44965-2 (ebk)

DOI: 10.1201/9781003449652

Typeset in CMR10 font
by KnowledgeWorks Global Ltd.

To all students
who like Calculus

Contents

Preface

Here we are, at the beginning of this journey entitled *Differentiable Calculus in Several Variables: A Learning-by-Doing Approach*. The main focus of this book is the functions – more precisely, the functions of several variables. In some occurrences, the number of variables will be kept to two or three; this is done for the clarity of the exposition without sacrificing the rigour and generality required when stating the theoretical results.

The intention of this book is to bind theoretical notions with practice questions and, thus, to engage the reader in a more dynamic process of learning. I have called this plan of action a *Learning-by-Doing Approach*. Such a school of thought can be traced back to Antiquity, in the fourth century BCE, when Aristotle wrote in his *Nicomachean Ethics* that

> *For the things we have to learn before we can do them, we learn by doing them.*

In modern times, Learning-by-Doing has been documented by research studies in education as an effective route that provides students with hands-on participation in the instruction strategy.

This book stems from a one-semester Calculus course, which I taught for many years to various groups of students at the University College Dublin, the largest university in Ireland. The delivery of the course took place pre-, during- and post-pandemic period in a face-to-face, online and hybrid mode. This diversity of settings allowed me to identify the difficulties the undergraduate students encountered, and I am happy to share them with the reader in the present work. The goal of this book is to immerse the reader in the subtleties of Calculus through an active perspective.

Let me next briefly describe the content of this book.

Chapter 1 is an introduction to vectors, convergent sequence in $\mathbb{R}^m$ $(m \geq 2)$ and fundamental notions in topology: open, closed and compact sets.

Chapter 2 discusses the basics of functions, their domain, range and graphs. Also, in this chapter, the reader is introduced to the level sets of a function of several variables. As an application of these concepts, we look closer at quadrics in $\mathbb{R}^3$ and at parametrized curves as a special type of functions.

Chapter 3 presents the notions of limit and continuity. This is what distinguishes the study of functions of several variables from single variable ones. Indeed, in the case of one-variable functions, the limit concept entails the study of the behaviour of the function to the left and to the right of the point in question on the real line. In contrast, the limit of a function of several variables becomes more intricate, due to the infinitely many paths of approaching a point in its domain of definition. Special attention was paid to the study of functions defined piecewise.

Chapter 4 discusses partial derivatives, tangent planes and differentiable functions. These concepts are introduced to the reader starting from the one real variable setting and then underlining the different aspects that emerge when moving to several variables. The length and breadth of relationships between the existence of partial derivatives, continuity and differentiability under specific assumptions on the function are described from various angles.

Chapter 5 deals with the Chain Rule, implicit differentiation and Mean Value Theorem in a several variable setting.

Chapter 6 presents the directional derivative, its relationship with the gradient vector and various geometric properties.

Chapter 7 extends the Chain Rule to the case of second-order derivatives and introduces one particular differential quantity named the Laplace operator. The study is further continued to higher order derivatives in Chapter 8. As an application, the reader is introduced to the Taylor Approximation Theorem and its consequences.

Chapter 9 presents the Implicit Function Theorem and justifies the implicit differentiation introduced in Chapter 5.

Chapters 10 and 11 provide the reader with knowledge on maxima and minima, constraint optimization and the Lagrange Multipliers Method. The basics on linear algebra required for a smooth reading of this book are recalled in the Appendix.

As a prerequisite, a good knowledge of one-variable Calculus is assumed. In order to steer through the proofs and to solve the exercises in the text at a reasonable pace, the reader must be familiar with the following topics from one-variable Calculus:

- Basics on general set theory; infimum and supremum of a set of real numbers

- Sequences of real numbers and their convergence; subsequences; Squeeze Theorem

- Sided limits for functions of one real variable

- Continuity and Extreme Value Theorem for functions of one real variable

- Differentiable functions and rules of differentiation: definition of a differentiable function of one real variable; product and quotient rule; chain rule; L'Hôpital's rule

- Mean Value Theorem for functions of one real variable

As already mentioned, the focus is on Learning-by-Doing. The theory is illustrated through many relevant examples, followed by a large number of exercises; these are not intended to build a routine in learning Calculus. The requirements of exercises are rendered by action verbs: find, show, verify, check, construct. In this way, the reader is taken out from the ordinary routine of computing and calculating. Instead, the reader is compelled to analyze and organize his analytical skills. Rather than placing the exercises in bulk at the end of each chapter, I chose to include sets of practice questions after each relevant section in the text where the related theoretical concepts are introduced. In this way, the reader has the possibility to check his understanding, work on the new concepts and gain confidence during the learning activity. As the theory unfolds, the exercises become more complex; sometimes they span over several topics. Hints have been added in order to guide the reader in the process.

This book can be used by students to independently learn Several Variables Calculus or in parallel with other related courses formally delivered at the undergraduate level. It can also be used by tutors and instructors to select specific topics and exercises in order to design their course content or for assessment. Finally, it can be used by anyone who needs to refresh their knowledge in this field.

Marius Ghergu
Dublin, Ireland

Foreword to the Student

This textbook provides a well-rounded material which enables the student to acquire a good knowledge of the Calculus of Several Variables. For an efficient and thorough study, I recommend that you follow the chapters in the order they are presented. This is because the content in each chapter relies on the material introduced previously in the book.

As it is customary, this textbook contains both theory and exercises. Since the theory is essential in grasping the topics, I recommend you to devote enough time to read it carefully. The theoretical concepts are introduced from particular situations to the most general ones. The proofs of the theorems in the text are carefully detailed and I insisted on explaining all subtleties that lie behind any difficult argument. On many occasions, the proofs are far from obvious and for a deep understanding of Several Variable Calculus, I advise you to dwell on them for some time or go over the arguments developed in these proofs after a while. The study of several variable functions relies more than in the single variable case on the geometric intuition and Linear Algebra. This is somehow natural, as one needs to understand, justify and interpret the interaction of variables through a function and its graph. To support your learning, many examples are provided each time a new topic is presented. Also, the needed results in Linear Algebra are included in the Appendix at the end of this textbook.

There are plenty of exercises included in the text. These are not meant to enhance your routine calculation ability but to help you build up and strengthen your analytical skills. While some exercises are drawn from the proofs in the book, others require you to connect topics encountered in several chapters of the material. The exercises are placed at the end of each relevant section and not at the end of each chapter. In this way, they form an individual unit so that your progress will be gauged step by step. It is recommended to start solving the exercises only after you understand the examples provided in the section.

Full solutions to all exercises are provided in Chapter 12 of this book. I took great care in explaining the steps to solve them and I believe you will find it beneficial to your learning.

Finally, I hope you will find this book a useful and enjoyable resource to learn and deepen your knowledge of Calculus of Several Variables.

Marius Ghergu

1

Vectors and Sets in $\mathbb{R}^m$

This chapter introduces vectors in $\mathbb{R}^m$ and their algebraic operations: addition, subtraction, scalar and cross product. These are next used to obtain various relationships between lines and planes in $\mathbb{R}^3$. Next, the focus is moved on sets and sequences in $\mathbb{R}^m$. An introduction to point set topology is also provided, which contains an account on open, closed and compact sets of $\mathbb{R}^m$.

1.1   Vectors in $\mathbb{R}^m$

The set $\mathbb{R}$ of real numbers is the underlying set for all notions (such as sequences and functions) in the one-variable Calculus. Geometrically, we may think of $\mathbb{R}$ as the collection of all points on a horizontal line on which we established a reference point called the origin. The origin corresponds to zero, and all positive real numbers are represented on the right of it, while the negative real numbers are represented on the left of the origin.

The set $\mathbb{R}^2$ consists of all pairs (a, b) of real numbers a, b. To represent the element $(a, b) \in \mathbb{R}^2$, we need two orthogonal lines having in common their reference point, that is, the origin which we denote by O. The orthogonal lines are one horizontal, called the x-axis, and one vertical, called the y-axis. The element $(a, b) \in \mathbb{R}^2$ is thus represented as a unique point in the xy-system of axes in which a units were measured on the x-axis and b units were measured on the y-axis.

If $A(a_1, b_1)$ and $B(a_2, b_2)$ are two points in the xy-system of axes, then the oriented segment with starting point at A and terminal point at B will be called the *vector* $\overrightarrow{AB}$. What distinguishes a vector from a line segment is that the former has a direction. In this way, the vectors $\overrightarrow{AB}$ and $\overrightarrow{BA}$ are distinct, while the segment lines $[AB]$ and $[BA]$ are not.

Further, given the two points $A(a_1, b_1)$ and $B(a_2, b_2)$, we call the pair $(a_2 - a_1, b_2 - b_1)$ the *components* of the vector $\overrightarrow{AB}$. Thus, if $P(a_2 - a_1, b_2 - b_1)$, then the vectors $\overrightarrow{AB}$ and $\overrightarrow{OP}$ have the same components, and we say that they are *equal* and write $\overrightarrow{AB} = \overrightarrow{OP}$. Geometrically, this amounts to say that $AOPB$ is a parallelogram; see Figure 1.1.

DOI: 10.1201/9781003449652-1

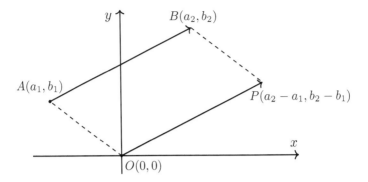

Figure 1.1
The vectors $\overrightarrow{AB}$ and $\overrightarrow{OP}$ in $\mathbb{R}^2$.

In a similar way, one can define vectors in $\mathbb{R}^3$. Let us start by saying that $\mathbb{R}^3$ is the set of triples $\mathbb{R}^3 = \{(a, b, c) : a, b, c \in \mathbb{R}\}$ and that for any points $A(a_1, b_1, c_1)$ and $A(a_2, b_2, c_2)$, the vector $\overrightarrow{AB}$ is the oriented line segment with starting point at A and terminal point at B. As before, the triple $(a_2 - a_1, b_2 - b_1, c_2 - c_1)$ is called the components of $\overrightarrow{AB}$. To represent the vector $\overrightarrow{AB}$, we need a three-axis system; see Figure 1.2.

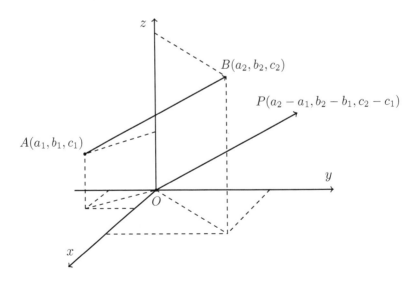

Figure 1.2
Vectors in $\mathbb{R}^3$.

The above notion of vectors can be extended to the case of $\mathbb{R}^m$ even if the geometrical intuition cannot go beyond the cases $m = 2, 3$. For an integer $m \geq 2$, we denote by $\mathbb{R}^m$ the set of all m-tuples of real numbers, namely

$$\mathbb{R}^m = \{(a_1, a_2, \ldots, a_m) : a_1, a_2, \ldots, a_m \in \mathbb{R}\}.$$

Thus,

$$\mathbb{R}^m \text{ is the Cartesian product } \underbrace{\mathbb{R} \times \mathbb{R} \times \cdots \times \mathbb{R}}_{m}.$$

In analogy to the case $m = 1$, the elements in $\mathbb{R}^m$ are called points. The vector in $\mathbb{R}^m$ with starting point at $A(a_1, a_2, \ldots, a_m)$ and terminal point at $B(b_1, b_2, \ldots, b_m)$ has components $(b_1 - a_1, b_2 - a_2, \ldots, b_m - a_m)$. In this text book we shall mostly use bold letters to denote vectors; for instance, $\overrightarrow{AB}$ will be denoted by $\mathbf{u}$ and on many occasions we shall write $\mathbf{u} = (b_1 - a_1, b_2 - a_2, \ldots, b_m - a_m)$. Doing so, we understand that the vector $\mathbf{u}$ has the components given by the m-tuple $(b_1 - a_1, b_2 - a_2, \ldots, b_m - a_m)$. Thus, if $P(b_1 - a_1, b_2 - a_2, \ldots, b_m - a_m)$, then $\mathbf{u} = \overrightarrow{AB} = \overrightarrow{OP}$. The vector with components $(0, 0, \ldots, 0) \in \mathbb{R}^m$ will be called the *zero vector* and will be denoted $\mathbf{0}$ (in bold script).

Given two vectors $\mathbf{u} = (u_1, u_2, \ldots, u_m)$ and $\mathbf{v} = (v_1, v_2, \ldots, v_m)$ in $\mathbb{R}^m$, we define the *addition* $\mathbf{u} + \mathbf{v}$ by

$$\mathbf{u} + \mathbf{v} = (u_1 + v_1, u_2 + v_2, \ldots, u_m + v_m).$$

If $\lambda \in \mathbb{R}$, we also define $\lambda\mathbf{u} = (\lambda u_1, \lambda u_2, \ldots, \lambda u_m)$. The real number λ is frequently called a *scalar*, and $\lambda\mathbf{u}$ is called the *multiplication* of $\mathbf{u}$ with the scalar λ. In this way, the vector $-\mathbf{v} = (-1)\mathbf{v}$ equals $(-v_1, -v_2, \ldots, -v_m)$, and thus we may compute the *subtraction* of $\mathbf{u}$ and $\mathbf{v}$ as follows:

$$\mathbf{u} - \mathbf{v} = \mathbf{u} + (-1)\mathbf{v} = (u_1 - v_1, u_2 - v_2, \ldots, u_m - v_m).$$

Geometrically, the sum $\mathbf{u} + \mathbf{v}$ and the difference $\mathbf{u} - \mathbf{v}$ are calculated by the so-called *parallelogram law*. If $\mathbf{u} = \overrightarrow{OA}$ and $\mathbf{v} = \overrightarrow{OB}$ are two vectors having the origin O as the starting point (see Figure 1.3), then $\mathbf{u} + \mathbf{v} = \overrightarrow{OC}$ and $\mathbf{u} - \mathbf{v} = \overrightarrow{BA}$.

If $\mathbf{u} = (u_1, u_2, \ldots, u_m)$ is a vector in $\mathbb{R}^m$, then the quantity

$$\|\mathbf{u}\| = \sqrt{u_1^2 + u_2^2 + \cdots + u_m^2}$$

is called the *length* (*or the norm*) of the vector $\mathbf{u}$. A vectors $\mathbf{u}$ with length 1 is called a *unit vector*. The vectors $\mathbf{e}_i$ ($1 \leq i \leq m$) whose components are equal to 1 on the i-th position and zero elsewhere are called the *standard unit vectors*. If $m = 2, 3$, we denote $\mathbf{e}_1 = \mathbf{i}$, $\mathbf{e}_2 = \mathbf{j}$ and $\mathbf{e}_3 = \mathbf{k}$. Thus, any vector $\mathbf{u} = (u_1, u_2, \ldots, u_m) \in \mathbb{R}^m$ can be uniquely written as a linear combination of standard unit vectors by

$$\mathbf{u} = u_1\mathbf{e}_1 + u_2\mathbf{e}_2 + \cdots + u_m\mathbf{e}_m.$$

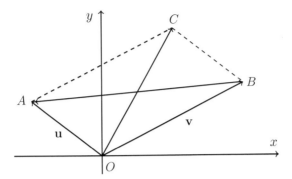

Figure 1.3
The parallelogram law for adding and subtracting vectors.

If $\mathbf{u} = (u_1, u_2, \ldots, u_m)$ and $\mathbf{v} = (v_1, v_2, \ldots, v_m)$, then their *scalar (or dot) product* is given by

$$\mathbf{u} \bullet \mathbf{v} = u_1 v_1 + u_2 v_2 + \cdots + u_m v_m. \qquad (1.1)$$

We shall use the bullet notation $\bullet$ for the scalar product to distinguish it from the standard product of real numbers. As the name suggests, the scalar product of two vectors is indeed a real number, not a vector. The vectors $\mathbf{u}$ and $\mathbf{v}$ are called *orthogonal* if $\mathbf{u} \bullet \mathbf{v} = 0$ and we write $\mathbf{u} \perp \mathbf{v}$. Two nonzero vectors $\mathbf{u}$ and $\mathbf{v}$ in $\mathbb{R}^m$ are called *parallel* if there exists a scalar $\lambda \in \mathbb{R}$, such that $\mathbf{u} = \lambda \mathbf{v}$.

We can further define the *angle* between two nonzero vectors $\mathbf{u}$ and $\mathbf{v}$ as the unique number $\theta \in [0, \pi]$ such that

$$\cos \theta = \frac{\mathbf{u} \bullet \mathbf{v}}{\|\mathbf{u}\| \cdot \|\mathbf{v}\|}. \qquad (1.2)$$

Such a notion is more meaningful in $\mathbb{R}^3$ where θ is in fact the angle formed by the support lines of the two vectors $\mathbf{u}$ and $\mathbf{v}$.

Example 1.1 *Let* $A(2, -1, 0)$, $B(3, 0, -2)$, $C(1, 1, -1)$ *and* $D(6, -3, -2)$.

    (i) *Find the components of* $\mathbf{u} = \overrightarrow{AB}$ *and* $\mathbf{v} = \overrightarrow{BC}$;

    (ii) *Find the length of the vectors* $\overrightarrow{AB}$ *and* $\overrightarrow{BC}$;

    (iii) *Check that the vectors* $\mathbf{w} = \overrightarrow{AD}$ *and* $\mathbf{v} = \overrightarrow{BC}$ *are parallel*;

    (iv) *Find the angle between the vectors* $\overrightarrow{AB}$ *and* $\overrightarrow{BC}$.

Solution. (i) We have $\mathbf{u} = (1, 1, -2)$ and $\mathbf{v} = (-2, 1, 1)$.

(ii) $\|\overrightarrow{AB}\| = \|\mathbf{u}\| = \sqrt{(-2)^2 + 1^2 + 1^2} = \sqrt{6}$ and similarly $\|\overrightarrow{BC}\| = \|\mathbf{v}\| = \sqrt{6}$.

(iii) We have $\mathbf{w} = \overrightarrow{AD} = (4, -2, -2)$ and $\mathbf{v} = \overrightarrow{BC} = (-2, 1, 1)$. Using the standard unit vectors, we have $\mathbf{v} = -2\mathbf{i} + \mathbf{j} + \mathbf{k}$ and $\mathbf{w} = 4\mathbf{i} - 2\mathbf{j} - 2\mathbf{k}$. Since $\mathbf{w} = -2\mathbf{v}$, one has that $\mathbf{w}$ and $\mathbf{v}$ are parallel.

(iv) Using formula (1.2), we have $\cos\theta = \frac{\mathbf{u} \bullet \mathbf{v}}{\|\mathbf{u}\| \cdot \|\mathbf{v}\|} = -\frac{3}{6} = -\frac{1}{2}$, so $\theta = 120^0$.

The following properties of the scalar product follow directly from (1.1):

Theorem 1.2 (Properties of the scalar product)

Let $\mathbf{u}$, $\mathbf{v}$ *and* $\mathbf{w}$ *be three vectors in* $\mathbb{R}^m$. *Then:*

(i) $\mathbf{u} \bullet (\mathbf{v} + \mathbf{w}) = \mathbf{u} \bullet \mathbf{v} + \mathbf{u} \bullet \mathbf{w}$;

(ii) $(\mathbf{u} + \mathbf{v}) \bullet \mathbf{w} = \mathbf{u} \bullet \mathbf{w} + \mathbf{v} \bullet \mathbf{w}$;

(iii) $\mathbf{u} \bullet \mathbf{v} = \mathbf{v} \bullet \mathbf{u}$;

(iv) $\mathbf{u} \bullet \mathbf{u} = \|\mathbf{u}\|^2 \geq 0$ *and* $\mathbf{u} \bullet \mathbf{u} = 0$ *if and only if* $\mathbf{u} = \mathbf{0}$.

Less obvious is the next property known as the *Cauchy-Schwarz inequality*, which has extensive applications in analysis and beyond:

$$|\mathbf{u} \bullet \mathbf{v}| \leq \|\mathbf{u}\| \cdot \|\mathbf{v}\| \quad \text{for all vectors } \mathbf{u}, \mathbf{v} \text{ in } \mathbb{R}^m. \tag{1.3}$$

Before we prove inequality (1.3), one has to clarify the term *proof*, at least to explain its meaning in this book.

Proving a mathematical statement is similar to the task undertaken by an attorney in court who uses *evidence* and *logical reasoning*. Our evidence in mathematics consists of definitions, theorems and postulates, while the logical reasoning amounts to a sequence of *deductive arguments* which yield the claim.

In our case, one has to combine the properties (i)–(iv) in Theorem 1.2 to derive the inequality (1.3) through a sequence of deductive arguments. Indeed, let us start from property (iv), which holds not only for $\mathbf{u}$, $\mathbf{v}$, but for a large class of vectors in $\mathbb{R}^m$. Precisely, for all $t \in \mathbb{R}$, the vector $t\mathbf{u} + \mathbf{v}$ must also satisfy (iv) from which we deduce

$$E(t) := \|t\mathbf{u} + \mathbf{v}\|^2 = (t\mathbf{u} + \mathbf{v}) \bullet (t\mathbf{u} + \mathbf{v}) \geq 0 \quad \text{for all } t \in \mathbb{R}. \tag{1.4}$$

Making use of properties (i)–(iii), one may expand $E(t)$ as

$$E(t) = t^2(\mathbf{u} \bullet \mathbf{u}) + 2t(\mathbf{u} \bullet \mathbf{v}) + \mathbf{v} \bullet \mathbf{v} \geq 0 \quad \text{for all } t \in \mathbb{R}.$$

By virtue of property (iv), the above statement is equivalent to

$$E(t) = t^2\|\mathbf{u}\|^2 + 2t(\mathbf{u} \bullet \mathbf{v}) + \|\mathbf{v}\|^2 \geq 0 \quad \text{for all } t \in \mathbb{R}. \tag{1.5}$$

Here comes another turn in the sequence of our deductive arguments: the quantity $E(t)$ on the right-hand side of (1.5) is a quadratic function in $t \in \mathbb{R}$ which, according to the above inequality, is nonnegative. Thus, its graph lies on or above the horizontal axis, which amounts to say that the quadratic equation $E(t) = 0$ has either two equal real roots or two complex roots. Hence, the discriminant of the quadratic equation is less than or equal to zero. This yields $(\mathbf{u} \bullet \mathbf{v})^2 \leq \|\mathbf{u}\|^2 \cdot \|\mathbf{v}\|^2$. Finally, taking the square roots on both sides, one deduces (1.3). Nice and elegant! The equality case in (1.5) is proposed for discussion in Exercise 4 of this section.

Besides the scalar product, we define the cross product of two vectors in $\mathbb{R}^3$ as follows.

Definition 1.3 (The cross product of two vectors in $\mathbb{R}^3$)
Let $\mathbf{u} = (u_1, u_2, u_3)$ and $\mathbf{v} = (v_1, v_2, v_3)$ be two vectors in $\mathbb{R}^3$.
The cross product of $\mathbf{u}$ and $\mathbf{v}$ is given by

$$\mathbf{u} \times \mathbf{v} = \begin{vmatrix} \mathbf{i} & \mathbf{j} & \mathbf{k} \\ u_1 & u_2 & u_3 \\ v_1 & v_2 & v_3 \end{vmatrix} = \begin{vmatrix} u_2 & u_3 \\ v_2 & v_3 \end{vmatrix} \mathbf{i} - \begin{vmatrix} u_1 & u_3 \\ v_1 & v_3 \end{vmatrix} \mathbf{j} + \begin{vmatrix} u_1 & u_2 \\ v_1 & v_2 \end{vmatrix} \mathbf{k}$$

$$= (u_2 v_3 - u_3 v_2)\mathbf{i} - (u_1 v_3 - u_3 v_1)\mathbf{j} + (u_1 v_2 - u_2 v_1)\mathbf{k}.$$

The main properties of determinants required in this book are given in the Appendix. The result below summarizes the features of the cross product.

Theorem 1.4 (Properties of the cross product)
Let $\mathbf{u} = (u_1, u_2, u_3)$, $\mathbf{v} = (v_1, v_2, v_3)$ and $\mathbf{w} = (w_1, w_2, w_3)$ be three vectors in $\mathbb{R}^3$. Then:

(i) $\mathbf{u} \times \mathbf{v} = -\mathbf{v} \times \mathbf{u}$;

(ii) $\mathbf{u} \times \mathbf{v}$ *is orthogonal to both $\mathbf{v}$ and $\mathbf{u}$;*

Its direction is determined by the right-hand rule[1];

(iii) $\mathbf{u} \times \mathbf{v} \neq \mathbf{0}$ *if and only if $\mathbf{v}$ and $\mathbf{u}$ are linearly independent;*

(iv) $\|\mathbf{u} \times \mathbf{v}\| = \|\mathbf{u}\| \cdot \|\mathbf{v}\| \sin \theta$, *where θ is the angle between the vectors $\mathbf{u}$ and $\mathbf{v}$;*

(v) $(\mathbf{u} \times \mathbf{v}) \bullet w = \begin{vmatrix} u_1 & u_2 & u_3 \\ v_1 & v_2 & v_3 \\ w_1 & w_2 & w_3 \end{vmatrix}$;

(vi) $(\mathbf{u} \times \mathbf{v}) \bullet \mathbf{w} = (\mathbf{v} \times \mathbf{w}) \bullet \mathbf{u} = (\mathbf{w} \times \mathbf{u}) \bullet \mathbf{v}$.

---

[1] If the fingers of your right-hand curl in the direction of the rotation from $\mathbf{u}$ to $\mathbf{v}$, then your thumb points in the direction of $\mathbf{u} \times \mathbf{v}$.

Exercises

Exercise 1. Verify, by direct calculation, the properties (i)–(iv) in Theorem 1.2 related to the scalar product.

Exercise 2. Using the properties of the scalar product (i)–(iv) and the Cauchy-Schwarz inequality (1.3), check the following statements:

(i) For any vector $\mathbf{u}$ in $\mathbb{R}^m$, we have $\|\mathbf{u}\| \geq 0$ and $\|\mathbf{u}\| = 0$ if and only if $\mathbf{u} = \mathbf{0}$;

(ii) For any vector $\mathbf{u}$ in $\mathbb{R}^m$ and any $\lambda \in \mathbb{R}$, we have $\|\lambda \mathbf{u}\| = |\lambda| \|\mathbf{u}\|$;

(iii) (Triangle inequality) For any vectors $\mathbf{u}, \mathbf{v}$ in $\mathbb{R}^m$ and any $\lambda \in \mathbb{R}$, we have
$$\|\mathbf{u} + \mathbf{v}\| \leq \|\mathbf{u}\| + \|\mathbf{v}\|.$$

Hint: To prove (iii) square both sides of the above inequality, expand the resulting brackets and use (1.3).

Exercise 3. Let $\mathbf{u}, \mathbf{v}$ be two vectors in $\mathbb{R}^m$. Using the definition of the norm of a vector, check that:

(i) $\big| \|\mathbf{u}\| - \|\mathbf{v}\| \big| \leq \|\mathbf{u} - \mathbf{v}\|$;

(ii) $\|\mathbf{u} + \mathbf{v}\|^2 + \|\mathbf{u} - \mathbf{v}\|^2 = 2(\|\mathbf{u}\|^2 + \|\mathbf{v}\|^2)$;

(iii) $\|\mathbf{u} + \mathbf{v}\|^2 - \|\mathbf{u} - \mathbf{v}\|^2 = 4(\mathbf{u} \bullet \mathbf{v})$.

Exercise 4. Discuss the equality case in (1.3) as follows.
Let $\mathbf{u}, \mathbf{v}$ be two nonzero vectors in $\mathbb{R}^m$ such that $\big| \mathbf{u} \bullet \mathbf{v} \big| = \|\mathbf{u}\| \cdot \|\mathbf{v}\|$.

(i) Replacing $\mathbf{v}$ by $-\mathbf{v}$, deduce that it is enough to assume $\mathbf{u} \bullet \mathbf{v} = \|\mathbf{u}\| \cdot \|\mathbf{v}\|$;

(ii) Let $t = -\dfrac{\|\mathbf{v}\|}{\|\mathbf{u}\|}$. Using (1.5) deduce that $E(t) = 0$;

(iii) Using (1.4) deduce $t\mathbf{u} + \mathbf{v} = \mathbf{0}$ and then $\mathbf{u} = \lambda \mathbf{v}$ for some scalar $\lambda \in \mathbb{R}$.

Exercise 5. Prove Theorem 1.4 about the properties of the cross product by expanding the expressions using Definition 1.3.

Exercise 6. Let $\mathbf{u}$ and $\mathbf{v}$ be two vectors in $\mathbb{R}^3$. Using a direct calculation or the property (iv) in Theorem 1.4 together with (1.2), check that
$$\|\mathbf{u} \times \mathbf{v}\|^2 = \|\mathbf{u}\|^2 \|\mathbf{v}\|^2 - (\mathbf{u} \bullet \mathbf{v})^2.$$

Exercise 7. Let A, B and C be three points in $\mathbb{R}^3$, which are not collinear. Denote $\mathbf{u} = \overrightarrow{BA}$ and $\mathbf{v} = \overrightarrow{BC}$. Using the geometric properties of the cross product, deduce that:

(i) $\mathrm{dist}(A, BC) = \dfrac{\|\mathbf{u} \times \mathbf{v}\|}{\|\mathbf{v}\|}$;   (ii) $\mathrm{dist}(C, AB) = \dfrac{\|\mathbf{u} \times \mathbf{v}\|}{\|\mathbf{u}\|}$.

Exercise 8. Let $\mathbf{u}$, $\mathbf{v}$ and $\mathbf{w}$ be three vectors in $\mathbb{R}^3$ with starting point at $\mathbf{0} = (0,0,0)$.

(i) Show that the volume of the parallelepiped in $\mathbb{R}^3$ constructed with $\mathbf{u}$, $\mathbf{v}$ and $\mathbf{w}$ as adjacent sides is $V = |(\mathbf{u} \times \mathbf{v}) \bullet \mathbf{w}|$;

(ii) Deduce that $\mathbf{u}$, $\mathbf{v}$ and $\mathbf{w}$ are *coplanar* (that is, they lie in the same plane) if and only if $(\mathbf{u} \times \mathbf{v}) \bullet \mathbf{w} = 0$.

Exercise 9. Let $\mathbf{u}$, $\mathbf{v}$ and $\mathbf{w}$ be three unit vectors in $\mathbb{R}^3$ with starting point at $\mathbf{0}$.

(i)   Assume $\|(\mathbf{u} \times \mathbf{v}) \times \mathbf{w}\| = 1$. Show that the three vectors are coplanar and $\mathbf{u} \perp \mathbf{v}$;

(ii) Assume now $\|(\mathbf{u} \times \mathbf{v}) \times \mathbf{w}\| = \|(\mathbf{v} \times \mathbf{w}) \times \mathbf{u}\| = 1$. Using part (i) above, show that $\mathbf{u} \perp \mathbf{v}$ and either $\mathbf{w} = \mathbf{u}$ or $\mathbf{w} = -\mathbf{u}$;

(iii) Are there unit vectors $\mathbf{u}$, $\mathbf{v}$ and $\mathbf{w}$ such that

$$\|(\mathbf{u} \times \mathbf{v}) \times \mathbf{w}\| = \|(\mathbf{v} \times \mathbf{w}) \times \mathbf{u}\| = \|(\mathbf{w} \times \mathbf{u}) \times \mathbf{v}\| = 1?$$

1.2   Lines and Planes in $\mathbb{R}^3$

With the help of vectors and their corresponding operations, we can describe lines and planes in $\mathbb{R}^3$. The equation of a line in $\mathbb{R}^2$ has the form $ax + by = c$. Somehow contrary to intuition, the equation $ax + by + cz = d$ defines in $\mathbb{R}^3$ a plane instead of a line. To find the equation of a line one needs its direction, given by a nonzero vector $\mathbf{u}$ and a point $P(x_0, y_0, z_0)$ through which it passes. Precisely, we have:

Theorem 1.5 *Let* $\mathbf{u} \in \mathbb{R}^3$ *be a nonzero vector. The equation of the line* ℓ *that passes through* $P(x_0, y_0, z_0)$ *and has direction* $\mathbf{u}$ *is*

$$(x, y, z) = (x_0, y_0, z_0) + t\mathbf{u}, \quad t \in \mathbb{R}. \tag{1.6}$$

Proof Let $Q(x, y, z)$ be a point on the line ℓ (see Figure 1.4).

Then, the vectors $\overrightarrow{PQ}$ and $\mathbf{u}$ are parallel, so $\overrightarrow{PQ} = t\mathbf{u}$ for some $t \in \mathbb{R}$, which yields the equation (1.6).   □

We call the equality (1.6) the *parametric equation* of the line ℓ since the components of any point (x, y, z) on the line depend on the scalar $t \in \mathbb{R}$. Turning to the equation of a plane in $\mathbb{R}^3$ which was already announced above, we have:

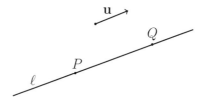

Figure 1.4
The line ℓ passing through P and having direction **u**.

Theorem 1.6 *Let* $\mathbf{n} = (a, b, c)$ *be a nonzero vector in* $\mathbb{R}^3$. *The equation of the plane which passes through* $P(x_0, y_0, z_0)$ *and is orthogonal to* $\mathbf{n}$ *is*

$$\sigma : a(x - x_0) + b(y - y_0) + c(z - z_0) = 0. \tag{1.7}$$

Proof Indeed, if $Q(x, y, z)$ is another point in the plane (see Figure 1.5), then the vectors $\overrightarrow{PQ}$ and $\mathbf{n}$ are orthogonal, so $\mathbf{n} \bullet \overrightarrow{PQ} = 0$.

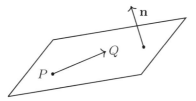

Figure 1.5
The plane σ which passes through P and is orthogonal to $\mathbf{n}$.

This means $\mathbf{n} \bullet \left((x, y, z) - (x_0, y_0, z_0)\right) = 0$, which is equivalent to (1.7). □
  As an immediate consequence of the above theorem, we have that given the plane of equation $ax + by + cz = d$, then the vector $\mathbf{n} = (a, b, c)$ is orthogonal to this plane.

Definition 1.7 *The angle of two planes in* $\mathbb{R}^3$ *is the angle* $\theta \in [0, \pi/2]$ *formed by the orthogonal vectors to the two planes.*

In this way, two planes are parallel (resp. orthogonal) if their orthogonal vectors are parallel (resp. orthogonal).
  Further relations between two planes can be derived through the relationship between their orthogonal vectors. These properties are stated below.

Corollary 1.8 *Consider the planes*

$$\sigma_1 : a_1 x + b_1 y + c_1 z = d_1 \quad and \quad \sigma_2 : a_2 x + b_2 y + c_2 z = d_2.$$

(i) *The planes σ_1 and σ_2 are parallel if and only if their corresponding orthogonal vectors $\mathbf{n}_1 = (a_1, b_1, c_1)$ and $\mathbf{n}_2 = (a_2, b_2, c_2)$ are parallel, that is,*

$$(a_1, b_1, c_1) = \lambda(a_2, b_2, c_2) \quad \text{for some } \lambda \in \mathbb{R}.$$

(ii) *The planes σ_1 and σ_2 are orthogonal if and only if $\mathbf{n}_1$ and $\mathbf{n}_2$ are orthogonal, that is,*

$$a_1 a_2 + b_1 b_2 + c_1 c_2 = 0.$$

(iii) *The angle $\theta \in [0, \pi/2]$ between the two planes σ_1 and σ_2 satisfies*

$$\cos \theta = \frac{|\mathbf{n}_1 \bullet \mathbf{n}_2|}{\|\mathbf{n}_1\| \cdot \|\mathbf{n}_2\|}.$$

Observe that the formula for $\cos \theta$ in Theorem 1.6(iii) is sign sensitive. We must include the absolute value of the dot product $\mathbf{n}_1 \bullet \mathbf{n}_2$, since $\cos \theta \geq 0$ (the angle θ lies in the range $0 \leq \theta \leq \pi/2$).

In light of Theorem 1.6, one may regard a line in $\mathbb{R}^3$ as the intersection of two planes as the following example illustrates.

Example 1.9 *Consider the planes of equation*

$$\sigma_1 : 2x - 3y + z = 1 \quad \text{and} \quad \sigma_2 : x + 2y - z = 6.$$

(i) *Find the normal vectors $\mathbf{n}_1$ and $\mathbf{n}_2$ to the above planes;*

(ii) *Find the vector $\mathbf{n} = \mathbf{n}_1 \times \mathbf{n}_2$;*

(iii) *Find the parametric equation of the line ℓ of the intersection between σ_1 and σ_2.*

Solution. (i) By Theorem 1.6, we have $\mathbf{n}_1 = (2, -3, 1)$ and $\mathbf{n}_2 = (1, 2, -1)$.

$$\text{(ii) } \mathbf{n} = \mathbf{n}_1 \times \mathbf{n}_2 = \begin{vmatrix} \mathbf{i} & \mathbf{j} & \mathbf{k} \\ 2 & -3 & 1 \\ 1 & 2 & -1 \end{vmatrix} = \mathbf{i} + 3\mathbf{j} + 7\mathbf{k}.$$

(iii) The line ℓ is orthogonal to both $\mathbf{n}_1$ and $\mathbf{n}_2$, so its direction is given by $\mathbf{n} = (1, 3, 7)$. It remains to find a point that lies on ℓ. Taking for instance $z = 1$ in the equation of the two planes, one finds $x = 3$ and $y = 2$. Hence, $P(3, 2, 1)$ lies on the line ℓ whose direction is $\mathbf{n}$. By Theorem 1.5, the parametric equation of ℓ is

$$(x, y, z) = (3, 2, 1) + t(1, 3, 7), \quad t \in \mathbb{R},$$

which we may also write

$$\frac{x - 3}{1} = \frac{y - 2}{3} = \frac{z - 1}{7} = t, \quad t \in \mathbb{R}.$$

Example 1.10 *Consider the plane of equation* $\sigma : 2x - y + 4z = 3$ *and the line of parametric equation* $\ell : (x, y, z) = (1 + t, 2 - 3t, 3 - 2t),\ t \in \mathbb{R}$.

   (i) *Find the intersection point between the line* ℓ *and the plane* σ;

   (ii) *Find the angle between the line* ℓ *and the plane* σ;

   (iii) *Find the equation of the plane* τ *which passes through* $P(2, 0, 1)$ *and is orthogonal to* ℓ.

Solution. (i) The intersection between the line ℓ and the plane σ is found by imposing $(1 + t, 2 - 3t, 3 - 2t) \in \sigma$. Hence,

$$2(1 + t) - (2 - 3t) + 4(3 - 2t) = 3 \Longrightarrow 12 - 3t = 3 \Longrightarrow t = 3.$$

Thus, the intersection point is $Q(4, -7, -3)$.

  (ii) The direction of ℓ is given by the vector $\mathbf{u} = (1, -3, -2)$ and the angle α between the line ℓ and the plane σ is $\frac{\pi}{2} - \theta$, where θ is the angle between $\mathbf{u}$ and the orthogonal vector $\mathbf{n} = (2, -1, 4)$ to the plane σ. With the formula in Corollary 1.8(iii) we find

$$\cos \theta = \frac{|\mathbf{u} \bullet \mathbf{n}|}{\|\mathbf{u}\| \cdot \|\mathbf{n}\|} = \frac{3}{7\sqrt{6}}.$$

Thus, $\alpha = \frac{\pi}{2} - \arccos \frac{3}{7\sqrt{6}}$.

  (iii) The required plane τ is orthogonal to $\mathbf{u}$ and contains $P(2, 0, 1)$. Hence, the vectors $\mathbf{u} = (1, -3, -2)$ and $(x - 2, y, z - 1)$ are orthogonal. Thus

$$\tau : (1, -3, -2) \bullet (x - 2, y - 0, z - 1) = 0 \Longrightarrow \tau : x - 3y - 2z = 0.$$

Example 1.11 *Let* $A(-1, 0, 2)$, $B(2, 1, 1)$ *and* $C(0, -1, 0)$.

   (i) *Write the parametric equations of the lines* AB *and* AC;

   (ii) *Write the equation of the plane* σ *that contains* A, B *and* C;

   (iii) *Find the equation of the plane which is orthogonal to* σ *and contains the line* AC.

Solution. (i) The directions of the lines AB and AC are $\mathbf{u} = (3, 1, -1)$ and $\mathbf{v} = (1, -1, -2)$. The parametric equations of the lines AB and AC are now

$$AB : (x, y, z) = (-1, 0, 2) + t(3, 1, -1), \quad t \in \mathbb{R}$$

and

$$AC : (x, y, z) = (-1, 0, 2) + t(1, -1, -2), \quad t \in \mathbb{R}.$$

  (ii) A vector that is orthogonal to $\mathbf{u}$ and $\mathbf{v}$ is

$$\mathbf{w} = \mathbf{u} \times \mathbf{v} = \begin{vmatrix} \mathbf{i} & \mathbf{j} & \mathbf{k} \\ 3 & 1 & -1 \\ 1 & -1 & -2 \end{vmatrix} = -3\mathbf{i} + 5\mathbf{j} - 4\mathbf{k}.$$

Thus, σ is the plane which passes through $A(-1, 0, 2)$ and is orthogonal to **w**. We have $-3(x+1) + 5(y-0) - 4(z-2) = 0$, that is, $-3x + 5y - 4z = -5$.

(iii) Let $\tau : ax + by + cz = d$ be the equation of the required plane. A vector that is normal to the plane τ is $\mathbf{n} = (a, b, c)$. We first impose that A and C belong to the plane τ which yields $-a + 2c = d$ and $d = -b$. Finally, we impose that $\mathbf{n}$ and $\mathbf{w}$ are orthogonal which means $\mathbf{n} \bullet \mathbf{w} = 0$, that is, $-3a + 5b - 4c = 0$. From the above equalities we find $a = 7c$, $b = 5c$ and $d = -5c$ and the desired equation is $\tau : 7x + 5y + z = -5$.

Before we move further, let us recap all the results on the lines and planes presented in this chapter.

Important

(A) Given a nonzero vector $\mathbf{u} = (u_1, u_2, u_3) \in \mathbb{R}^3$ and a point $P(x_0, y_0, z_0)$, the parametric equation of the line passing through P and having direction $\mathbf{u}$ is

$$(x, y, z) = (x_0, y_0, z_0) + t(u_1, u_2, u_3), \quad t \in \mathbb{R}.$$

If none of the components of $\mathbf{u}$ are zero, we may write this equation as
$$\frac{x - x_0}{u_1} = \frac{y - y_0}{u_2} = \frac{z - z_0}{u_3}.$$

(B) The equation of a plane in $\mathbb{R}^3$ is $ax + by + cz = d$, where $(a, b, c) \neq \mathbf{0}$. A normal vector to the plane is $\mathbf{n} = (a, b, c)$.

(C) The angle $\theta \in [0, \pi/2]$ between by the planes

$$a_1 x + b_1 y + c_1 = d_1,$$

$$a_2 x + b_2 y + c_2 = d_2,$$

is uniquely determined by $\cos \theta = \dfrac{|\mathbf{n}_1 \bullet \mathbf{n}_2|}{\|\mathbf{n}_1\| \cdot \|\mathbf{n}_2\|}$. In particular:

(i) The planes are parallel if and only if

$$(a_1, b_1, c_1) = \lambda(a_2, b_2, c_2) \quad \text{for some } \lambda \in \mathbb{R}.$$

(ii) The planes are orthogonal if and only if their normal vectors $\mathbf{n}_1 = (a_1, b_1, c_1)$ and $\mathbf{n}_2 = (a_2, b_2, c_2)$ are orthogonal, that is,
$$a_1 a_2 + b_1 b_2 + c_1 c_2 = 0.$$

Exercises

Exercise 10. Find the parametric equation of the line ℓ which passes through $(1, 0, -1)$ and is orthogonal to the plane $2x - 3y + z = 11$.

Exercise 11. Find the equation of the plane that contains the intersection line ℓ of the planes $x + 2y = 1$, $x + y - z = 3$ and is orthogonal to the plane $2x - 3y + 2z = 5$.

Exercise 12. Find the equation of a plane that passes through $A(1, -1, -1)$, $B(-1, -6, 2)$ and is orthogonal to the plane $2x + 3y - z = 1$.

Exercise 13. For what values of $m \in \mathbb{R}$ do the planes $y + z = 3$ and $-x + my + z = 9$ intersect each other in an angle of measure $\pi/6$?

Exercise 14. For what values of $p \in \mathbb{R}$, $p \neq 0$ does the line of equation

$$\frac{x - 1}{-1} = \frac{y + 1}{1} = \frac{z}{p}$$

make an angle of measure $\pi/3$ with the plane $z = x - 5$?

1.3   Points or Vectors?

Recall that the elements of $\mathbb{R}^m$ are called points, in analogy to the set of real numbers $\mathbb{R}$. We may establish a one-to-one correspondence between points and the components of vectors in $\mathbb{R}^m$. Each vector can be identified with a point P in $\mathbb{R}^m$ through its components and conversely, to each point P in $\mathbb{R}^m$ we assign a unique vector $\mathbf{u}$ in $\mathbb{R}^m$ having the origin $\mathbf{0}$ as a starting point and P as a terminal point; that is, $\mathbf{u} = \overrightarrow{OP}$. Thus, a point $\mathbf{a} = (a_1, a_2, \ldots, a_m)$ may designate an element of the set $\mathbb{R}^m$ but may also represent the components of a vector in $\mathbb{R}^m$. When it comes to sets in $\mathbb{R}^m$, their elements will be called points (and not vectors). We shall use the notation $\mathbf{a} = (a_1, a_2, \ldots, a_m)$ for points in $\mathbb{R}^m$ and only when the direction is relevant in the context we shall understand that $\mathbf{a} = (a_1, a_2, \ldots, a_m)$ refers to the components of the vector $\mathbf{a}$.

Given two points $\mathbf{a} = (a_1, a_2, \ldots, a_m)$ and $\mathbf{b} = (b_1, b_2, \ldots, b_m)$ in $\mathbb{R}^m$, then

$$\|\mathbf{a} - \mathbf{b}\| = \sqrt{(a_1 - b_1)^2 + (a_2 - b_2)^2 + \cdots + (a_m - b_m)^2}$$

is called the *Euclidean distance* between $\mathbf{a}$ and $\mathbf{b}$.

1.4 Convergent Sequences in $\mathbb{R}^m$

Before we proceed further, we have to clarify the notion of convergence of a sequence in $\mathbb{R}^m$. Given a sequence $\{\mathbf{a}_n\}$ in $\mathbb{R}^m$, its general term is $\mathbf{a}_n \in \mathbb{R}^m$ which is written

$$\mathbf{a}_n = (a_{1,n}, a_{2,n}, \ldots, a_{m,n}) \quad \text{where} \quad a_{i,n} \in \mathbb{R}, \ 1 \le i \le m. \tag{1.8}$$

If

$$1 \le k_1 < k_2 < \cdots < k_n < \ldots$$

is an increasing sequence of positive integers, then $\{\mathbf{a}_{k_n}\}$ is called a *subsequence* of $\{\mathbf{a}_n\}$. In slightly different terms, a subsequence of $\{\mathbf{a}_n\}$ is a sequence whose terms are chosen from those of $\{\mathbf{a}_n\}$ by preserving the order in which they appear.

Definition 1.12 *Let $\{\mathbf{a}_n\}$ be a sequence in $\mathbb{R}^m$.*

(i) *We say that the sequence $\{\mathbf{a}_n\} \subset \mathbb{R}^m$ is bounded if there exists a positive real number $M > 0$ such that $\|\mathbf{a}_n\| \le M$ for all $n \ge 1$;*

(ii) *We say that the sequence $\{\mathbf{a}_n\} \subset \mathbb{R}^m$ converges to $\mathbf{a} \in \mathbb{R}^m$ if*

$$\|\mathbf{a}_n - \mathbf{a}\| \to 0 \quad as \quad n \to \infty.$$

In this case, we call $\mathbf{a}$ the limit of $\{\mathbf{a}_n\}$ and write

$$\lim_{n \to \infty} \mathbf{a}_n = \mathbf{a} \quad or \quad \mathbf{a}_n \to \mathbf{a} \ as \ n \to \infty.$$

For instance, the sequence $\{\mathbf{a}_n\} \subset \mathbb{R}^2$ given by $\mathbf{a}_n = \left(\frac{2+n}{n^2}, \cos(3n)\right)$ is bounded. Indeed, for all $n \ge 1$ one has

$$\|\mathbf{a}_n\| = \sqrt{\left(\frac{2+n}{n^2}\right)^2 + \cos^2(3n)} \le \sqrt{\left(\frac{2n+n}{n^2}\right)^2 + 1} = \sqrt{\frac{9}{n^2} + 1} \le \sqrt{10}.$$

Definition 1.12(ii) allows us to translate the convergence of a sequence $\{\mathbf{a}_n\} \subset \mathbb{R}^m$ into the convergence of sequences of real numbers. If $\mathbf{a}_n$ is given by (1.8) and $\mathbf{a} = (a_1, a_2, \ldots, a_m) \in \mathbb{R}^m$, then $\{\mathbf{a}_n\}$ converges to $\mathbf{a} \in \mathbb{R}^m$ if and only if $\{a_{i,n}\}$ converges, as a sequence of real numbers, to a_i for all $1 \le i \le m$. Indeed, this comes from the estimate

$$|a_{i,n} - a_i| \le \|\mathbf{a}_n - \mathbf{a}\| \le m \cdot \max_{1 \le j \le m} |a_{j,n} - a_j| \quad \text{for all } 1 \le i \le m. \tag{1.9}$$

Using this fact, the next theorem which is the $\varepsilon - \delta$ criterion for convergence follows naturally.

Theorem 1.13 ($\varepsilon - \delta$ criterion for convergent sequences in $\mathbb{R}^m$)

Let $\{\mathbf{a}_n\}$ *be a sequence* $\mathbb{R}^m$. *The following statements are equivalent:*

(i) $\{\mathbf{a}_n\}$ *converges to* $\mathbf{a} \in \mathbb{R}^m$ *in the sense of Definition 1.12;*

(ii) *For any* $\varepsilon > 0$ *there exists an integer* $N_\varepsilon \geq 1$ *such that*

$$\|\mathbf{a}_n - \mathbf{a}\| < \varepsilon \quad \text{whenever } n \geq N_\varepsilon. \tag{1.10}$$

Theorem 1.13 claims the equivalence between two mathematical statements formulated in (i) and (ii) above. This means that the proof of Theorem 1.13 has two parts. Firstly, the statement (i) is assumed to be true and along a sequence of logical arguments we show that statement (ii) holds. This part of the proof is denoted (i)$\Longrightarrow$(ii). Secondly, we assume that statement (ii) holds and prove that this entails the validity of statement (i). We label this part as (ii)$\Longrightarrow$(i).

Proof Let $\mathbf{a}_n$ be given by (1.8) and let $\mathbf{a} = (a_1, a_2, \ldots, a_m)$.

(i)$\Longrightarrow$(ii). If $\{\mathbf{a}_n\}$ converges to $\mathbf{a}$, then Definition 1.12(ii) yields $\|\mathbf{a}_n - \mathbf{a}\| \to 0$ as $n \to \infty$. Now, (1.9) implies

$$|a_{i,n} - a_i| \leq \|\mathbf{a}_n - \mathbf{a}\| \quad \text{for all } 1 \leq i \leq m.$$

This further yields

$$|a_{i,n} - a_i| \to 0 \quad \text{as } n \to \infty \quad \text{for all } 1 \leq i \leq m.$$

Thus, the sequence of real numbers $\{a_{i,n}\}$ converges to a_i for all $1 \leq i \leq m$. We may thus use the $\varepsilon - \delta$ characterization of the convergence for sequences of real numbers. This means that for any $\varepsilon > 0$ there exists an integer $N_\varepsilon \geq 1$ such that

$$|a_{i,n} - a_i| \leq \frac{\varepsilon}{m} \quad \text{for all } n \geq N_\varepsilon. \tag{1.11}$$

At this stage, we should point out that the standard $\varepsilon - \delta$ criterion does not contain any denominator m in (1.11) and this has been chosen for aesthetic reasons (put simply, we want to cancel the factor m in (1.9)).

Because we work with a finite number of sequences of real numbers $\{a_{n,i}\}$, we may assume that (1.11) holds for all $1 \leq i \leq m$. Thus, (1.11) and (1.9) imply (1.10) which proves the claim in (ii).

(ii)$\Longrightarrow$(i). Let $\varepsilon > 0$. By (ii), one may find an integer $N_\varepsilon \geq 1$ such that (1.10) holds. Then, the estimate (1.9) yields

$$|a_{i,n} - a_i| < \varepsilon \quad \text{for all } n \geq N_\varepsilon \text{ and } 1 \leq i \leq m.$$

This shows that $\{a_{i,n}\}$ converges to a_i for all $1 \leq i \leq m$, that is, $|a_{n,i} - a_i| \to 0$ as $n \to \infty$. Now, (1.9) yields $\|\mathbf{a}_n - \mathbf{a}\| \to 0$ as $n \to \infty$, that is, the statement (i) holds. $\qquad\square$

Theorem 1.14 (Properties of convergent sequences in $\mathbb{R}^m$)

Let $\{\mathbf{a}_n\}, \{\mathbf{b}_n\} \subset \mathbb{R}^m$ *be two convergent sequences to* $\mathbf{a} \in \mathbb{R}^m$ *and* $\mathbf{b} \in \mathbb{R}^m$ *respectively. Then:*

(i) $\lim\limits_{n\to\infty} \left(\mathbf{a}_n \pm \mathbf{b}_n\right) = \mathbf{a} \pm \mathbf{b}$;

(ii) $\lim\limits_{n\to\infty} \lambda\mathbf{a}_n = \lambda\mathbf{a}$ *for all* $\lambda \in \mathbb{R}$;

(iii) $\lim\limits_{n\to\infty} \mathbf{a}_n \bullet \mathbf{b}_n = \mathbf{a} \bullet \mathbf{b}$;

(iv) $\lim\limits_{n\to\infty} \|\mathbf{a}_n\| = \|\mathbf{a}\|$.

Exercises

Exercise 15. Using Definition 1.12 check the statements (i)–(iv) in Theorem 1.14.

Hint: For part (iv) use Exercise 3(i).

Exercise 16. Prove (using logical arguments) or disprove (by constructing a counterxample) the following statement:

If $\{\mathbf{a}_n\}$ is a sequence in $\mathbb{R}^m$ such that $\{\|\mathbf{a}_n\|\}$ is convergent, then $\{\mathbf{a}_n\}$ is also convergent.

1.5   Sets in $\mathbb{R}^m$

In this section we describe some specific properties of sets which are peculiar to the Several Variables Calculus. The relevant sets on the real line are the open and closed intervals $(a, b) = \{x \in \mathbb{R} : a < x < b\}$ and $[a, b] = \{x \in \mathbb{R} : a \leq x \leq b\}$ respectively. A natural extension of the intervals to the plane $\mathbb{R}^2$ are the rectangles

$$(a, b) \times (c, d) = \{(x, y) \in \mathbb{R}^2 : a < x < b, c < y < d\},$$
$$[a, b] \times [c, d] = \{(x, y) \in \mathbb{R}^2 : a \leq x \leq b, c \leq y \leq d\}.$$

However, this is not enough to describe specific properties of sets in $\mathbb{R}^2$ and in general in $\mathbb{R}^m$.

Remark 1.15 For a fixed point $\mathbf{a} \in \mathbb{R}^m$ and a positive real number $r > 0$, the condition $\|\mathbf{x} - \mathbf{a}\| = r$ describes the set of all points $\mathbf{x} \in \mathbb{R}^m$ whose Euclidean distance to $\mathbf{a}$ is r. In $\mathbb{R}^2$ this is a circle centred at $\mathbf{a}$ and having the radius r; in $\mathbb{R}^3$ the same condition describes the sphere centred at $\mathbf{a}$ and having radius r and so on.

Definition 1.16 *Let* $\mathbf{a} \in \mathbb{R}^m$ *and* $r > 0$.

(i) *The set* $B_r(\mathbf{a}) := \{\mathbf{x} \in \mathbb{R}^m : \|\mathbf{x} - \mathbf{a}\| < r\}$ *is called the open ball centred at* $\mathbf{a}$ *and having the radius* r;

(ii) *The set* $\overline{B}_r(\mathbf{a}) := \{\mathbf{x} \in \mathbb{R}^m : \|\mathbf{x} - \mathbf{a}\| \leq r\}$ *is called the closed ball centred at* $\mathbf{a}$ *and having the radius* r.

The difference between the open ball $B_r(\mathbf{a})$ and the closed ball $\overline{B}_r(\mathbf{a})$ is that the later contains the points on the circle $\|\mathbf{x} - \mathbf{a}\| = r$ while the former does not. In dimension $m = 2$, the open and closed balls are called *open and closed disc* respectively. These are illustrated in Figure 1.6.

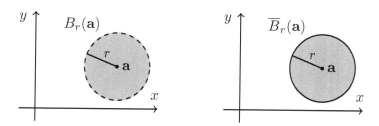

Figure 1.6
The open disc (left) and the closed disc (right) in $\mathbb{R}^2$.

Definition 1.17 *Let* $D \subset \mathbb{R}^m$ *and* $\mathbf{a} \in \mathbb{R}^m$.

(i) $\mathbf{a}$ *is called an isolated point of* D *if there exists* $r > 0$ *such that* $B_r(\mathbf{a}) \cap D = \{\mathbf{a}\}$;

(ii) $\mathbf{a}$ *is called an interior point of* D *if there exists* $r > 0$ *such that* $B_r(\mathbf{a}) \subset D$;

(iii) $\mathbf{a}$ *is called an exterior point of* D *if* $\mathbf{a}$ *is an interior point of the complement set* $D^c = \mathbb{R}^m \setminus D$;

(iv) $\mathbf{a}$ *is called a boundary point of* D *if* $\mathbf{a}$ *is neither an interior nor an exterior point of* D.

In Figure 1.7, $\mathbf{a}$ is an interior point, $\mathbf{b}$ is an exterior point while $\mathbf{c}$ is a boundary point of D. Let us note that another way of characterizing boundary points is the following: $\mathbf{a}$ is a boundary point of D if and only if any open ball centred at $\mathbf{a}$ contains points in D as well as points in the complement D^c.

Definition 1.18 *Let* $D \subset \mathbb{R}^m$.

(i) *The interior of* D *is the set denoted* D° *which contains all interior points of* D. *We say that* D *is open if* $D = D^\circ$;

(ii) *The boundary of* D *is the set denoted* ∂D *which contains all boundary points of* D;

(iii) *The closure of* D *is the set* $\overline{D} = D^\circ \cup \partial D$. *We say that* D *is closed if* $D = \overline{D}$.

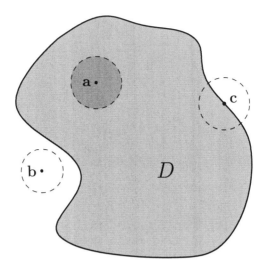

Figure 1.7
Interior, exterior and boundary points of a set $D \subset \mathbb{R}^2$.

Observe that an element $\mathbf{a} \in \overline{D}$ can either be an interior point or a boundary point of D, so $\overline{D} = D^\circ \cup \partial D$. Also, we have $D^\circ \subset D \subset \overline{D}$ and a set D is closed if and only if its complement D^c is open.

A characterization of the closure $\overline{D}$ through convergent sequences is provided below.

Theorem 1.19 *Let $D \subset \mathbb{R}^m$. Then $\mathbf{a} \in \overline{D}$ if and only if there exists a sequence $\{\mathbf{a}_n\} \subset D$ such that $\{\mathbf{a}_n\}$ converges to $\mathbf{a}$.*

Proof Since $\overline{D} = D^\circ \cup \partial D$, we have to discuss two cases.
Case 1: $\mathbf{a} \in \partial D$. Thus, $\mathbf{a}$ is neither an interior nor an exterior point of D. Hence, every ball $B_r(\mathbf{a})$, $r > 0$, contains points in D and points in the complement D^c. Taking $r = 1/n$, it follows that for every $n \geq 1$ there exists $\mathbf{a}_n \in D \cap B_{1/n}(\mathbf{a})$. Thus, the sequence $\{\mathbf{a}_n\} \subset D$ satisfies

$$\|\mathbf{a}_n - \mathbf{a}\| \leq \frac{1}{n} \quad \text{for all } n \geq 1. \tag{1.12}$$

This shows that $\mathbf{a}_n \to \mathbf{a}$ as $n \to \infty$ which concludes the proof in this part.
Case 2: $\mathbf{a} \in D^\circ$. Then, there exists $r > 0$ such that $B_r(\mathbf{a}) \subset D$. It is enough to construct a sequence $\{\mathbf{a}_n\}$ with the property (1.12). Indeed, let $N \geq 1$ be large enough so that $\frac{1}{N} < r$. Then $\frac{1}{n} < r$ for all $n \geq N$. For any $n \geq N$, we have $\overline{B}_{1/n}(\mathbf{a}) \subset B_r(\mathbf{a}) \subset D$. For any $n \geq N$ take $\mathbf{a}_n \in \partial B_{1/n}(\mathbf{a})$. This means that $\mathbf{a}_n \in D$ and $\|\mathbf{a}_n - \mathbf{a}\| = \frac{1}{n}$. Thus the sequence $\{\mathbf{a}_n\}_{n \geq N}$ satisfies (1.12) for all $n \geq N$ and then $\{\mathbf{a}_n\}$ converges to $\mathbf{a}$ as $n \to \infty$. $\square$

Example 1.20 *Let*

$$D = (-2,3] \times [-1,3) \quad and \quad E = \{(x,y) \in \mathbb{R}^2 : -2 \leq x < 2,\ -1 < y \leq |x+1|\}.$$

Find D°, $\overline{D}$, ∂D *and* E°, $\overline{E}$, ∂E.

Solution. The two sets D and E are depicted in Figure 1.8.

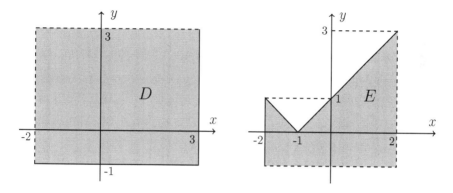

Figure 1.8
The sets $D, E \subset \mathbb{R}^2$.

We have $D^{\circ} = (-2,3) \times (-1,3)$ and $\overline{D} = [-2,3] \times [-1,3]$. In set notation, we have

$$D^{\circ} = \{(x,y) \in \mathbb{R}^2 : -2 < x < 3,\ -1 < y < 3\},$$
$$\overline{D} = \{(x,y) \in \mathbb{R}^2 : -2 \leq x \leq 3,\ -1 \leq y \leq 3\}.$$

The boundary of D consists of the union of all four segments that define the rectangle in Figure 1.8. Thus, $\partial D = \big(\{-2,3\} \times [-1,3]\big) \cup \big([-2,3] \times \{-1,3\}\big)$.
In set notation, we have

$$E^{\circ} = \{(x,y) \in \mathbb{R}^2 : -2 < x < 2,\ -1 < y < |x+1|\},$$
$$\overline{E} = \{(x,y) \in \mathbb{R}^2 : -2 \leq x \leq 2,\ -1 \leq y \leq |x+1|\}.$$

The boundary of E consists of the union of the five segment lines that define the pentagon in Figure 1.8. Hence, $\partial E = \big(\{-2\} \times [-1,1]\big) \cup \big(\{2\} \times [-1,3]\big) \cup \{(x,y) \in \mathbb{R}^2 : x \in [-2,2], y = -1 \text{ or } y = |x+1|\}$.

Definition 1.21 *Let* $D \subset \mathbb{R}^m$.

(i) *We say that* D *is bounded if there exists* $M > 0$ *such that* $\|\mathbf{a}\| \leq M$ *for all* $\mathbf{a} \in D$;

(ii) *We say that* D *is compact if* D *is closed and bounded.*

Condition $\|\mathbf{a}\| \leq M$ for all $\mathbf{a} \in D$ means $\mathbf{a} \in \overline{B}_M(\mathbf{0})$ for all $\mathbf{a} \in D$. Thus, $D \subset \mathbb{R}^m$ is bounded if and only if there exists a closed ball centred at the origin that contains D. Ellipses, discs and rectangles in the plane are bounded while half planes and strips in the plane are not.

Later on in Chapter 10 we shall make use of the following characterization of compact sets.

Theorem 1.22 *A set $D \subset \mathbb{R}^m$ is compact if and only if any sequence in D has a convergent subsequence to an element in D.*

Exercises

Exercise 17. Using Definition 1.18(i), check that the following sets are open:

   (i) $A = \{(x, y) \in \mathbb{R}^2 : -1 < y < 2\}$;

  (ii) $B = \{(x, y) \in \mathbb{R}^2 : y > x\}$;

  (ii) $C = \{(x_1, x_2, \ldots, x_m) \in \mathbb{R}^m : x_m > 0\}$.

Hint: Show that each of the above set is open by finding a suitable ball around every point contained in the set.

Exercise 18. Let $\mathbf{a}, \mathbf{b} \in \mathbb{R}^m$ and $r, \rho > 0$ be such that $\|\mathbf{a} - \mathbf{b}\| + \rho < r$.

   (i)   Show that $\overline{B}_\rho(\mathbf{b}) \subset B_r(\mathbf{a})$;

  (ii)   Deduce in particular that $B_r(\mathbf{a})$ is an open set in $\mathbb{R}^m$.

Exercise 19. Verify that:

   (i)   if $G_1, G_2 \subset \mathbb{R}^m$ are open sets, then $G_1 \cup G_2$ and $G_1 \cap G_2$ are also open sets;

  (ii)   if $D_1, D_2 \subset \mathbb{R}^m$ are closed sets, then $D_1 \cup D_2$ and $D_1 \cap D_2$ are also closed sets.

Exercise 20. Using Definition 1.18, check that:

   (i)   A set $D \subset \mathbb{R}^m$ is open if and only if D does not contain any boundary point;

  (ii)   A set $D \subset \mathbb{R}^m$ is closed if and only if D contains all its boundary points.

Exercise 21. Let $D \subset \mathbb{R}^m$ be a nonempty set. Prove that:

  (i) $(\overline{D})^c = (D^c)^\circ$;   (ii) $(D^\circ)^c = \overline{D^c}$;   (iii) $\partial D = \partial D^c$.

Exercise 22. Find the interior, closure and the boundary of the following sets. It is useful to graph each set in your analysis:

(i) $A = [-2, 6] \times (1, 8)$;

(ii) $B = (-\infty, 1] \times (-1, \infty)$;

(iii) $C = \{(x, y) \in \mathbb{R}^2 : -1 \leq x < 2, x^2 \leq y < 5\}$;

(iv) $D = \{(x, y) \in \mathbb{R}^2 : xy^2 \neq 1\}$;

(v) $E = \{(x_1, x_2, x_3) \in \mathbb{R}^3 : x_1^2 + x_2^2 < x_3^2 \leq 1\}$.

Exercise 23. Verify the following statements:

(i) If $D \subset \mathbb{R}^m$ is bounded, then $\overline{D}$ is also bounded;

(ii) If $D \subset \mathbb{R}^m$ is bounded, then $\overline{D}$ and ∂D are compact sets.

Hint: Use Theorem 1.19.

Exercise 24. Give example of an unbounded set $D \subset \mathbb{R}^m$ such that ∂D is compact.

Exercise 25. Let $D_1, D_2 \subset \mathbb{R}^m$ be compact sets. Check that $D_1 \cup D_2$ and $D_1 \cap D_2$ are also compact.

2

Functions of Several Variables

In this chapter the notion of a function between two sets $A \subset \mathbb{R}^m$ and $B \subset \mathbb{R}^p$ is presented. Domain, codomain, level sets and graph of a function of several variables are defined and studied. Particular attention is paid to quadrics and curves.

2.1   Functions, Domains and Codomains

Given two nonempty sets A and B, a function from A to B is a rule that assigns to each element $x \in A$ a unique element $y \in B$. If f denotes the assigning rule, we write $f : A \to B$ and $y = f(x)$. Note that A and B can be any nonempty sets (of people, objects, etc). A is called the *domain* of f while B is called the *codomain* of f. Also $f(A) = \{f(x) : x \in A\}$ is called the *range* of f. As it can be easily seen, the range of f is a subset of the codomain B.

If $A \subset \mathbb{R}^m$, $m \geq 2$ and $B = \mathbb{R}$, we shall call $f : A \to \mathbb{R}$ a function of several real variables. For instance, $f : \mathbb{R}^2 \to \mathbb{R}$, $f(x, y) = xe^{-y}$ is a function of two variables and we can compute the values of f at various points $(x, y) \in \mathbb{R}^2$, such as

$$f(1, 1) = 1 \cdot e^{-1} = e^{-1}, \quad f(-2, 0) = -2e^0 = -2$$

and so on.

If $A \subset \mathbb{R}^m$ and $B = \mathbb{R}^p$, with $m \geq 1$ and $p \geq 2$, then any function $F : A \to \mathbb{R}^p$ will be called a *vector valued function*. This means that given any $\mathbf{x} \in A$, the outcome $F(\mathbf{x})$ is a vector in $\mathbb{R}^p$. Thus, we are entitled to write $F(\mathbf{x}) = (f_1(\mathbf{x}), f_2(\mathbf{x}), \ldots, f_p(\mathbf{x}))$ and call $f_1, f_2, \ldots, f_p : A \to \mathbb{R}$ the components of F. Take as an illustration,

$$F : \mathbb{R}^2 \to \mathbb{R}^3, \quad F(x, y) = (x^2 + y, ye^x, \cos(xy)).$$

Then F maps vectors in $\mathbb{R}^2$ into vectors in $\mathbb{R}^3$; for instance $F(-2, 0) = (4, 0, 1)$ and so on. Throughout this book we shall be concerned with the study of functions $f : D \subset \mathbb{R}^m \to \mathbb{R}$ and their vector valued counterparts $F : D \subset \mathbb{R}^m \to \mathbb{R}^p$. The domain D of definition of f is taken as the largest set in $\mathbb{R}^m$ so that f makes sense, that is, $f(\mathbf{x})$ can be computed at all $\mathbf{x} \in D$. Similarly,

DOI: 10.1201/9781003449652-2

the domain D for a vector valued function $F = (f_1, f_2, \ldots, f_p)$ is the largest set in $\mathbb{R}^m$ on which all its components $f_1, f_2, \ldots, f_p$ exist.

As a rule of thumb, when establishing the domain of definition of a function one has to take into account that division by zero, taking square roots and logarithms of negative numbers are not allowed.

Example 2.1 *Find the domain of definition of the functions:*

(i) $f(x, y) = \sqrt{1 - \dfrac{x^2}{y}};$ (ii) $F(x, y) = \left(\ln(1 - x^2 - y^2), \sqrt{xy} \right).$

Solution. (i) We impose $y \neq 0$ and $1 - \frac{x^2}{y} \geq 0$. This means $y \neq 0$ and $\frac{x^2}{y} \leq 1$. Notice that

$$\frac{x^2}{y} \leq 1 \Longleftrightarrow y < 0 \quad \text{OR} \quad \begin{cases} y > 0, \\ y \geq x^2. \end{cases}$$

Condition $y < 0$ represents the lower half-plane. Conditions $y > 0$ and $y \geq x^2$ represent the interior and the boundary of the parabola $y = x^2$ except the origin $(0, 0)$. In set notation,

$$D = \{(x, y) \in \mathbb{R}^2 \mid y < 0 \text{ OR } (y > 0, y \geq x^2)\}.$$

(ii) We impose $1 - x^2 - y^2 > 0$ and $xy \geq 0$. This means

$$x^2 + y^2 < 1 \quad \text{and} \quad (x \geq 0, y \geq 0) \quad \text{or} \quad (x \leq 0, y \leq 0).$$

Thus, the domain of definition is the union of two sectors of the unit disc that lie in the first and third quadrant of the plane; see Figure 2.1. In set notation, we have

$$D = \left\{ (x, y) \in \mathbb{R}^2 \mid x^2 + y^2 < 1 \text{ AND } \left[(x \geq 0, y \geq 0) \text{ OR } (x \leq 0, y \leq 0) \right] \right\}.$$

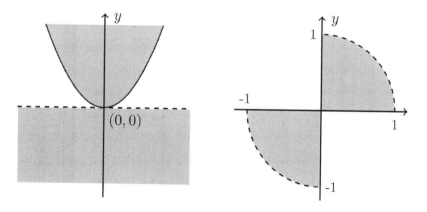

Figure 2.1
The domain D of f (left) and F (right).

As in the one-variable case, if $A \subset \mathbb{R}^m$, $B \subset \mathbb{R}^p$, $m, p \geq 1$, a function $F : A \to B$ is called:

- *injective (or one-to-one)* if whenever $\mathbf{x}, \mathbf{y} \in A$, $\mathbf{x} \neq \mathbf{y}$, then $F(\mathbf{x}) \neq F(\mathbf{y})$;

- *surjective (or onto)* if for every $\mathbf{u} \in B$ there exists $\mathbf{x} \in A$ such that $F(\mathbf{x}) = \mathbf{u}$;

- *bijective* if it is both injective and surjective.

From the above, we easily derive that $F : A \to B$ is bijective if and only if for any $\mathbf{u} \in B$ there exists a *unique* $\mathbf{x} \in A$ such that $F(\mathbf{x}) = \mathbf{u}$.

A relevant example in this sense is the *stereographic projection*. For the sake of presentation, let us place ourselves in the three dimensional space and denote by $\mathbb{S}^2$ the unit sphere (following the notations from Chapter 1, $\mathbb{S}^2 = \partial B_1(0)$, where $B_1(0) \subset \mathbb{R}^3$ is the unit ball in $\mathbb{R}^3$). We shall call $N(0, 0, 1)$ the north pole of the sphere $\mathbb{S}^2$. For any point $P(x, y, z) \in \mathbb{S}^2 \setminus \{N\}$ denote by $Q(X, Y, 0)$ the intersection of the line NP with the xy plane; see Figure 2.2.

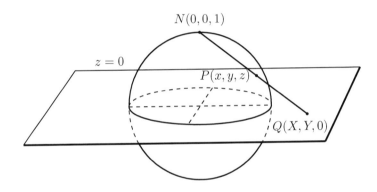

Figure 2.2
The stereographic projection on $\mathbb{R}^3$.

We now define

$$\Phi : \mathbb{S}^2 \setminus \{N\} \to \mathbb{R}^2 \quad \text{by} \quad \Phi(x, y, z) = (X, Y).$$

To determine X and Y we use the fact that triangles NTR and NOS in Figure 2.3 are similar, so

$$\frac{TR}{OS} = \frac{NT}{NO} \implies \frac{x}{X} = \frac{1 - z}{1} \implies X = \frac{x}{1 - z}.$$

Next, triangles NTU and NOV are similar and this yields $Y = \frac{y}{1-z}$.

Hence

$$\Phi : \mathbb{S}^2 \setminus \{N\} \to \mathbb{R}^2, \quad \Phi(x, y, z) = \left(\frac{x}{1 - z}, \frac{y}{1 - z} \right). \tag{2.1}$$

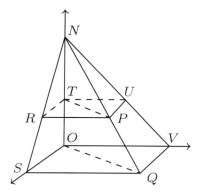

Figure 2.3
The similar triangles $NTR \sim NOS$ and $NTU \sim NOV$.

The stereographic projection is bijective. Indeed, if $(X, Y) \in \mathbb{R}^2$, then $\Phi(x, y, z) = (X, Y)$ implies

$$x = X(1 - z) \quad \text{and} \quad y = Y(1 - z).$$

From $x^2 + y^2 + z^2 = 1$ we deduce $z = \frac{X^2 + Y^2 - 1}{X^2 + Y^2 + 1}$ and then

$$(x, y, z) = \left(\frac{2X}{X^2 + Y^2 + 1}, \frac{2Y}{X^2 + Y^2 + 1}, \frac{X^2 + Y^2 - 1}{X^2 + Y^2 + 1} \right).$$

Thus, for all $(X, Y, Z) \in \mathbb{R}^2$ the equation $\Phi(x, y, z) = (X, Y)$ has a unique solution $(x, y, z) \in \mathbb{S}^2 \setminus \{N\}$, which concludes that Φ is bijective. Its inverse $\Phi^{-1} : \mathbb{R}^2 \to \mathbb{S}^2 \setminus \{N\}$ is thus given by

$$\Phi^{-1}(X, Y) = \left(\frac{2X}{X^2 + Y^2 + 1}, \frac{2Y}{X^2 + Y^2 + 1}, \frac{X^2 + Y^2 - 1}{X^2 + Y^2 + 1} \right).$$

Exercises

Exercise 1. For each of the function $f(x, y)$ below describe its domain D in set notation and sketch it in the xy plane:

(i) $f(x, y) = \dfrac{1}{3x^2 + y^2};$

(ii) $f(x, y) = \dfrac{1}{\sqrt{16 - x^2 - y^2}};$

(iii) $f(x, y) = \dfrac{8}{\sqrt{x^2 + y^2 - 4}}$;

(vi) $f(x, y) = \ln\left(1 - \dfrac{e^x}{y}\right)$;

(iv) $f(x, y) = \sqrt{y - x}$;

(v) $f(x, y) = \ln(x^2 - y^2)$;

(vii) $f(x, y) = \sqrt{1 + xy^2}$.

Exercise 2. Determine the domain D of the following functions. Write D in set notation and describe it geometrically.

(i) $F(x, y) = \left(\ln(x - \sqrt{y}), \ln(y - \sqrt{x})\right)$;

(ii) $F(x, y) = \left(\dfrac{x}{y}, \dfrac{y^2}{y + z}, \dfrac{z^3}{x + y + z}\right)$.

Exercise 3. Let $\mathbb{S}^2$ be the unit sphere in $\mathbb{R}^3$ and $S(0, 0, -1) \in \mathbb{S}$ be its south pole.

(i) Determine the stereographic projection $\Psi : \mathbb{S}^2 \setminus \{S\} \to \mathbb{R}^2$ with respect to the south pole S;

(ii) If Φ is the stereographic projection in $\mathbb{S}^2$ with respect to the north pole defined by (2.1), show that

$$(\Phi \circ \Psi^{-1})(X, Y) = \left(\frac{X}{X^2 + Y^2}, \frac{Y}{X^2 + Y^2}\right)$$

for all $(X, Y) \in \mathbb{R}^2 \setminus \{(0, 0)\}$.

Exercise 4. Let $\mathbb{S}^{m-1}$ denote the unit sphere in $\mathbb{R}^{m-1}$, $m \geq 2$ and $N(0, 0, \ldots, 0, 1) \in \mathbb{S}^{m-1}$ be its north pole. Check that the stereographic projection in $\mathbb{R}^m$ given by $\Phi : \mathbb{S}^{m-1} \setminus \{N\} \to \mathbb{R}^{m-1}$,

$$\Phi(x_1, x_2, \ldots, x_m) = \left(\frac{x_1}{1 - x_m}, \frac{x_2}{1 - x_m}, \ldots, \frac{x_{m-1}}{1 - x_m}\right)$$

is bijective.

Exercise 5. Find all values of $k \in \mathbb{R}$ such that

$$F : \mathbb{R}^m \to \mathbb{R}^m, \quad F(\mathbf{x}) = \begin{cases} \mathbf{x}\|\mathbf{x}\|^k & \text{if } \mathbf{x} \neq 0 \\ 0 & \text{if } \mathbf{x} = 0 \end{cases} \quad \text{is bijective.}$$

2.2   The Graph of a Function

Definition 2.2 (The graph of a function)

Let $F : D \subset \mathbb{R}^m \to \mathbb{R}^p$, $m, p \geq 1$, be a *function of m variables (possibly vector valued)*. The set $G(f) = \{(\mathbf{x}, F(\mathbf{x})) : \mathbf{x} \in D\} \subset \mathbb{R}^m \times \mathbb{R}^p$ is called the *graph of F*.

According to the above definition, the graph of a function $F : D \subset \mathbb{R}^m \to \mathbb{R}^p$ is a subset of $\mathbb{R}^{m+p}$. If $m + p \leq 3$, we can visualize its graph as a subset in the plane or in the three dimensional space. However, the condition $m + p \leq 3$ leaves us with a small number of cases: $(m, p) = (1, 1), (1, 2), (2, 1)$.

Under the mere condition of continuity, the range of a function $F : D \subset \mathbb{R}^2 \to \mathbb{R}^3$ is called a *surface*. The continuity of functions of several variables will be discussed in detail in Chapter 3. Here we shall content ourselves with the fact that all *elementary functions*, that is, polynomial, rational, trigonometric, exponential and logarithmic functions in several variables are continuous.

Example 2.3 *Let $r > 0$, $D = [0, 2\pi) \times [0, \pi] \subset \mathbb{R}^2$ and $F : D \to \mathbb{R}^3$ be given by*

$$F(\varphi, \theta) = (r \cos \varphi \sin \theta, r \sin \varphi \sin \theta, r \cos \theta).$$

Then the range of F is the sphere in $\mathbb{R}^3$ of radius r centred at the origin $O(0, 0, 0)$.

Solution. The triple (r, φ, θ) is called the spherical coordinates; see Figure 2.4. Letting

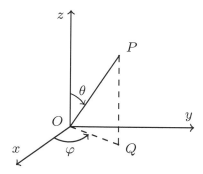

Figure 2.4
The angles φ and θ in the spherical coordinates (2.2).

$$x = r \cos \varphi \sin \theta,$$
$$y = r \sin \varphi \sin \theta, \qquad (2.2)$$
$$z = r \cos \theta,$$

we have $F(\varphi, \theta) = (x, y, z)$ and $x^2 + y^2 + z^2 = r^2$. This shows that the range of F is contained on the sphere of radius r centred at the origin O.

Conversely, for each triple (x, y, z) of the sphere of radius r centred at the origin, we can find $(\varphi, \theta) \in D$ such that $F(\varphi, \theta) = (x, y, z)$. Indeed, as shown in Figure 2.4, let $P(x, y, z)$ be any point on the sphere and $Q(x, y, 0)$ be its projection on the xy plane. Then, φ is the angle between the semiline Ox and

OQ measured anticlockwise in the xy plane, while θ is the angle made by OP and the semiline Oz.

Let $f : D \subset \mathbb{R}^2 \to \mathbb{R}$ be a function of two variables. Then, its graph $G(f)$ is a surface in $\mathbb{R}^3$ since $G(f)$ is the range of the function

$$F : D \to \mathbb{R}^3, \quad F(x, y, z) = (x, y, f(x, y)).$$

Of course, we need to impose F is continuous which amounts to f is continuous (again, continuity is postponed until next chapter). Figure 2.5 presents the graph of such a function $f : D \subset \mathbb{R}^2 \to \mathbb{R}$. A helpful way to gain information

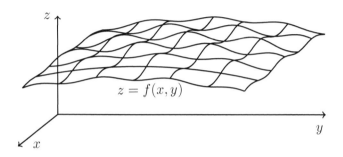

Figure 2.5
A surface in $\mathbb{R}^3$.

about this surface is to determine the shape of its intersection with planes that are parallel to the coordinate planes xz and yz. Such intersections are called the *traces* of the surface $z = f(x, y)$. Precisely, let $(a, b) \in D$. Then, $x \mapsto f(x, b)$ and $y \mapsto f(a, y)$ are one-variable functions whose corresponding graphs $z = f(x, b)$ and $z = f(a, x)$ yield these traces of the surface. They create a curved mesh of the surface as illustrate in Figure 2.5.

Example 2.4 *Using the traces of the surface, sketch the graph of $f(x, y) = x^2$.*

Solution. The trace of the graph in the plane $y = c$ is the parabola $z = x^2$ for all values $c \in \mathbb{R}$. Thus, the graph of f consists of infinitely many copies of the standard parabola $z = x^2$ placed along the y axis; see Figure 2.6.

The vertical line test that determines whether a shape in $\mathbb{R}^2$ is the graph of a function works further for functions of two variables. Indeed, if $f(x, y)$ is a function of two variables, then any vertical line must intersect the traces, and thus the whole surface $z = f(x, y)$, in at most one point. As a consequence, the unit sphere in $\mathbb{R}^3$ is not the graph of a function, since it is intersected by the z axis in exactly two points namely the south and north pole of the sphere. However, the upper and lower unit semisphere are graphs of a function

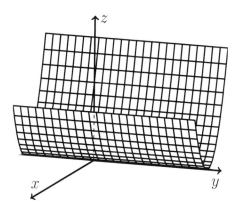

Figure 2.6
The graph of $f(x, y) = x^2$.

as we can easily convince ourselves by taking $f(x, y) = \sqrt{1 - x^2 - y^2}$ and $g(x, y) = -\sqrt{1 - x^2 - y^2}$ respectively.

Example 2.5 *The graph* $y = e^{-x^2}$ *is rotated about the x axis. What is the equation of the obtained surface?*

Solution. Let (x, y, z) be a point on the surface. Then, $y = e^{-x^2}$ and the distance from (x, y, z) to $(x, 0, 0)$ is e^{-x^2}. Thus, $\sqrt{y^2 + z^2} = e^{-x^2}$ which yields the equation of the surface $y^2 + z^2 = e^{-2x^2}$.

Example 2.6 *The graph* $y = x^3$ *is rotated about the y axis. What is the equation of the obtained surface?*

Solution. As before, let (x, y, z) be a point on the surface. Then, $y = x^3$, or $x = y^{1/3}$ and the distance from (x, y, z) to $(0, y, 0)$ is $y^{1/3}$. Thus, $\sqrt{x^2 + z^2} = y^{1/3}$ which yields the equation of the surface $x^2 + z^2 = y^{2/3}$.

Exercises

Exercise 6. Using the traces of the surface, sketch the grapf of the following functions: (i) $f(x, y) = y^2$;   (ii) $f(x, y) = y^2 - 2xy - 3y$.

Exercise 7. Let $a > 0$ be a real number and $g : [0, a] \to \mathbb{R}$ be a continuous function such that $g \not\equiv 0$. The graph of g is rotated about x axis in the plane.

What is the equation of the resulting surface in $\mathbb{R}^3$? Is this surface the graph of a function $f : D \subset \mathbb{R}^2 \to \mathbb{R}$?

Exercise 8. Find the equation of the surface in $\mathbb{R}^3$ described by all points $P(x, y, z)$ for which the distance from P to the y axis is half the distance from P to the origin $O(0, 0, 0)$.

Exercise 9. Find the equation of the surface in $\mathbb{R}^3$ described by all points $P(x, y, z)$ for which the distance from P to the x axis equals the distance from P to the plane $x = -1$.

2.3   Level Sets

Another relevant concept in the study of functions of several variables is that of level sets.

Definition 2.7 *Let $f : D \subset \mathbb{R}^m \to \mathbb{R}$ be a function of m variables and $c \in \mathbb{R}$. The level set of f at level c is given by*

$$L(f, c) = \{\mathbf{x} \in D : f(\mathbf{x}) = c\}.$$

Note that $L(f, c)$ is always a subset of the domain D of f. It may be empty if the equation $f(\mathbf{x}) = c$ has no solutions, that is, if c is not in the range of f.

If $D \subset \mathbb{R}^2$, that is, f is a function of two variables, then $L(f, c)$ is called *level curve*. Similarly, if $D \subset \mathbb{R}^3$, then $L(f, c)$ is called *level surface*.

Example 2.8 *Find the level curves of $f : \mathbb{R}^2 \to \mathbb{R}$, $f(x, y) = \sqrt{x^2 + y^2}$.*

Solution. To find $L(f, c)$ we solve

$$f(x, y) = c \implies L(f, c) = \begin{cases} \emptyset & \text{if } c < 0 \\ \{(0, 0)\} & \text{if } c = 0 \\ \{x^2 + y^2 = c^2\} & \text{if } c > 0. \end{cases}$$

Thus, the level curve $L(f, c)$ is either empty, reduces to the origin, or is the circle of radius $\sqrt{c}$ centred at the origin.

Example 2.9 *Let*

$$f(x, y) = \ln\left(\frac{x}{y} - \frac{y}{x}\right)$$

(i) *Determine the domain of definition of f;*

(ii) *Find the level curves of f.*

Solution. (i) We impose $x, y \neq 0$ and $\frac{x}{y} - \frac{y}{x} > 0$. This latter condition yields

$$\frac{x}{y} - \frac{y}{x} = \frac{x^2 - y^2}{xy} > 0 \implies \begin{cases} xy > 0 \\ x^2 > y^2 \end{cases} \quad OR \quad \begin{cases} xy < 0 \\ x^2 < y^2. \end{cases}$$

The domain is depicted in Figure 2.7.

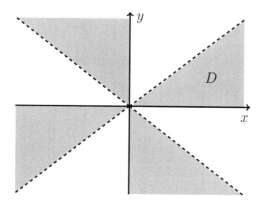

Figure 2.7
The domain of $f(x,y) = \ln\left(\frac{x}{y} - \frac{y}{x}\right)$.

(ii) To find $L(f,c)$ we solve

$$f(x,y) = c \implies \frac{x^2 - y^2}{xy} = e^c \implies x^2 - e^c xy - y^2 = 0.$$

Divide the last equation by y^2 (note that $y \neq 0$, so this is allowed!) to deduce that $t = \frac{x}{y}$ satisfies

$$t^2 - e^c t - 1 = 0 \implies t_{1,2} = \frac{e^c \pm \sqrt{e^{2c} + 4}}{2}.$$

Observe that

$$t_1 = \frac{e^c + \sqrt{e^{2c} + 4}}{2} > 1 \quad \text{and} \quad t_2 = \frac{e^c - \sqrt{e^{2c} + 4}}{2} \in (-1, 0)$$

so the lines $x = t_1 y$ and $x = t_2 y$ lie in the domain D. Thus, the level curve $L(f,c)$ consists of the two lines of equation $x = t_1 y$ and $x = t_2 y$.

Example 2.10 Let $f(x, y, z) = x^4 - 2x^2 + y^2 + 2z^2 - 6y + 4z$.

(i) *Using the completion of squares, determine the values of $c \in \mathbb{R}$ for which the level surface $L(f,c)$ is nonempty;*

(ii) *Verify that all nonempty level surfaces of f are bounded.*

Solution. (i) Using the completion of squares, we have

$$f(x, y, z) = x^4 - 2x^2 + y^2 - 6y + 2z^2 + 4z$$
$$= (x^4 - 2x^2) + (y^2 - 6y) + 2(z^2 + 2z)$$
$$= (x^2 - 1)^2 + (y - 3)^2 + 2(z + 1)^2 - 12.$$

Now, $(x, y, z) \in L(f, c)$ if and only if $f(x, y, z) = c$ which means

$$(x^2 - 1)^2 + (y - 3)^2 + 2(z + 1)^2 = c + 12. \tag{2.3}$$

If $c < -12$, the above equation has no real solutions, since the left-hand side is a sum of perfect squares, hence a nonnegative quantity, while the right-hand side is negative. Assume now $c \geq -12$.

If $c = -12$, then (2.3) becomes

$$(x^2 - 1)^2 + (y - 3)^2 + 2(z + 1)^2 = 0.$$

We have a sum of three perfect squares on the left-hand side which equals zero. This yields each of these perfect squares is zero, so $x^2 - 1 = 0$, $y - 3 = 0$ and $z + 1 = 0$. Hence, $L(f, -12) = \{(1, 3, -1), (-1, 3, -1)\}$.

If $c > -12$, there are also solutions of (2.3). Indeed, let us take $x = 1$, $z = -1$ for which (2.3) becomes $(y - 3)^2 = c + 12$ with $y = 3 \pm \sqrt{c + 12}$. Hence,

$$(1, 3 + \sqrt{c + 12}, -1) \in L(f, c) \quad \text{for all} \quad c > -12.$$

Thus, the level surface $L(f, c)$ is nonempty if and only if $c \geq -12$.

Remark that we were not required to determine the whole surface set $L(f, c)$, just to show it is nonempty. For $c = -12$, it is easy to find the whole level surface $L(f, -12)$ and this reduces to a finite set of points. However, for $c > -12$ this is not an easy task; the level surface $L(f, c)$ contains infinitely many points and to anticipate the topic of the next section it is called ellipsoid.

(ii) The level surface $L(f, c)$ is nonempty if and only if $c \geq -12$. In light of the Definition 1.21, we have to show that there exists $M > 0$ such that

$$\|(x, y, z)\| \leq M \quad \text{for all} \quad (x, y, z) \in L(f, c).$$

This is equivalence to the existence of $A > 0$ such that

$$|x| \leq A, \quad |y| \leq A, \quad |z| \leq A \quad \text{for all } (x, y, z) \in L(f, c).$$

Indeed, if such a condition holds, then

$$\|(x, y, z)\| = \sqrt{x^2 + y^2 + z^2} \leq A\sqrt{3} \quad \text{for all} \quad (x, y, z) \in L(f, c).$$

The above condition is further equivalent to the existence of constants a_1, a_2, a_3, b_1, b_2, $b_2 \in \mathbb{R}$ such that

$$x \in [a_1, b_1], \quad y \in [a_2, b_2], \quad z \in [a_3, b_3] \quad \text{for all } (x, y, z) \in L(f, c). \tag{2.4}$$

Let us proceed from (2.3) which implies

$$(x^2 - 1)^2 \le c + 12, \quad (y - 3)^2 \le c + 12, \quad 2(z + 1)^2 \le c + 12. \qquad (2.5)$$

We start with the second inequality in (2.5) and we have

$$(y - 3)^2 \le c + 12 \implies \sqrt{(y - 3)^2} \le \sqrt{c + 12}$$
$$\implies |y - 3| \le \sqrt{c + 12}$$
$$\implies -\sqrt{c + 12} \le y - 3 \le \sqrt{c + 12}$$
$$\implies 3 - \sqrt{c + 12} \le y \le 3 + \sqrt{c + 12}.$$

This shows that $y \in [3 - \sqrt{c + 12}, 3 + \sqrt{c + 12}]$. From the third inequality in (2.5) we find

$$2(z + 1)^2 \le c + 12 \implies \sqrt{(z + 1)^2} \le \sqrt{\frac{c + 12}{2}}$$
$$\implies |z + 1| \le \sqrt{\frac{c + 12}{2}}$$
$$\implies -\sqrt{\frac{c + 12}{2}} \le z + 1 \le \sqrt{\frac{c + 12}{2}}$$
$$\implies -1 - \sqrt{\frac{c + 12}{2}} \le z \le -1 + \sqrt{\frac{c + 12}{2}}.$$

Hence $z \in \left[-1 - \sqrt{\frac{c+12}{2}}, -1 + \sqrt{\frac{c+12}{2}} \right]$.

We left the analysis of the first inequality in (2.5) at the end of our discussion. Similar to the above approach, we have

$$(x^2 - 1)^2 \le c + 12 \implies \sqrt{(x^2 - 1)^2} \le \sqrt{c + 12}$$
$$\implies |x^2 - 1| \le \sqrt{c + 12}$$
$$\implies -\sqrt{c + 12} \le x^2 - 1 \le \sqrt{c + 12}$$
$$\implies 1 - \sqrt{c + 12} \le x^2 \le 1 + \sqrt{c + 12}.$$

Since $x^2 \ge 0$ we deduce

$$d \le x^2 \le 1 + \sqrt{c + 12}, \quad \text{where} \quad d = \max\{0, 1 - \sqrt{c + 12}\}.$$

Taking the square roots as previously, we find $\sqrt{d} \le |x| \le \sqrt{1 + \sqrt{c + 12}}$. In particular, x satisfies $|x| \le \sqrt{1 + \sqrt{c + 12}}$ which yields

$$x \in \left[-\sqrt{1 + \sqrt{c + 12}}, \sqrt{1 + \sqrt{c + 12}} \right].$$

We have shown that $(x, y, z) \in L(f, c)$ satisfies (2.4), thus $L(f, c)$ is bounded whenever $c \ge -12$.

Exercises

Exercise 10. Find the level sets for each of the following functions:

(i) $f(x,y) = \begin{cases} \dfrac{xy}{x^2+y^2} & \text{if } (x,y) \neq (0,0) \\ 0 & \text{if } (x,y) = (0,0); \end{cases}$

(ii) $f(x,y,z) = -3$;

(iii) $f(x,y,z) = x + 2y - 7z$;

(iv) $f(x,y) = e^{x-2y}$;

(v) $f(x,y) = \cos(xy)$;

(vi) $f(x,y,z) = x^2 + z^2$.

Exercise 11. Let $f(x,y) = \sqrt{2 - \dfrac{\sin^2 x}{y}}$.

(i) Sketch the domain of definition D of f in the plane;

(ii) Find the level curve of f at level $c = 1$.

Exercise 12. Let $f(x,y) = \sqrt{2x^2 + y^2} - x$.
The level curve of f at level $c > 0$ is a circle of radius 1. Find c.

Exercise 13. Let $f(x,y,z) = x^2 + ay^2 + z^2 - 2by - 4z + 7$. Find $a, b \in \mathbb{R}$ if the level surface of f at level $c = 17$ is a sphere that passes through $(-1, 2, 3)$.

Exercise 14. (i) Find the functions $f : \mathbb{R}^2 \to \mathbb{R}$ such that for all $c > 0$ the level curve of f at level c is the circle centred at $(-2, 3)$ having radius $2c$;
(ii) Find the functions $f : \mathbb{R}^3 \to \mathbb{R}$ such that for all $c > 0$ the level surface of f at level c is the sphere centred at $(1, 0, -1)$ having radius $2\sqrt{c}$.

Exercise 15. The level curves of $f : \mathbb{R}^2 \to \mathbb{R}$ are circles.

(i) Does it follow from here that f is a polynomial function?

(ii) If $g : \mathbb{R} \to \mathbb{R}$ is a one-variable function, what are the level sets of $g \circ f : \mathbb{R}^2 \to \mathbb{R}$?

Exercise 16. Let $f(x,y,z) = x^2 + 5y^2 + 2z^2 - 2xz + 2yz + 4y$.

(i) Find $c \in \mathbb{R}$ for which the level surface $L(f,c)$ is nonempty;

(ii) Find $c \in \mathbb{R}$ for which the level surface $L(f,c)$ reduces to a single element.

Exercise 17. Let $f(x,y,z) = 5x^2 + 2y^4 + y^2z^2 + 2xy^2 + 2y^3z - 4x + 2$.

(i) Using the completion of squares method, check that the level surfaces $L(f,0)$ and $L(f,1)$ are empty;

(ii) Check that the level surface $L(f,3)$ is nonempty and unbounded.

Hint: try to show that $L(f,3)$ contains points of the form $(x,0,z)$ for some specific x and for all $z \in \mathbb{R}$.

Exercise 18. Give example of a function $f : \mathbb{R}^2 \to \mathbb{R}$ such that all its level curves are unbounded sets.

Exercise 19. Let $f : [0,1] \to \mathbb{R}$ be a one-variable function. In the xyz orthogonal system, the graph of $z = f(y)$ is rotated about the z axis. Check that the new object is the graph of a function $g(x,y)$. Find g in terms of f and its level curves.

2.4 Quadric Surfaces

We have seen in Chapter 1 that the graphical representation of a first order polynomial equation

$$ax + by + cz = d \tag{2.6}$$

is a plane. There is nothing special about this case: as soon as we know one point contained in the plane and a normal vector (remember, for (2.6) this is given by $\mathbf{n} = (a,b,c)$) then the plane is completely determined.

In this section we take the discussion one step further and consider second order polynomial equations in three variables:

$$Ax^2 + By^2 + Cz^2 + Dxy + Eyz + Fxz + Gx + Hy + Iz + J = 0. \tag{2.7}$$

The graphical representation of such equation is called a *quadric surface*. This situation is not unfamiliar to us: in Section 2.2 we looked at the case where all coefficients in (2.7) are zero except for $A = B = C = 1$ and $J = -1$; this corresponds to the *unit sphere*. In fact, regrouping terms in (2.7) and using the completion of squares, one can reduce (2.7) to either

$$Ax^2 + By^2 + Iz = 0 \quad \text{or} \quad Ax^2 + By^2 + Cz^2 + J = 0. \tag{2.8}$$

We shall see below how this is done through a specific example. If $C = 0$ or $I = 0$, the above equations yield *cylindrical surfaces* in $\mathbb{R}^3$, that is, the horizontal traces of the surface are always the same. Otherwise, assuming that $(A,B,C) \neq (0,0,0)$, the reduced equations (2.8) can be written as

$$\frac{z}{c} = \pm\frac{x^2}{a^2} \pm \frac{y^2}{b^2} \quad \text{or} \quad \pm\frac{x^2}{a^2} \pm \frac{y^2}{b^2} \pm \frac{z^2}{c^2} = 1 \quad \text{or} \quad \pm\frac{x^2}{a^2} \pm \frac{y^2}{b^2} \pm \frac{z^2}{c^2} = 0. \tag{2.9}$$

Table 2.1 summarizes the six quadrics that follow from (2.9). These are pictured in Figures 2.8 and 2.9.

TABLE 2.1

The six important quadrics.

Quadric	Equation	Figure
Ellipsoid	$\dfrac{x^2}{a^2} + \dfrac{y^2}{b^2} + \dfrac{z^2}{c^2} = 1$	Figure 2.8 (a)
Double cone	$\dfrac{z^2}{c^2} = \dfrac{x^2}{a^2} + \dfrac{y^2}{b^2}$	Figure 2.8 (b)
Hyperboloid of one sheet	$\dfrac{x^2}{a^2} + \dfrac{y^2}{b^2} - \dfrac{z^2}{c^2} = 1$	Figure 2.8 (c)
Hyperboloid of two sheets	$\dfrac{x^2}{a^2} - \dfrac{y^2}{b^2} - \dfrac{z^2}{c^2} = 1$	Figure 2.8 (d)
Elliptic paraboloid	$\dfrac{z}{c} = \dfrac{x^2}{a^2} + \dfrac{y^2}{b^2}$	Figure 2.9 (e)
Hyperbolic paraboloid	$\dfrac{z}{c} = \dfrac{x^2}{a^2} - \dfrac{y^2}{b^2}$	Figure 2.9 (f)

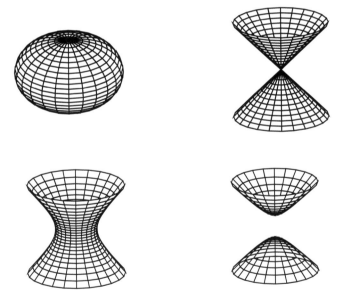

Figure 2.8

From top left: The quadrics (a)–(d).

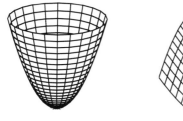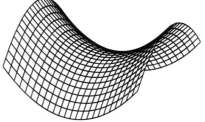

Figure 2.9
From left: The quadrics (e)–(f).

Example 2.11 *Using the completion of squares and then a linear change of variables, match the quadric*

$$x^2 + 5y^2 - 3z^2 + 2xy - 4yz + 2x + 2y = 3$$

with a standard one from the above table.

Solution. The above equation can be written as

$$(x + y + 1)^2 + (2y - z)^2 - 4z^2 = 4.$$

Let $X = x + y + 1$, $Y = 2y - z$ and $Z = z$, so that

$$x = X - \frac{Y + Z}{2} - 1, \quad y = \frac{Y + Z}{2}, \quad z = Z.$$

Our original quadric is equivalent, in the new XYZ system to

$$\frac{X^2}{2^2} + \frac{Y^2}{2^2} - Z^2 = 1$$

which is a hyperboloid of one sheet.

Exercises

Exercise 20. Using the completion of squares method, arrange the following second order equations in variables x, y and z in order to match them with one of the quadrics described in Table 2.1:

    (i) $3x^2 + 2y^2 + z^2 - 4xy + 2yz + 2x - 5 = 0$;

    (ii) $x^2 + 3y^2 - 2xy - 4y - z + 2 = 0$;

(iii) $-x^2 + 3y^2 + 4z^2 + 4xz + 6y + 3 = 0$;

(iv) $3x^2 - 2y^2 + z^2 - 2xy + 4xz + 2y - 6 = 0$;

(v) $-x^2 + y^2 - 3z^2 - 4xz - 2yz - 2x = 0$;

(vi) $x^2 + y^2 + z^2 - xy - yz - zx = 7$.

Exercise 21. Describe the level sets of each of the following functions. Observe that these level sets are given by equations of the quadrics discussed above.

(i) $f(x, y, z) = \dfrac{x^2 + y^2}{1 + z^2}$;

(iii) $f(x, y, z) = \dfrac{x}{y^2 + z^2}$;

(ii) $f(x, y, z) = \dfrac{x^2 - y^2}{1 + z^2}$;

(iv) $f(x, y, z) = \dfrac{1 - y^2}{x^2 + z^2}$.

2.5   Curves

In this section we consider functions $\gamma : I \to \mathbb{R}^p$ where $I \subset \mathbb{R}$ is an interval on the real line. Observe that γ is a vector valued function of one variable, so we may write $\gamma = (\gamma_1, \gamma_2, \ldots, \gamma_p)$. These functions provide the right setting for defining the notion of a *curve* in $\mathbb{R}^p$. In such circumstances we shall use the notation γ instead of f or F, the way it has been used so far in this chapter.

Definition 2.12 *Let $I \subset \mathbb{R}$ be an interval. A function $\gamma : I \to \mathbb{R}^p$, $\gamma = (\gamma_1, \gamma_2, \ldots, \gamma_p)$ is called a parametrized curve in $\mathbb{R}^p$ if each component $\gamma_j : I \to \mathbb{R}$, $1 \le j \le p$, is a continuous function of one variable. The set $\Gamma = \{\gamma(t) : t \in I\}$ is called the trace of the curve.*

The input variable in I is denoted by t and it is called the *parameter* for the parametrized curve γ. In vectorial notation, we have

$$\gamma(t) = \gamma_1(t)\mathbf{e}_1 + \gamma_2(t)\mathbf{e}_2 + \cdots + \gamma_p(t)\mathbf{e}_p.$$

Example 2.13 *Let $\mathbf{a}, \mathbf{b} \in \mathbb{R}^p$ be two distinct points in $\mathbb{R}^p$. Then $\gamma : [0, 1] \to \mathbb{R}^p$, $\gamma(t) = t\mathbf{a} + (1 - t)\mathbf{b}$ is a parametrized curve whose trace is the closed segment $[\mathbf{a}, \mathbf{b}]$.*

Example 2.14 *Let $\gamma : [0, 2\pi] \to \mathbb{R}^2$, $\gamma(t) = (\cos t, \sin t)$. Then γ is a parametrized curve whose trace is the unit circle in the plane.*

As it can be easily seen, the parametrization of the unit circle is not unique. For instance, $\gamma : [0, 1] \to \mathbb{R}^2$, $\gamma(t) = (\sin(2\pi t), \cos(2\pi t))$ is another parametrization.

Example 2.15 *Let $\gamma : [0, 7\pi] \to \mathbb{R}^3$, $\gamma(t) = (\cos t, \sin t, t)$. Then, γ is a parametrized curve in space whose trace is called helix.*

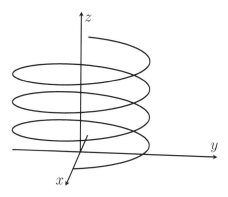

Figure 2.10
The helix.

The helix is a spiral staircase which lies on the cylinder $x^2 + y^2 = 1$, see
Figure 2.10.

Example 2.16 *Find the parametrized curve of the intersection between the
cone $x^2 = y^2 + z^2$ and the plane $z - x = 1$.*

Solution. Plugging $z = x + 1$ in the equation of the cone we find $x^2 = y^2 + (x+1)^2$. From here we deduce $x = -\frac{y^2+1}{2}$ and then $z = \frac{1-y^2}{2}$. Letting
$y = t$ we find the parametrized curve $\gamma : \mathbb{R} \to \mathbb{R}$, $\gamma(t) = \left(-\frac{t^2+1}{2}, t, \frac{1-t^2}{2}\right)$.

Definition 2.17 (Differentiable and regular curves)
Let $\gamma : I \to \mathbb{R}^p$, $\gamma = (\gamma_1, \gamma_2, \ldots, \gamma_p)$ be a parametrized curve.

(i) *The curve γ is called differentiable if each γ_j, $1 \le j \le p$, is
differentiable on I. We denote by γ' or $\frac{d\gamma}{dt}$ the vector $(\gamma_1', \gamma_2', \ldots, \gamma_p')$;*

(ii) *The curve γ is called continuously differentiable if it is differentiable on I and each $\gamma_i' : I \to \mathbb{R}$, $1 \le j \le p$, is continuous;*

(ii) *The curve γ is called regular if it is continuously differentiable
and $\gamma'(t) \ne \mathbf{0}$ for all $t \in I$. In such a case, $\gamma'(t)$ is called the tangent
vector to γ at $t \in I$.*

Definition 2.18 (Tangent line to a curve)
Let $\gamma : I \to \mathbb{R}^p$ be a regular curve and $\tau \in I$. The tangent line to γ at
$P = \gamma(\tau)$ is the line that passes through P and has the direction $\gamma'(\tau)$.

Example 2.19 *Find the parametric equation of the tangent line to the helix
$\gamma : \mathbb{R} \to \mathbb{R}^3$, $\gamma(t) = (\cos t, -2\sin t, t)$ at $(-1, 0, \pi)$.*

Solution. The point corresponding to $(-1, 0, \pi)$ is $\gamma(\pi)$ and the direction of the tangent line is given by $\gamma'(\pi) = (0, 2, 1)$. Thus, the parametric equation of the tangent line is

$$(x, y, z) = (-1, 0, \pi) + t(0, 2, 1) = (-1, 2t, t + \pi), \quad t \in \mathbb{R}.$$

Exercises

Exercise 22. In each of the following cases, find the parametrized curve of intersection between the given surfaces:

(i)   The hyperboloid $x^2 - y^2 - z^2 = 2$ and the plane $x = y - 2$;

(ii) The hyperbolic paraboloid $y = z^2 - x^2$ and the cylinder $x^2 + z^2 = 4$;

(iii) The ellipsoid $x^2 + y^2 + 2z^2 = 6$ and the cone $x = -\sqrt{y^2 + z^2}$.

Exercise 23. Let $\gamma : \mathbb{R} \to \mathbb{R}^3$, $\gamma(t) = (\cos t, t^2 + 2, e^t - 1)$. At which point on the curve is the tangent line parallel to z axis?

Exercise 24. Let $\gamma, \eta : \mathbb{R} \to \mathbb{R}^3$ be given by

$$\gamma(t) = \big(\ln(t^2 + 1), te^t, 3t \big) \quad \text{and} \quad \eta(t) = \big(t^2 - t, -e^{-t}, 3 - t \big).$$

Find $t \in \mathbb{R}$ such that the tangent vectors to $\gamma(t)$ and $\eta(t)$ are orthogonal.

Exercise 25. Let $\gamma, \eta : \mathbb{R} \to \mathbb{R}^3$ be two curves given by

$$\gamma(t) = (t, t^2, 2t - 1) \quad \text{and} \quad \eta(s) = (2s - 1, 2 - s^2, s).$$

(i) Determine the intersection point between the two curves;

(ii) Find the angle of the tangent lines to the two curves at the intersection point.

Exercise 26. (i) Let $\gamma : [0, 1] \to \mathbb{R}^2$ be given by $\gamma(t) = (t, t^2)$. Find $c \in (0, 1)$ such that

$$\gamma(1) - \gamma(0) = \gamma'(c).$$

(ii) Let $\eta : [0, 1] \to \mathbb{R}^2$ be given by $\eta(t) = (t^2, t^3)$. Are there points $c \in (0, 1)$ such that

$$\eta(1) - \eta(0) = \eta'(c)?$$

Exercise 27. Let $\gamma, \eta : I \to \mathbb{R}^p$ be two differentiable curves. Verify the following properties of differentiation:

(i) $(\gamma + \eta)'(t) = \gamma'(t) + \eta'(t)$;

(ii) $(c\gamma)'(t) = c\gamma'(t)$, for any constant $c \in \mathbb{R}$;

(iii) for any differentiable function $f : I \to \mathbb{R}$ one has

$$(f\gamma)'(t) = f(t)\gamma'(t) + f'(t)\gamma(t);$$

(iv) $(\gamma \bullet \eta)'(t) = \gamma'(t) \bullet \eta(t) + \gamma(t) \bullet \eta'(t).$

Exercise 28. The trace of the regular curve $\gamma : I \to \mathbb{R}^3$ lies on the unit sphere. Check that for each $t \in I$ the vectors $\gamma(t)$ and $\gamma'(t)$ are orthogonal. Hint: Since the trace lies on the unit sphere, one has $\|\gamma(t)\| = 1$ for all $t \in I$. Differentiate $\|\gamma(t)\|^2 = 1$ using Exercise 27(iv).

Exercise 29. Let $\gamma = (\gamma_1, \gamma_2) : I \to \mathbb{R}$ be a differentiable curve such that $\gamma_2'(t) \neq 0$ for all $t \in I$. If all the tangent lines to the trace of γ pass through the origin, show that $\gamma_1 = c\gamma_2$ for some $c \in \mathbb{R}$.

3

Limits and Continuity

This chapter discusses the notion of limit for functions of several variables. First, the nonexistence of the limit at a point is presented by the Two-path test. Next, criteria for the existence of the limit are investigated in which the Squeeze Principle plays a crucial role. The concept of the limit of a multi-variable function is further used to define the continuity and the continuous extension of a several variable function.

3.1 Limit of a Function

Throughout this chapter, unless otherwise stated, $D \subset \mathbb{R}^m$ denotes an open set. By analogy with one real variable, we can state the definition of the limit as follows.

Definition 3.1 *Let $f : D \to \mathbb{R}$ be a function of m variables and $\mathbf{a} \in \overline{D}$. We say that f has limit ℓ at $\mathbf{a}$ if for any sequence $\{\mathbf{a}_n\} \subset D \setminus \{\mathbf{a}\}$ which converges to $\mathbf{a}$, we have*

$$\lim_{n \to \infty} f(\mathbf{a}_n) = \ell.$$

We write

$$\lim_{\mathbf{x} \to \mathbf{a}} f(\mathbf{x}) = \ell, \quad or \quad f(\mathbf{x}) \to \ell \quad as \quad \mathbf{x} \to \mathbf{a}.$$

In resemblance to Theorem 1.13 we can establish the following $\varepsilon - \delta$ characterization of the limit.

Theorem 3.2 $(\varepsilon - \delta$ criterion for the limit of a function$)$
Let $f : D \to \mathbb{R}$ be a function of m variables and $\mathbf{a} \in \overline{D}$. The following statements are equivalent:

(i) *f has limit ℓ at $\mathbf{a}$ in the sense of Definition 3.1;*

(ii) *For any $\varepsilon > 0$, there exists $\delta > 0$ such that*

$$\|f(\mathbf{x}) - \ell\| < \varepsilon \quad whenever \quad \mathbf{x} \in D \setminus \{\mathbf{a}\}, \|\mathbf{x} - \mathbf{a}\| < \delta. \quad (3.1)$$

Proof The proof of Theorem 3.2 is achieved by showing that the two implications (i)$\Longrightarrow$(ii) and (ii)$\Longrightarrow$(i) hold.

DOI: 10.1201/9781003449652-3

(i)$\Longrightarrow$(ii). In proving Theorem 3.2 we shall use a different way of thinking to that we employed in Section 1.1 for the Cauchy-Schwarz inequality (1.3). Our approach relies on the so-called *proof by contradiction* method. This consists in assuming the conclusion of Theorem 3.2(ii) to be false and then, by a sequence of logical arguments, we arrive at contradicting something which was assumed to be true. In essence, this lies in the fact that mathematical statements can either be true or false. The fact that $\sqrt{2}$ is irrational, or there is no least positive rational number are standard mathematical results which can be demonstrated using the proof by contradiction.

To proceed with, we first assume that the statement in (ii) does not hold. Then, there exists $\varepsilon > 0$ such that for all $\delta > 0$ one can find $\mathbf{x}_\delta \in D \setminus \{\mathbf{a}\}$ such that

$$\|\mathbf{x}_\delta - \mathbf{a}\| < \delta \quad \text{and} \quad |f(\mathbf{x}_\delta) - \ell| \geq \varepsilon.$$

Note that in this setting $\varepsilon > 0$ is fixed but $\delta > 0$ can be taken arbitrarily small. So, let $\delta = \frac{1}{n}$, $n \geq 1$. Thus, for all $n \geq 1$ there exists $\mathbf{x}_n \in D \setminus \{\mathbf{a}\}$ with

$$\|\mathbf{x}_n - \mathbf{a}\| < \frac{1}{n} \quad \text{and} \quad |f(\mathbf{x}_n) - \ell| \geq \varepsilon. \tag{3.2}$$

The first estimate in (3.2) yields $\mathbf{x}_n \to \mathbf{a}$ as $n \to \infty$. Thus, by our hypothesis (i) it follows that $f(\mathbf{x}_n) \to \ell$ as $n \to \infty$ which contradicts the second estimate in (3.2). This means that (ii) holds true.

(ii)$\Longrightarrow$(i). Take $\varepsilon > 0$ and let $\{\mathbf{x}_n\} \subset D \setminus \{\mathbf{a}\}$ be a sequence that converges to $\mathbf{a}$. By Theorem 1.13, one may find $N_\delta \geq 1$ such that

$$\|\mathbf{x}_n - \mathbf{a}\| < \delta \quad \text{whenever } n \geq N_\delta.$$

Then, from the statement (ii) we find

$$|f(\mathbf{x}_n) - \ell| < \varepsilon \quad \text{for all } n \geq N_\delta.$$

By Theorem 1.13, the above statement yields $f(\mathbf{x}_n) \to \ell$ as $n \geq 0$. In light of Definition 3.1, this means that f has limit ℓ at $\mathbf{a}$ and concludes our proof. $\square$

3.2 Two-Path Test for the Nonexistence of a Limit

Here comes the main difference between single and multivariable calculus. For functions of one real variable, the existence of the limit at a point is equivalent to the fact that the two sided limits (to the left and to the right at that point) exist and are equal. Instead, for functions of several variables $f : D \to \mathbb{R}$ there are infinitely many ways of approaching a point $\mathbf{a} \in D$ as Figure 3.1 suggests.

The fact that a point $\mathbf{a} \in \overline{D}$ can be approached through many paths allows us to derive the following criterion for the nonexistence of the limit.

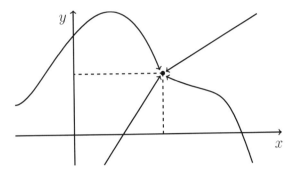

Figure 3.1
Infinitely many paths of approaching a point in $\mathbb{R}^2$.

Theorem 3.3 (Two-Path Test for the nonexistence of a limit)
  Let $f : D \subset \mathbb{R}^m \to \mathbb{R}$ *be a function of m variables and* $\mathbf{a} \in \overline{D}$. *If* $f(\mathbf{x})$
approaches two different values as $\mathbf{x}$ *approaches* $\mathbf{a}$ *along two different paths in*
D, *then* f *has no limit at* $\mathbf{a}$.

Proof As in the proof of Theorem 3.2, we argue by contradiction. Thus, we
assume that f has limit at $\mathbf{a}$. By the assumption in Theorem 3.3, there exist
two curves $\gamma, \overline{\gamma} \subset D$ that pass through $\mathbf{a}$ along which $f(\mathbf{x})$ approaches two
distinct values, say ℓ and $\overline{\ell}$. Thus, one may find two sequence $\{\mathbf{a}_n\} \subset \gamma \setminus \{\mathbf{a}\}$
and $\{\overline{\mathbf{a}}_n\} \subset \overline{\gamma} \setminus \{\mathbf{a}\}$ that converge to $\mathbf{a}$ and such that

$$\lim_{n \to \infty} f(\mathbf{a}_n) = \ell \quad \text{and} \quad \lim_{n \to \infty} f(\overline{\mathbf{a}}_n) = \overline{\ell}.$$

This contradicts exactly the Definition 3.1 of the limit at $\mathbf{a}$ and thus implies
that our assumption is false. Hence, f has no limit at $\mathbf{a}$.   □

Example 3.4 *Let*

$$f : \mathbb{R}^2 \to \mathbb{R}, \quad f(x,y) = \begin{cases} \dfrac{xy}{x^2 + y^2} & \text{if } (x,y) \neq (0,0) \\ 0 & \text{if } (x,y) = (0,0). \end{cases}$$

  (i) *Show that for all* $(a,b) \in \mathbb{R}^2$ *the functions* $f(a,\cdot) : \mathbb{R} \to \mathbb{R}$ *and*
$f(\cdot,b) : \mathbb{R} \to \mathbb{R}$ *are continuous functions of one real variable;*

  (ii) *Show that* f *has no limit at* $(0,0)$ *in the sense of Definition 3.1.*

Remark 3.5 This example shows that the function f is separately continu-
ous, that is, if we fix $a, b \in \mathbb{R}$, then $f(a, \cdot)$ and $f(\cdot, b)$ are continuous. However,
f is not jointly (or global) continuous as a function of two real variables.

Solution. (i) Observe that for any $a, b \in \mathbb{R}$, the functions $f(a, \cdot), f(\cdot, b) : \mathbb{R} \to \mathbb{R}$ are given by

$$f(a, y) = \begin{cases} 0 & \text{if } a = 0 \\ \frac{ay}{a^2+y^2} & \text{if } a \neq 0 \end{cases} \quad \text{and} \quad f(x, b) = \begin{cases} 0 & \text{if } b = 0 \\ \frac{bx}{x^2+b^2} & \text{if } b \neq 0. \end{cases}$$

Thus $f(a, \cdot), f(\cdot, b)$ are continuous on $\mathbb{R}$.

(ii) First we approach $(0,0)$ along the x axis. By taking $y = 0$ one has

$$f(x, y) = f(x, 0) = \frac{x \cdot 0}{x^2 + 0^2} = 0.$$

Hence

$$f(x, y) \to 0 \quad \text{as} \quad (x, y) \to (0,0) \quad \text{along the line } y = 0.$$

Next, we approach $(0,0)$ along the line $y = x$ which gives

$$f(x, y) = f(x, x) = \frac{x^2}{x^2 + x^2} = \frac{1}{2}.$$

Thus

$$f(x, y) \to \frac{1}{2} \quad \text{as} \quad (x, y) \to (0,0) \quad \text{along the line } y = x.$$

Hence, $f(x, y)$ approaches two different values as (x, y) approaches $(0,0)$ along two different paths. By Theorem 3.3, it follows that $f(x, y)$ has no limit at $(0,0)$.

Remark 3.6 To show that f has no limit at $(0,0)$ we could use the polar coordinates for the pair (x, y) namely $x = r \cos \theta, y = r \sin \theta$. Then

$$f(x, y) = f(r \cos \theta, r \sin \theta) = \frac{r^2 \cos \theta \sin \theta}{r^2 \cos^2 \theta + r^2 \sin^2 \theta} = \cos \theta \sin \theta = \frac{1}{2} \sin(2\theta).$$

This shows that $f(x, y)$ is constant along any ray that emanates from the origin and this constant depends on the angle between the ray and the x axis. Hence, f has no limit at the origin.

Example 3.7 *Let*

$$f : \mathbb{R}^2 \to \mathbb{R}, \quad f(x, y) = \begin{cases} \frac{y \sin(x - y)}{x^2 + y^4} & \text{if } (x, y) \neq (0,0) \\ 0 & \text{if } (x, y) = (0,0). \end{cases}$$

(i) *Show that the function f has a limit at $(0,0)$ along any line that passes through $(0,0)$.*

(ii) *Does f have limit at $(0,0)$ (in the sense of Definition 3.1)?*

Solution. (i) Any line passing through $(0,0)$ has the equation $y = mx$, $m \in \mathbb{R}$. Along such a line, we have

$$f(x,y) = f(x, mx) = \frac{mx \sin(1-m)x}{x^2 + m^4 x^4}.$$

If $m = 1$, then $f(x,y) = f(x,x) = 0 \to 0$ as $(x,y) \to (0,0)$.

If $m \neq 1$, using the fact that $(\sin t)/t \to 1$ as $t \to 0$, we deduce

$$f(x,y) = f(x, mx) = \frac{\sin(1-m)x}{(1-m)x} \cdot \frac{m(1-m)}{1 + m^4 x^2} \to m(1-m)$$

as $(x,y) \to (0,0)$, along the line $y = mx$.

(ii) Since along the lines $y = mx$ the function $f(x,y)$ approaches different values (depending on m), it follows that f has no limit at $(0,0)$.

Example 3.8 *Let*

$$f : \mathbb{R}^2 \to \mathbb{R}, \quad f(x,y) = \begin{cases} \dfrac{x^2 y}{x^4 + 3y^2} & \text{if } (x,y) \neq (0,0) \\ 0 & \text{if } (x,y) = (0,0). \end{cases}$$

Does f has a limit at $(0,0)$?

Solution. Note first that along any line $y = mx$, $m \in \mathbb{R}$, we have

$$f(x,y) = f(x, mx) = \frac{mx^3}{x^4 + 3m^2 x^2} = \frac{mx}{x^2 + 3m^2} \to 0 \quad \text{as } (x,y) \to (0,0).$$

Can we draw the conclusion that f has limit zero as $(x,y) \to (0,0)$? The answer is certainly NO, because in light of Theorem 3.3, we have not exhausted *all paths* that approach $(0,0)$. Then, how to determine a suitable path through which $f(x,y)$ tends to a nonzero value? The idea is to select a curve along which x and y bring the same contribution to the denominator of $f(x,y)$. Precisely, let us investigate the behaviour of $f(x,y)$ along the curve $x^4 = y^2$; as you see, we dropped the constant factor 3 at the denominator of $f(x,y)$. Solving $x^4 = y^2$ we are led to $y = \pm x^2$, so let us see what is the behaviour of $f(x,y)$ along the curve $y = x^2$ (we could have taken $y = -x^2$ as well!). Along $y = x^2$, we have

$$f(x,y) = f(x, x^2) = \frac{x^2 \cdot x^2}{x^4 + 3x^4} = \frac{1}{4}.$$

Thus,

$$f(x,y) \to \frac{1}{4} \quad \text{as} \quad (x,y) \to (0,0) \text{ along the curve } y = x^2.$$

We have seen that $f(x,y)$ approaches 0 along any straight line $y = mx$ and approaches $\frac{1}{4}$ along the parabola $y = x^2$ as $(x,y) \to (0,0)$. By Theorem 3.3, it follows that $f(x,y)$ has no limit at $(0,0)$.

Remark 3.9 Consider the parabola $y = mx^2$; we now approach $(0,0)$ along parabolas rather than straight lines. We find

$$f(x,y) = f(x, mx^2) = \frac{mx^4}{x^4 + m^2 x^4} = \frac{m}{1 + m^2} \to \frac{m}{1 + m^2} \quad \text{as } (x,y) \to (0,0).$$

Observe that these limits vary with the choice of m, so by the two-path test, $f(x,y)$ has no limit at $(0,0)$.

Example 3.10 *Show that the function*

$$f : \mathbb{R}^2 \to \mathbb{R}, \quad f(x,y) = \begin{cases} \dfrac{x(y+1)}{7(x-1)^6 + (y+1)^2} & \text{if } (x,y) \neq (1,-1) \\ 0 & \text{if } (x,y) = (1,-1), \end{cases}$$

has no limit at $(1,-1)$.

Solution. Following the idea in Example 3.8 we look for paths that pass through $(1,-1)$ such that $(x-1)^6 = (y+1)^2$, that is, $y = -1 \pm (x-1)^3$. Let us only focus on the path $y = -1 + (x-1)^3$ along which we have:

$$f(x,y) = f(x, -1 + (x-1)^3) = \frac{x(x-1)^3}{7(x-1)^6 + (x-1)^6} = \frac{x}{8(x-1)^3}.$$

From here we find

$$f(x,y) = \frac{x}{8(x-1)^3} \to +\infty$$

as $(x,y) \to (1,-1)$ along the curve $y = -1 + (x-1)^3$, $x > 1$. Further,

$$f(x,y) = \frac{x}{8(x-1)^3} \to -\infty$$

as $(x,y) \to (1,-1)$ along the curve $y = -1 + (x-1)^3$, $x < 1$. In light of Theorem 3.3 it follows that $f(x,y)$ has no limit at $(1,-1)$.

Exercises

Exercise 1. Some limits can be computed by reducing them to the limit of a single variable function. Using this fact compute the limits below:

(i) $\displaystyle\lim_{(x,y)\to(0,0)} \frac{e^{xy} - 1}{xy}$;

(iii) $\displaystyle\lim_{(x,y)\to(0,0)} \frac{\cos(xe^y) - 1}{x^2 e^{2y}}$;

(ii) $\displaystyle\lim_{(x,y)\to(0,0)} \frac{\ln(x^2 + y^2 + 1)}{x^2 + y^2}$;

(iv) $\displaystyle\lim_{(x,y)\to(0,0)} \frac{\sin(x^2 + y^2)}{x^2 + y^2}$.

Exercise 2. Using Example 3.8 and Remark 3.9 construct a function $f : \mathbb{R}^2 \to \mathbb{R}$ with the properties (i)–(iii) below.

(i) $f(x,y)$ tends to zero as $(x,y) \to (0,0)$ along each line $y = mx$, $m \in \mathbb{R}$;

(ii) $f(x,y)$ tends to zero as $(x,y) \to (0,0)$ along each parabola $y = mx^2$, $m \in \mathbb{R}$;

(iii) $f(x,y)$ tends to a nonzero value as $(x,y) \to (0,0)$ along $y = x^3$.

Does f have limit at $(0,0)$?

Exercise 3. Verify that the function

$$f : \mathbb{R}^2 \to \mathbb{R}, \quad f(x,y) = \begin{cases} \dfrac{3xy^2 \cos y}{x^2 + y^4} & \text{if } (x,y) \neq (0,0) \\ 0 & \text{if } (x,y) = (0,0), \end{cases}$$

has no limit at $(0,0)$.

Exercise 4. Let $f : \mathbb{R}^2 \to \mathbb{R}$ be given by

$$f(x,y) = \begin{cases} \dfrac{(x+1)^2 \sin(y-2)}{(x+1)^4 + (y-2)^2} & \text{if } (x,y) \neq (-1,2) \\ 0 & \text{if } (x,y) = (-1,2). \end{cases}$$

(i) Find two paths that pass through $(-1,2)$ along which f tends to two distinct values;

(ii) Deduce that f has no limit at $(-1,2)$.

Exercise 5. Let $A \subset \mathbb{R}^m$ be a nonempty open set and

$$f : \mathbb{R}^m \to \mathbb{R} \quad \text{be given by} \quad f(\mathbf{x}) = \begin{cases} 1 & \text{if } \mathbf{x} \in A \\ 0 & \text{if } \mathbf{x} \notin A. \end{cases}$$

(i) Check that f has limit at points $\mathbf{a} \in \mathbb{R}^m \setminus \partial A$ and has no limit at points $\mathbf{a} \in \partial A$;

(ii) Construct a counterexample to illustrate that (i) above fails to hold if A is not open.

3.3 Functions with Limit at a Point

In Section 3.2 we discussed the Two-Path Test method for the nonexistence of the limit. In the current section we shall discuss the existence of the limit of a function. The following theorem which is the Squeeze Theorem in several variables is the key approach to our study.

Theorem 3.11 (Squeeze Theorem)

Let $f, g, h : D \subset \mathbb{R}^m \to \mathbb{R}$ *be functions of m variables and* $\mathbf{a} \in \overline{D}$ *be such that:*

(i) $h(\mathbf{x}) \leq f(\mathbf{x}) \leq g(\mathbf{x})$ *for all* $\mathbf{x} \in D \setminus \{\mathbf{a}\}$;

(ii) *there exists* $\lim\limits_{\mathbf{x} \to \mathbf{a}} g(\mathbf{x}) = \lim\limits_{\mathbf{x} \to \mathbf{a}} h(\mathbf{x}) = \ell$.

Then $\lim\limits_{\mathbf{x} \to \mathbf{a}} f(\mathbf{x}) = \ell$.

Proof We want to deduce the conclusion of Theorem 3.11 by using the Definition 3.1 of the limit of a function. To this aim, let $\{\mathbf{a}_n\} \subset D \setminus \{\mathbf{a}\}$ be a convergent sequence to $\mathbf{a}$. Then, property (i) above states that $h(\mathbf{a}_n) \leq f(\mathbf{a}_n) \leq g(\mathbf{a}_n)$ for all $n \geq 1$. Also, property (ii) above shows that $\lim\limits_{n \to \infty} g(\mathbf{a}_n) = \lim\limits_{n \to \infty} h(\mathbf{a}_n) = \ell$. We may then apply the Squeeze Principle for the real valued sequences $\{f(\mathbf{a}_n)\}$, $\{g(\mathbf{a}_n)\}$ and $\{h(\mathbf{a}_n)\}$ to deduce $\lim\limits_{n \to \infty} f(\mathbf{a}_n) = \ell$. Since the sequence $\{\mathbf{a}_n\}$ that converges to $\mathbf{a}$ was arbitrarily chosen, it follows from Definition 3.1 that $\lim\limits_{\mathbf{x} \to \mathbf{a}} f(\mathbf{x}) = \ell$. □

Particularly important to our approach in the study of the limit for functions of several variables is the consequence below.

Corollary 3.12 Let $f, g : D \subset \mathbb{R}^m \to \mathbb{R}$ *be functions of m variables and* $\mathbf{a} \in \overline{D}$ *be such that:*

(i) $|f(\mathbf{x})| \leq g(\mathbf{x})$ *for all* $\mathbf{x} \in D \setminus \{\mathbf{a}\}$;

(ii) $\lim\limits_{\mathbf{x} \to \mathbf{a}} g(\mathbf{x}) = 0$.

Then $\lim\limits_{\mathbf{x} \to \mathbf{a}} f(\mathbf{x}) = 0$.

Proof Observe first that condition (i) above yields $-g(\mathbf{x}) \leq f(\mathbf{x}) \leq g(\mathbf{x})$ for all $x \in D \setminus \{\mathbf{a}\}$. We may now apply Theorem 3.11 with $h(\mathbf{x}) = -g(\mathbf{x})$ and the conclusion follows. □

Remark 3.13 Suppose we want to show that the limit of $f(\mathbf{x})$ as $\mathbf{x} \to \mathbf{a}$ is ℓ. We want to use Corollary 3.12 and we estimate $|f(\mathbf{x}) - \ell| \leq g(\mathbf{x})$ where the function $g(\mathbf{x})$ has limit 0 as $\mathbf{x} \to \mathbf{a}$. Very useful in such approach are the following inequalities:

$$\frac{A}{A+B} \leq 1 \quad \text{for all } A, B \geq 0, \ (A, B) \neq (0, 0), \tag{3.3}$$

$$|AB| \leq \frac{1}{2}\left(A^2 + B^2\right) \quad \text{for all } A, B \in \mathbb{R}. \tag{3.4}$$

Indeed, inequality (3.3) is obvious while inequality (3.4) is equivalent to

$$A^2 + B^2 \geq 2|AB| \iff (|A| - |B|)^2 \geq 0 \quad (\text{TRUE!}).$$

Example 3.14 *Let*

$$f : \mathbb{R}^2 \to \mathbb{R}, \quad f(x,y) = \begin{cases} \dfrac{5x^2y}{x^2 + y^2} & \text{if } (x,y) \neq (0,0), \\ 0 & \text{if } (x,y) = (0,0). \end{cases}$$

Show that $\lim\limits_{(x,y)\to(0,0)} f(x,y) = 0.$

Solution 1. We use inequality (3.3) as follows:

$$|f(x,y)| = \left| \frac{5x^2y}{x^2 + y^2} \right| = \underbrace{\frac{x^2}{x^2 + y^2}}_{\leq 1} \cdot 5|y| \leq 5|y| \to 0 \text{ as } (x,y) \to (0,0).$$

Hence, by taking $g(x,y) = 5|y|$ we fulfil the conditions in Corollary 3.12, namely:

$$\begin{cases} |f(x,y)| \leq g(x,y) & \text{for all } (x,y) \in \mathbb{R}^2 \setminus \{(0,0)\}, \\ \lim\limits_{(x,y)\to(0,0)} g(x,y) = 0. \end{cases} \tag{3.5}$$

By Corollary 3.12 it follows that $\lim\limits_{(x,y)\to(0,0)} f(x,y) = 0.$

Solution 2. By inequality (3.4) in Remark 3.13, we have

$$|xy| \leq \frac{1}{2}(x^2 + y^2).$$

Thus,

$$|f(x,y)| = \left| \frac{5x^2y}{x^2 + y^2} \right| = \underbrace{\frac{|xy|}{x^2 + y^2}}_{\leq 1/2} \cdot 5|x| \leq \frac{5}{2}|x| \to 0 \text{ as } (x,y) \to (0,0).$$

Hence, letting now $g(x,y) = \frac{5}{2}|x|$, we have that f and g fulfil again the conditions (3.5) above. By Corollary 3.12 the conclusion follows.

Remark 3.15 The above example shows that our main goal is to eliminate the denominator of $f(x,y)$ by inequalities of type (3.3) or (3.4) and then apply Corollary 3.12. Inequalities (3.3) or (3.4) suffice to achieve this goal. The following examples illustrate this fact.

Example 3.16 *Let*

$$f : \mathbb{R}^2 \to \mathbb{R}, \quad f(x,y) = \begin{cases} \dfrac{x^2 y^4}{x^4 + 8y^6} & \text{if } (x,y) \neq (0,0) \\ 0 & \text{if } (x,y) = (0,0). \end{cases}$$

Show that $\displaystyle\lim_{(x,y)\to(0,0)} f(x,y) = 0.$

Solution. We cannot use the inequality (3.3) in Remark 3.13 (why?) but inequality (3.4) for $A = x^2$ and $B = |y|^3$ produces

$$x^2 |y|^3 \leq \frac{1}{2}(x^4 + y^6) \quad \text{for all } (x,y) \in \mathbb{R}^2 \setminus \{(0,0)\}.$$

Now we estimate

$$|f(x,y)| = \frac{x^2 |y|^3}{x^4 + 8y^6} \cdot |y| \leq \underbrace{\frac{x^2 |y|^3}{x^4 + y^6}}_{\leq 1/2} \cdot |y| \leq \frac{|y|}{2} \to 0 \text{ as } (x,y) \to (0,0).$$

Letting $g(x,y) = \frac{|y|}{2}$, we have that f and g satisfy the conditions in Corollary 3.12 which allows us to conclude.

Example 3.17 *Let* $f : \mathbb{R}^2 \setminus \{(0,0)\} \to \mathbb{R}$, $f(x,y) = \dfrac{x^2(y+2) + y^2(3x+2)}{x^2 + y^2}$.
Show that $\displaystyle\lim_{(x,y)\to(0,0)} f(x,y) = 2.$

Solution. We want to follow the argument presented in Remark 3.13 and we first estimate

$$|f(x,y) - 2| = \left| \frac{x^2 y + 3xy^2}{x^2 + y^2} \right| \leq \frac{x^2}{x^2 + y^2} \cdot |y| + \frac{y^2}{x^2 + y^2} \cdot 3|x|.$$

By inequality (3.3) in Remark 3.13 one has

$$\frac{x^2}{x^2 + y^2} \leq 1 \quad \text{and} \quad \frac{y^2}{x^2 + y^2} \leq 1 \quad \text{for all } (x,y) \in \mathbb{R}^2 \setminus \{(0,0)\}.$$

Thus,

$$|f(x,y) - 2| \leq |y| + 3|x| \to 0 \text{ as } (x,y) \to (0,0).$$

Hence, by taking $g(x,y) = |y| + 3|x|$ it follows by Corollary 3.12 that $f(x,y) - 2$ tends to zero as $(x,y) \to (0,0)$, that is, $\displaystyle\lim_{(x,y)\to(0,0)} f(x,y) = 2.$

Important

(A) To show that a function $f(x, y)$ has no limit at (a, b):

    (i) try to find two paths approaching (a, b) along which f tends to two different values;

    (ii) if you are unsure whether or not f has a limit, experiment with a few paths approaching (a, b) as we did in Example 3.8.

(B) To show that a function $f(x, y)$ has limit ℓ at (a, b):

    (i) try first to use the substitution;

    (ii) if that does not work, try and estimate $|f(x, y) - \ell|$, that is, try and show that this is less than or equal to a quantity that tends to zero as $(x, y) \to (a, b)$. See the techniques in Example 3.17.

    The inequalities (3.3) or (3.4) are useful in this sense!

The result below states the properties of functions with the limit at a point.

Theorem 3.18 (Properties of functions with limit at a point)
Let $f, g : D \subset \mathbb{R}^m \to \mathbb{R}$ be two functions of m variables and $\mathbf{a} \in \overline{D}$. If $\lim\limits_{\mathbf{x} \to a} f(\mathbf{x}) = \ell$ and $\lim\limits_{\mathbf{x} \to a} g(\mathbf{x}) = L$, then:

    (i) $\lim\limits_{\mathbf{x} \to a} (f \pm g)(\mathbf{x}) = \ell \pm L$;

    (ii) $\lim\limits_{\mathbf{x} \to a} (cf)(\mathbf{x}) = c\ell$;

    (iii) $\lim\limits_{\mathbf{x} \to a} (f \cdot g)(\mathbf{x}) = \ell \cdot L$;

    (iv) $\lim\limits_{\mathbf{x} \to a} \max\{f, g\}(\mathbf{x}) = \max\{\ell, L\}$;

    (v) $\lim\limits_{\mathbf{x} \to a} \min\{f, g\}(\mathbf{x}) = \min\{\ell, L\}$;

    (vi) $\lim\limits_{\mathbf{x} \to a} |f|(\mathbf{x}) = |\ell|$;

    (vii) If $L \neq 0$, then $\lim\limits_{\mathbf{x} \to a} \dfrac{f}{g}(\mathbf{x}) = \dfrac{\ell}{L}$.

Proof Let $\{\mathbf{a}_n\} \subset D \setminus \{\mathbf{a}\} \subset \mathbb{R}^m$ be a sequence that converges to $\mathbf{a}$. By Definition 3.1, we have that $\lim\limits_{n \to \infty} f(\mathbf{a}_n) = \ell$ and $\lim\limits_{n \to \infty} g(\mathbf{a}_n) = L$.

By the properties of limit of sequences (see Theorem 1.14) it follows that

$$\lim_{n\to\infty} (f \pm g)(\mathbf{a}_n) = \lim_{n\to\infty} f(\mathbf{a}_n) \pm \lim_{n\to\infty} g(\mathbf{a}_n) = \ell \pm L,$$

$$\lim_{n\to\infty} (cf)(\mathbf{a}_n) = c \lim_{n\to\infty} f(\mathbf{a}_n) = c\ell,$$

$$\lim_{n\to\infty} (f \cdot g)(\mathbf{a}_n) = \left(\lim_{n\to\infty} f(\mathbf{a}_n) \right) \cdot \left(\lim_{n\to\infty} g(\mathbf{a}_n) \right) = \ell \cdot L,$$

$$\lim_{n\to\infty} \max\{f,g\}(\mathbf{a}_n) = \lim_{n\to\infty} \max\{f(\mathbf{a}_n), g(\mathbf{a}_n)\} = \max\{\ell, L\},$$

$$\lim_{n\to\infty} \min\{f,g\}(\mathbf{a}_n) = \lim_{n\to\infty} \min\{f(\mathbf{a}_n), g(\mathbf{a}_n)\} = \min\{\ell, L\},$$

$$\lim_{n\to\infty} |f|(\mathbf{a}_n) = \lim_{n\to\infty} |f(\mathbf{a}_n)| = |\ell|.$$

Further, if $L \neq 0$ we also have

$$\lim_{n\to\infty} \left(\frac{f}{g}\right)(\mathbf{a}_n) = \lim_{n\to\infty} \frac{f(\mathbf{a}_n)}{g(\mathbf{a}_n)} = \frac{\ell}{L}.$$

The conclusion follows now from Definition 3.1. □

Exercises

Exercise 6. Decide whether the following functions have limit at $(0,0)$:

(i) $f(x,y) = \dfrac{\sin(xy^2)}{x^2 + y^2}$;

(ii) $f(x,y) = \dfrac{x^3 + y^3}{x^2 + y^2}$;

(iii) $f(x,y) = \dfrac{x^2 \sin y}{x^4 + 2y^2}$;

(iv) $f(x,y) = \dfrac{\ln(x^2 + y^2 + 1)}{x^2 + 2y^2}$;

(v) $f(x,y) = \dfrac{\sin(xy)}{\sqrt{x^2 + y^2}}$;

(vi) $f(x,y) = \dfrac{x^2(y+1) + y^2(x+1)}{x^2 + y^2}$.

Exercise 7. Let $f : \mathbb{R}^2 \setminus \{(0,0)\} \to \mathbb{R}$, $f(x,y) = \dfrac{x^k y^3}{x^4 + 3y^6}$ where $k > 0$ is an integer.

(i)   Show that for $k = 1$ and $k = 2$ the function f has no limit at $(0,0)$;

(ii) Show that $\lim\limits_{(x,y)\to(0,0)} f(x,y) = 0$ for all $k \geq 3$.

Exercise 8. Let $f : \mathbb{R}^2 \setminus \{(0,0)\} \to \mathbb{R}$, $f(x,y) = \dfrac{x^4 y^5}{2x^6 + y^{12}}$.

(i)   Show that $2x^6 + y^{12} \geq 3x^4 y^4$ for all $x, y \in \mathbb{R}$;

(ii) Show that $\lim\limits_{(x,y)\to(0,0)} f(x,y) = 0$. Can we apply Remark 3.13?

Exercise 9. Check whether the following functions have limit at $(0,0,0)$:

(i) $f(x,y,z) = \dfrac{(x+y+z)^2}{x^2 + y^2 + z^2}$;   (ii) $f(x,y,z) = \dfrac{(x+y+z)^2 z}{x^2 + y^2 + z^2}$.

Exercise 10. Let $f : \mathbb{R}^2 \to \mathbb{R}$, $f(x,y) = \dfrac{x^a y^b}{x^4 + y^2}$.

For which integers $a \geq 0$ and $b \geq 0$ does $\lim\limits_{(x,y)\to(0,0)} f(x,y)$ exist?

Exercise 11. Find $c \in \mathbb{R}$ such that $f(x,y) = \dfrac{x(y+c)^2}{x^2 + (y+1)^2}$ has limit at $(0,-1)$.

Exercise 12. Let $f,g : D \subset \mathbb{R}^m \to \mathbb{R}$ and $\mathbf{a} \in D$ be such that

(i) g is *bounded*, that is $|g(\mathbf{x})| \leq M$ for all $\mathbf{x} \in D$, where $M > 0$ is a constant;

(ii) $\lim\limits_{\mathbf{x}\to\mathbf{a}} f(\mathbf{x}) = 0$.

Check that $\lim\limits_{\mathbf{x}\to\mathbf{a}} (fg)(\mathbf{x}) = 0$.

3.4   Continuous Functions

Definition 3.19 *Let $f : D \subset \mathbb{R}^m \to \mathbb{R}$ be a function of m variables and $\mathbf{a} \in D$. We say that f is continuous at $\mathbf{a}$ if $\lim\limits_{\mathbf{x}\to\mathbf{a}} f(\mathbf{x}) = f(\mathbf{a})$.*

We say that f is continuous on D if it is continuous at every point $\mathbf{a} \in D$.

To be continuous at a point $\mathbf{a}$ we need to fulfil three conditions:

- $\mathbf{a} \in D$ and not just $\mathbf{a} \in \overline{D}$ as we require in Definition 3.1. Thus, f must be defined at $\mathbf{a}$;

- f has limit at $\mathbf{a}$;

- the limit is precisely equal to $f(\mathbf{a})$.

Example 3.20 *The function $f : \mathbb{R}^2 \to \mathbb{R}$ given in Example 3.14 is continuous at $(0,0)$ because we showed that $\lim\limits_{(x,y)\to(0,0)} f(x,y) = 0 = f(0,0)$.*

Example 3.21 *The function $f : \mathbb{R}^m \to \mathbb{R}$ given by $f(\mathbf{x}) = \|\mathbf{x}\|$ is continuous on $\mathbb{R}^m$.*

Solution. Let $\mathbf{a} \in \mathbb{R}^m$. By Theorem 1.14(iv), it follows that for any sequence $\{\mathbf{a}_n\} \subset \mathbb{R}^m \setminus \{\mathbf{a}\}$, we have

$$\lim_{n\to\infty} f(\mathbf{a}_n) = \lim_{n\to\infty} \|\mathbf{a}_n\| = \|\mathbf{a}\|.$$

Hence, by Definition 3.1 one has $\lim\limits_{\mathbf{x}\to\mathbf{a}} f(\mathbf{x}) = \|\mathbf{a}\| = f(\mathbf{a})$, that is, f is continuous at $\mathbf{a}$. By the arbitrariness of $\mathbf{a}$ we conclude that f is continuous on $\mathbb{R}^m$.

Theorem 3.22 (Properties of continuous functions)
  Let $f, g : D \subset \mathbb{R}^m \to \mathbb{R}$ be two functions of m variables which are continuous at $\mathbf{a} \in D$. Then:

(i) *The functions*

$$f \pm g, cf, f \cdot g, |f|, \max\{f, g\}, \min\{f, g\}$$

are continuous at $\mathbf{a}$, where $c \in \mathbb{R}$.

(ii) *If $g(\mathbf{a}) \neq 0$, then $\dfrac{f}{g}$ is continuous at $\mathbf{a}$;*

(iii) *If $h : f(D) \to \mathbb{R}$ is a continuous function at $f(\mathbf{a})$, then $h \circ f : D \to \mathbb{R}$ is continuous at $\mathbf{a}$.*

Proof Properties (i) and (ii) follow from Theorem 3.18. To check (iii) let $\{\mathbf{a}_n\} \subset D \setminus \{\mathbf{a}\} \subset \mathbb{R}^m$ be a sequence that converges to $\mathbf{a}$ as $n \to \infty$. Since f is continuous at $\mathbf{a}$, one has that $\{f(\mathbf{a}_n)\} \subset \mathbb{R}$ converges to $f(\mathbf{a})$. Then, $h(f(\mathbf{a}_n))$ equals $h(f(\mathbf{a}))$ or forms a sequence that converges to $h(f(\mathbf{a}))$. In either case the sequence $\{h(f(\mathbf{a}_n))\}$ converges to $h(f(\mathbf{a}))$ which yields

$$\lim_{\mathbf{x}\to\mathbf{a}} (h \circ f)(\mathbf{x}) = h(f(\mathbf{a})) = (h \circ f)(\mathbf{a}),$$

that is, $h \circ f$ is continuous at $\mathbf{a}$.   □
   The property (iii) fails to hold if we assume h to have limit instead of being continuous at $f(\mathbf{a})$ (see Exercise 14).

Remark 3.23 All elementary functions (polynomials, rational, trigonometric, exponential and logarithmic functions) are continuous on their domain of definition. This means that we can compute their limit at a point by direct substitution.

For instance

$$\lim_{(x,y)\to(-1,2)} (x^2y - 3xy^3 - 4x) = (-1)^2 \cdot 2 - 3 \cdot (-1) \cdot 2^3 - 4 \cdot (-1) = 30,$$

because $f(x,y) = x^2y - 3xy^3 - 4x$ is a polynomial function, thus a continuous one, hence the limit at any point in the domain of definition (which is $\mathbb{R}^2$) is computed by direct substitution.

Example 3.24 *Find all points $(x,y) \in \mathbb{R}^2$ at which the function $f(x,y) = \ln(x^2 + y^2)$ is continuous.*

Solution. The function f is the composition of two elementary functions: the logarithmic function ln and the polynomial function $x^2 + y^2$. By Theorem 3.22(iii), it follows that f is continuous on its domain D of definition, which is the set of all points $(x,y) \in \mathbb{R}^2$ such that $x^2 + y^2 > 0$. Hence, f is continuous on $D = \mathbb{R}^2 \setminus \{(0,0)\}$.

One important property of continuous functions that we will use in this book is the following.

Theorem 3.25 *Let $f : D \subset \mathbb{R}^m \to \mathbb{R}$ be a function which is continuous at $\mathbf{a} \in D$ and such that $f(\mathbf{a}) > 0$. Then, there exists a ball $B_r(\mathbf{a}) \subset D$ and $\varepsilon > 0$ such that*

$$f(\mathbf{x}) > \varepsilon \quad \text{for all} \quad \mathbf{x} \in B_r(\mathbf{a}).$$

Proof Using Theorem 3.2, for all $\varepsilon > 0$ there exists $\delta > 0$ such that

$$|f(\mathbf{x}) - f(\mathbf{a})| < \varepsilon \quad \text{whenever } \mathbf{x} \in D, \|\mathbf{x} - \mathbf{a}\| < \delta.$$

Take $\varepsilon = f(\mathbf{a})/2 > 0$. Then, the above statement yields

$$|f(\mathbf{x}) - f(\mathbf{a})| < \frac{f(\mathbf{a})}{2} \quad \text{whenever } \mathbf{x} \in D \cap B_\delta(\mathbf{a}).$$

By taking $\delta > 0$ small enough, we may assume $B_\delta(\mathbf{a}) \subset D$ and thus, the above statements reads

$$-\frac{f(\mathbf{a})}{2} < f(\mathbf{x}) - f(\mathbf{a}) < \frac{f(\mathbf{a})}{2} \quad \text{whenever } \mathbf{x} \in B_\delta(\mathbf{a}) \subset D.$$

This implies

$$f(\mathbf{x}) > \varepsilon = \frac{f(\mathbf{a})}{2} \quad \text{whenever } \mathbf{x} \in B_\delta(\mathbf{a}) \subset D.$$

□

Exercises

Exercise 13. Let $f : \mathbb{R}^2 \to \mathbb{R}$, $f(x,y) = \begin{cases} 0 & \text{if } xy \neq 0 \\ 1 & \text{if } xy = 0. \end{cases}$

At which points in the plane is f continuous?

Exercise 14. Let $f : \mathbb{R}^2 \to \mathbb{R}$ and $h : \mathbb{R} \to \mathbb{R}$ be given by

$$f(x,y) = \begin{cases} 0 & \text{if } x \leq 0 \\ x & \text{if } x > 0 \end{cases} \quad \text{and} \quad h(t) = \begin{cases} 0 & \text{if } t = 0 \\ 1 & \text{if } t \neq 0. \end{cases}$$

(i) Compute $\lim\limits_{(x,y)\to(0,0)} f(x,y)$ and $\lim\limits_{t\to 0} h(t)$.

(ii) Compute $h \circ f$ and show that $\lim\limits_{(x,y)\to(0,0)} (h \circ f)(x,y)$ does not exist.

Does this fact contradict Theorem 3.22(iii)?

Exercise 15. Find all points in the plane at which the following functions are continuous:

(i) $f(x,y) = \sqrt{x^2 - xy + y^2}$;

(ii) $f(x,y) = \dfrac{\ln(x^2 + 1)}{2 + \cos(xy)}$;

(iii) $f(x,y) = \ln(x^2 - y^2)$.

Exercise 16. Prove that the level sets of a continuous function $f : D \subset \mathbb{R}^m \to \mathbb{R}$ are closed. Hint: Use Theorem 1.19.

Exercise 17. Let $f : \mathbb{R}^2 \to \mathbb{R}$ be a function of two variables. Prove or disprove (by constructing a counterexample) the statements below:

(i) If e^f is continuous at $(a,b) \in \mathbb{R}^2$, then f is also continuous at (a,b);

(ii) If $\cos(f)$ is continuous at $(a,b) \in \mathbb{R}^2$, then f is also continuous at (a,b).

Exercise 18. Using the approach in Theorem 3.25 prove the following statement for functions which have a positive limit at a point. If $f : D \subset \mathbb{R}^m \to \mathbb{R}$ and $\mathbf{a} \in D$ is such that $\lim\limits_{\mathbf{x}\to\mathbf{a}} f(\mathbf{x}) > 0$, then there exists a ball $B_r(\mathbf{a}) \subset D$ and $\varepsilon > 0$ such that

$$f(\mathbf{x}) > \varepsilon \quad \text{for all } \mathbf{x} \in B_r(\mathbf{a}) \setminus \{\mathbf{a}\}.$$

3.5 Continuous Extensions

Definition 3.26 *Let $f : D \setminus \{\mathbf{a}\} \subset \mathbb{R}^m \to \mathbb{R}$ be a function of m variables such that there exists*

$$\ell = \lim_{\mathbf{x}\to\mathbf{a}} f(\mathbf{x}) \in \mathbb{R}.$$

Then the function

$$\widetilde{f} : D \to \mathbb{R}, \quad \widetilde{f}(\mathbf{x}) = \begin{cases} f(\mathbf{x}) & \text{if } \mathbf{x} \in D \setminus \{\mathbf{a}\} \\ \ell & \text{if } \mathbf{x} = \mathbf{a} \end{cases}$$

is called the continuous extension of f at **a**.

Example 3.27 *Let* $D = \{(x, y) \in \mathbb{R}^2 : 0 < x^2 + y^2 < 1\}$ *and*

$$f : D \to \mathbb{R}, \quad f(x, y) = \frac{x^2 + y^2}{\sqrt{4 - 3(x^2 + y^2)} - 2}.$$

Find the continuous extension of f at $(0, 0)$.

Solution. To have a continuous extension at $(0, 0)$, there must exist

$$\ell = \lim_{(x,y) \to (0,0)} f(x, y) \in \mathbb{R}.$$

Hence, we compute

$$\ell = \lim_{(x,y) \to (0,0)} f(x, y) = \lim_{(x,y) \to (0,0)} \frac{x^2 + y^2}{\sqrt{4 - 3(x^2 + y^2)} - 2}$$

$$= \lim_{t = x^2 + y^2 \to 0} \frac{t}{\sqrt{4 - 3t} - 2}$$

$$= \lim_{t \to 0} \frac{t(\sqrt{4 - 3t} + 2)}{(\sqrt{4 - 3t} - 2)(\sqrt{4 - 3t} + 2)}$$

$$= \lim_{t \to 0} \frac{t(\sqrt{4 - 3t} + 2)}{(4 - 3t) - 4}$$

$$= \lim_{t \to 0} \frac{\sqrt{4 - 3t} + 2)}{-3} = -\frac{4}{3}.$$

Now, the continuous extension $\widetilde{f}$ of f at $(0, 0)$ is given by

$$\widetilde{f} : D \cup \{(0, 0)\} \to \mathbb{R}, \quad \widetilde{f}(x, y) = \begin{cases} \dfrac{x^2 + y^2}{\sqrt{4 - 3(x^2 + y^2)} - 2} & \text{if } (x, y) \in D \\ -\dfrac{4}{3} & \text{if } (x, y) = 0. \end{cases}$$

Example 3.28 *Let* $f(x, y) = e^{\frac{1}{x^2} - \frac{1}{y^2}}$

    (i) *Show that f has no limit at* $(0, 0)$;

    (ii) *Show that f has a continuous extension at any points* $(a, 0)$, $a \neq 0$;

    (iii) *Has f a continuous extension at* $(0, 1)$?

Solution. (i) Along the line $y = x$, we have $f(x,y) = f(x,x) = e^0 = 1 \to 1$ as $(x,y) \to (0,0)$. Along the line $y = 2x$, we have $f(x,y) = f(x,2x) = e^{\frac{3}{4x^2}} \to \infty$ as $(x,y) \to (0,0)$. Hence, f has no limit at $(0,0)$.

(ii) Let $a \in \mathbb{R}$, $a \neq 0$ and $r = \frac{|a|}{2} > 0$. If $(x,y) \in B_r(a,0)$, then $|x-a| < \frac{|a|}{2}$, so $a - \frac{|a|}{2} < x < a + \frac{|a|}{2}$. Looking at the cases $a > 0$ and $a < 0$ we see that $x^2 > \frac{a^2}{4}$. Then,

$$|f(x,y)| = e^{\frac{1}{x^2}} \cdot e^{-\frac{1}{y^2}} \le e^{\frac{4}{a^2}} \cdot e^{-\frac{1}{y^2}} \to 0 \qquad \text{as} \quad (x,y) \to (a,0),$$

since $e^{-\frac{1}{y^2}} \to 0$ as $y \to 0$. Thus, by Corollary 3.12 we deduce $\lim_{(x,y)\to(a,0)} f(x,y) = 0$ which shows that f has a continuous extension at $(a,0)$.

(iii) If f had a continuous extension at $(0,1)$, then $\lim_{(x,y)\to(0,1)} f(x,y)$ would exist as a real number. However, along the line $y = 1$, we have $f(x,y) = f(x,1) \to \infty$ as $(x,y) \to (0,1)$. Hence, f has no continuous extension at this point.

Exercises

Exercise 19. Check that the function

$$f : \mathbb{R}^2 \setminus \{(1,0)\} \to \mathbb{R}, \quad f(x,y) = \frac{(x-1)\sin^3 y}{(x-1)^2 + y^4}$$

has a continuous extension at $(1,0)$.

Exercise 20. Let $f : \mathbb{R}^2 \setminus \{(2,0)\} \to \mathbb{R}$, $f(x,y) = \dfrac{xy + ax + by + c}{\sqrt{(x-2)^2 + y^2}}$.

(i) Explain why f is continuous on $\mathbb{R}^2 \setminus \{(2,0)\}$;

(ii) Determine $a, b, c \in \mathbb{R}$ such that f has a continuous extension at $(2,0)$.

Exercise 21. Determine which of the following functions admit a continuous extension at the origin:

(i) $f : \mathbb{R}^2 \setminus \{(0,0)\} \to \mathbb{R}$, $f(x,y) = \dfrac{\sin(x^2 + y^4)}{\sqrt{x^4 + y^2}}$;

(ii) $f : \mathbb{R}^3 \setminus \{(0,0,0)\} \to \mathbb{R}$, $f(x,y,z) = \dfrac{xyz}{x^2 + y^2 + z^2}$;

(iii) $f : \mathbb{R}^m \setminus \{\mathbf{0}\} \to \mathbb{R}$, $f\mathbf{x}) = \dfrac{1}{\|\mathbf{x}\|} e^{-\frac{1}{\|\mathbf{x}\|}}$.

Exercise 22. Determine all real numbers $a > 0$ such that the function

$$f : \mathbb{R}^2 \setminus \{(0,0)\} \to \mathbb{R}, \quad f(x,y) = \frac{\sin\left(|x|^a \sin y\right)}{x^2 + y^2}$$

has a continuous extension at $(0,0)$.

Exercise 23. Let $k > 0$ be a real number and

$$D = \{(x,y) \in \mathbb{R}^2 : 0 < x^2 + y^2 < 1\}.$$

Assume that $f : D \to \mathbb{R}$ is a function that satisfies

$$|f(x,y) - 2| \leq \frac{x^2 |y|^k}{x^4 + y^4} \quad \text{for all } (x,y) \in D.$$

(i) Check that if $k = 4$, then any function with the above property has a continuous extension at $(0,0)$;

(ii) Assume $k = 2$. Construct two functions which satisfy the above condition of which one has a continuous extension at $(0,0)$ and the other has not.

4

Differentiable Functions

This chapter discusses the differentiability of functions of several variables. The chapter introduces first the partial derivatives, the equation of the tangent plane and the normal line to a surface in $\mathbb{R}^3$. Next, the focus is on the differentiability which is further extended to vector valued functions.

4.1 Partial Derivatives

Recall that if $I \subset \mathbb{R}$ is an open interval on the real line and $f : I \to \mathbb{R}$, then f is differentiable at $a \in I$ if the following limit exists as a real number:

$$\lim_{t \to 0} \frac{f(a+t) - f(a)}{t}. \tag{4.1}$$

The value of the above limit is called *the derivative of f at a* and it is denoted by $f'(a)$ or $\frac{df}{dx}(a)$.

Let now $D \subset \mathbb{R}^2$ be an open set, $f : D \to \mathbb{R}$ be a function of two variables and $(a, b) \in D$. The counterpart of (4.1) reads

$$\lim_{t \to 0} \frac{f(a+t, b) - f(a, b)}{t} \quad \text{and} \quad \lim_{t \to 0} \frac{f(a, b+t) - f(a, b)}{t}.$$

We shall call the above limits (if they exist as real numbers) the *partial derivatives of f at (a, b) with respect to x and y variable*. More precisely,

$$\frac{\partial f}{\partial x}(a, b) = \lim_{t \to 0} \frac{f(a+t, b) - f(a, b)}{t} \tag{4.2}$$

$$\frac{\partial f}{\partial y}(a, b) = \lim_{t \to 0} \frac{f(a, b+t) - f(a, b)}{t}. \tag{4.3}$$

Observe that

$$\frac{\partial f}{\partial x}(a, b) = \frac{df}{dt}(t, b)\Big|_{t=a} \quad \text{and} \quad \frac{\partial f}{\partial y}(a, b) = \frac{df}{dt}(a, t)\Big|_{t=b}. \tag{4.4}$$

Let $f : D \subset \mathbb{R}^2 \to \mathbb{R}$. The equalities (4.2)–(4.3) alone are not enough to quantify the differentiability of a function of two variables. Note that the existence of partial derivatives of f at (a, b) means that the one-variable functions

DOI: 10.1201/9781003449652-4

$f(a, \cdot)$ and $f(\cdot, b)$ are differentiable. Just us separate continuity does not imply jointly continuity of a function (see Remark 3.5), a similar conclusion applies in the case of differentiability. We shall come back to this matter in Section 4.3.

To present the above definition in the general context of m variables, recall that for $1 \leq j \leq m$, $\mathbf{e}_j$ denotes the standard unit vector in $\mathbb{R}^m$ with 1 on the j-th entry and 0 elsewhere. As in the previous chapter, D denotes an open set in $\mathbb{R}^m$ (or, depending the context, in $\mathbb{R}^2$).

Definition 4.1 *Let $f : D \subset \mathbb{R}^m \to \mathbb{R}$ be a function of m variables and $\mathbf{a} \in D$. The partial derivative of f with respect to its j-th variable at $\mathbf{a}$ (if exists, as a real number) is given by*

$$\frac{\partial f}{\partial x_j}(\mathbf{a}) = \lim_{t \to 0} \frac{f(\mathbf{a} + t\mathbf{e}_j) - f(\mathbf{a})}{t}. \tag{4.5}$$

The equalities (4.2)–(4.3) and more generally (4.5) give us a precise rule on how to compute these partial derivatives:

- To find $\frac{\partial f}{\partial x}$, regard y as a constant and differentiate $f(x, y)$ with respect to x;

- To find $\frac{\partial f}{\partial y}$, regard x as a constant and differentiate $f(x, y)$ with respect to y;

- More generally, if f is a function of m variables, to find $\frac{\partial f}{\partial x_j}$ we regard all other variables x_k, $1 \leq k \leq m$, $k \neq j$ as constants and differentiate f with respect to x_j.

Example 4.2 *Let $f(x, y) = x^2 + y^5 - 2xe^{3y}$. Find $\dfrac{\partial f}{\partial x}$ and $\dfrac{\partial f}{\partial y}$.*

Solution. Using the above rule, we find

$$\frac{\partial f}{\partial x}(x, y) = 2x - 2e^{3y} \quad \text{and} \quad \frac{\partial f}{\partial y}(x, y) = 5y^4 - 6xe^{3y}.$$

It is important to note that there is no relationship between the continuity (or the limit) of a function at a point and the existence of partial derivatives at that point, as the following examples suggest.

Example 4.3 *Let*

$$f : \mathbb{R}^2 \to \mathbb{R}, \quad f(x, y) = \begin{cases} \dfrac{xy}{x^2 + y^2} & \text{if } (x, y) \neq (0, 0) \\ 0 & \text{if } (x, y) = (0, 0). \end{cases}$$

Then, f has no limit at $(0,0)$ but $\frac{\partial f}{\partial x}(0,0)$ and $\frac{\partial f}{\partial y}(0,0)$ exist and are both zero.

Solution. We have seen in Example 3.4 that f has no limit at $(0,0)$. To compute the partial derivatives we use the Definition 4.1 (more precisely (4.2)–(4.3)) as follows

$$\frac{\partial f}{\partial x}(0,0) = \lim_{t \to 0} \frac{f(0+t,0) - f(0,0)}{t} = \lim_{t \to 0} \frac{0-0}{t} = 0.$$

Similarly,

$$\frac{\partial f}{\partial y}(0,0) = \lim_{t \to 0} \frac{f(0,0+t) - f(0,0)}{t} = \lim_{t \to 0} \frac{0-0}{t} = 0.$$

Let us also note that if $(x,y) \neq (0,0)$, then by the quotient rule we compute

$$\frac{\partial f}{\partial x}(x,y) = \frac{y(y^2 - x^2)}{(x^2 + y^2)^2} \quad \text{and} \quad \frac{\partial f}{\partial x}(x,y) = \frac{x(x^2 - y^2)}{(x^2 + y^2)^2}.$$

Example 4.4 *Let $f : \mathbb{R}^2 \to \mathbb{R}$, $f(x,y) = \sqrt{x^2 + y^2}$. Then, f is continuous at $(0,0)$ but $\frac{\partial f}{\partial x}(0,0)$ and $\frac{\partial f}{\partial y}(0,0)$ do not exist.*

Solution. From Example 3.21, we have that f is continuous on the whole $\mathbb{R}^2$. Further,

$$\lim_{t \to 0} \frac{f(0+t,0) - f(0,0)}{t} = \lim_{t \to 0} \frac{|t|}{t}$$

which does not exist (by taking sided limits as $t \to 0^+$ and as $t \to 0^-$). This shows that $\frac{\partial f}{\partial x}(0,0)$ does not exist and the same conclusion holds for $\frac{\partial f}{\partial y}(0,0)$.

Definition 4.5 *Let $f : D \subset \mathbb{R}^m \to \mathbb{R}$ be a function of m variables which has partial derivatives at $\mathbf{a} \in D$. The gradient of f at $\mathbf{a}$ is a vector in $\mathbb{R}^m$ defined by*

$$\nabla f(\mathbf{a}) := \left(\frac{\partial f}{\partial x_1}(\mathbf{a}), \frac{\partial f}{\partial x_2}(\mathbf{a}), \ldots, \frac{\partial f}{\partial x_m}(\mathbf{a}) \right).$$

The notation $\operatorname{grad} f$ is also used instead of ∇f.

For instance, if $f(x,y) = x^2 e^{-2y}$, then $\nabla f(x,y) = (2xe^{-2y}, -2x^2 e^{-2y})$.

Definition 4.6 *Let $f : D \subset \mathbb{R}^m \to \mathbb{R}$ be a function of m variables which has partial derivatives at $\mathbf{a} \in D$. Then $\mathbf{a}$ is called a critical point if $\nabla f(\mathbf{a}) = \mathbf{0}$; otherwise $\mathbf{a}$ is called a regular point of f.*

Theorem 4.7 *Let $f : D \subset \mathbb{R}^m \to \mathbb{R}$ be a function of m variables and $\mathbf{a} \in D$. Assume that:*

(i) *The partial derivatives $\dfrac{\partial f}{\partial x_j}$ exist on $D \setminus \{\mathbf{a}\}$, for all $1 \leq j \leq m$;*

(ii) *There exists a real number $M > 0$ such that*

$$\left| \frac{\partial f}{\partial x_j}(\mathbf{x}) \right| \leq M \quad \text{for all } x \in D \setminus \{\mathbf{a}\} \text{ and } 1 \leq j \leq m.$$

Then f is continuous at **a**.

Proof The strategy we adopt in proving this result will be encountered several times in this chapter and the reader is kindly advised to acquire it. For the clarity of the exposition, we shall assume $m = 2$, so $D \subset \mathbb{R}^2$ is an open set and the point at which we want to derive the continuity is $(a, b) \in D$. The main ingredient is the mighty Mean Value Theorem for one-variable functions.

To begin with, let us use the fact that D is open. Thus, there exists $r > 0$ such that the ball $B_r(a, b)$ is contained in D. This allows us to say that for any (x, y) in $B_r(a, b) \setminus \{(a, b)\}$, the segment line that joins (x, y) and (a, b) lies in the ball $B_r(a, b)$ (and thus in D). Let us proceed by estimating

$$|f(x, y) - f(a, b)| \leq |f(x, y) - f(a, y)| + |f(a, y) - f(a, b)|. \qquad (4.6)$$

We may assume $x \neq a$, $y \neq b$, otherwise there is no need to use the inequality (4.6) in the following argument. By assumption (i), the function $f(\cdot, y)$ is continuous on the closed interval with endpoints at a and x and differentiable on the interior of this interval. Thus, there exists $c_{x,y}$ between x and a such that

$$f(x, y) - f(a, y) = \frac{\partial f}{\partial x}(c_{x,y}, y)(x - a).$$

Similarly, there exists d_y between y and b such that

$$f(a, y) - f(a, b) = \frac{\partial f}{\partial y}(a, d_y)(y - b).$$

We now plug the above two equalities in (4.6). Using (ii) we deduce

$$|f(x, y) - f(a, b)| \leq M(|x - a| + |y - b|) \quad \text{for all } (x, y) \in B_r(a, b).$$

It remains now to apply Corollary 3.12 for $g(x, y) = M(|x - a| + |y - b|)$ which fulfils $g(x, y) \to 0$ as $(x, y) \to (a, b)$. Hence, $f(x, y) \to f(a, b)$ as $(x, y) \to (a, b)$, and thus f is continuous at (a, b). $\qquad \square$

Exercises

Exercise 1. Find the gradient of the following functions:

(i) $f(x, y) = e^x \cos(x^2 y)$;

(ii) $f(x, y, z) = \dfrac{e^{-3y}}{1 + x^2 z^4}$;

(iii) $f(x, y, z) = yz \ln(x^2 + z^2 + 1)$;

(iv) $f(x_1, x_2, x_3, x_4) = \dfrac{x_2 x_3}{1 + x_1^2 + x_4^2}$.

Exercise 2.

(i) Let $f(x, y) = \ln(x^2 + y^2)$.

Find (x, y) such that $\nabla f(x, y) = \left(\frac{2}{5}, -\frac{6}{5}\right)$;

(ii) Let $f(x, y, z) = x^2 y + \dfrac{z}{y}$.

Find (x, y, z) such that $\nabla f(x, y, z) = (2, -1, -1)$.

Exercise 3. Let $m \geq 2$. At which points in $\mathbb{R}^m$ does the norm function $f : \mathbb{R}^m \to \mathbb{R}$, $f(\mathbf{x}) = \|\mathbf{x}\|$ have partial derivatives?

Exercise 4. Let

$$f : \mathbb{R}^3 \to \mathbb{R}, \quad f(x, y, z) = \begin{cases} \dfrac{x^3 + y^4 + z^2}{x^2 + y^2 + z^2} & \text{if } (x, y, z) \neq (0, 0, 0) \\ 0 & \text{if } (x, y, z) = (0, 0, 0). \end{cases}$$

Using the definition of partial derivatives, verify that $\dfrac{\partial f}{\partial x}(0, 0, 0)$ and $\dfrac{\partial f}{\partial y}(0, 0, 0)$ exist but $\dfrac{\partial f}{\partial z}(0, 0, 0)$ does not.

Exercise 5. (i) Retake the argument in the proof of Theorem 4.7 and write down the details in the general case $m \geq 2$;

(ii) Verify that Theorem 4.7 does not hold in the case $m = 1$ by constructing a counterexample.

Exercise 6. Let $f : B_r(\mathbf{a}) \subset \mathbb{R}^m \to \mathbb{R}$, $m \geq 1$, be a function which admits partial derivative at every point in its domain of definition and such that $\nabla f(\mathbf{x}) = 0$ for all $\mathbf{x} \in B_r(\mathbf{a})$. Using the method of Theorem 4.7, check that f is constant.

Exercise 7. Let $f : \mathbb{R}^2 \to \mathbb{R}$ be a function which admits partial derivatives on $\mathbb{R}^2$. Using Exercise 6 above find f, knowing that for all $(x, y) \in \mathbb{R}^2$ one has

$$\frac{\partial f}{\partial x}(x, y) = 4 \quad \text{and} \quad \frac{\partial f}{\partial y}(x, y) = 5$$

and $f(0, 1) = 3$.

Exercise 8. Let $(a, b) \in \mathbb{R}^2$ and $f : \mathbb{R}^2 \to \mathbb{R}$ be a function with the properties:

(i) f admits partial derivatives at all points in $\mathbb{R}^2 \setminus \{(a, b)\}$;

(ii) There exist $\lim\limits_{(x,y) \to (a,b)} \dfrac{\partial f}{\partial x}(x, y)$ and $\lim\limits_{(x,y) \to (a,b)} \dfrac{\partial f}{\partial y}(x, y)$.

Deduce, with an argument similar to that in Theorem 4.7, that f can be extended to a continuous function at (a, b).

4.2   The Tangent Plane to a Surface

Geometrically, the derivative of a single variable function is the slope of the tangent line to its graph. Let us next discuss the geometric interpretation of the partial derivatives of a function $f : D \subset \mathbb{R}^2 \to \mathbb{R}$. Assume f admits partial derivatives at $(a, b) \in D$ and let $c = f(a, b)$. The graph of $z = f(x, y)$ is a surface $\mathcal{S}$ in $\mathbb{R}^3$ which intersects the plane $y = b$ along a curve $\mathcal{C}_1$. This means that $\mathcal{C}_1$ is the graph of $g(x) := f(x, b)$. The slope of its tangent T_1 at P is $g'(a) = \dfrac{\partial f}{\partial x}(a, b)$. Similarly, let the curve $\mathcal{C}_2$ be the intersection of $\mathcal{S}$ with the plane $x = a$; hence $\mathcal{C}_1$ is the graph of $h(y) := f(a, y)$. The slope of its tangent T_2 at P is $h'(b) = \dfrac{\partial f}{\partial y}(a, b)$.

In Figure 4.1, the curves $\mathcal{C}_1$ and $\mathcal{C}_2$ are pictured with a discontinuous line. The first observation we can make at this point is that:

Remark 4.8 The partial derivatives $\dfrac{\partial f}{\partial x}(a, b)$ and $\dfrac{\partial f}{\partial y}(a, b)$ give the rate of change of f at (a, b) in the x-direction and y-direction, respectively.

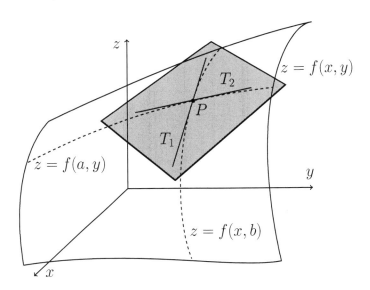

Figure 4.1
The tangent plane constructed with the tangent lines T_1 and T_2.

Definition 4.9 *The plane that contains the tangent lines T_1 and T_2 is called the tangent plane to the surface $\mathcal{S}$ at the point P. The line which is orthogonal to the tangent plane at P is called the normal line to the surface $\mathcal{S}$.*

Let us further explore the features of the tangent plane to a surface $\mathcal{S}$.

Theorem 4.10 *Suppose $f : D \subset \mathbb{R}^2 \to \mathbb{R}$ has continuous partial derivatives at $(a, b) \in D$ and let $c = f(a, b)$.*

(i) *The vector*

$$\mathbf{n} = \frac{\partial f}{\partial x}(a, b)\mathbf{i} + \frac{\partial f}{\partial y}(a, b)\mathbf{j} - \mathbf{k}$$

is orthogonal to the tangent plane to the surface $z = f(x, y)$ at the point $P(a, b, c)$.

(ii) *A parametric equation of the normal line to the tangent plane at $P(a, b, c)$ to the surface $z = f(x, y)$ is given by*

$$(x, y, z) = \big(a, b, f(a, b)\big) + t\Big(\frac{\partial f}{\partial x}(a, b), \frac{\partial f}{\partial y}(a, b), -1\Big), \ t \in \mathbb{R}.$$

(iii) *The equation of the tangent plane to the surface $z = f(x, y)$ at the point $P(a, b, c)$ is*

$$z - f(a, b) = \frac{\partial f}{\partial x}(a, b)(x - a) + \frac{\partial f}{\partial x}(a, b)(y - b).$$

Proof (i) Recall that $\frac{\partial f}{\partial x}(a, b)$ is the slope of the tangent line T_1 to the trace of the graph of f in the plane $y = b$. This tangent line has the direction

$$\mathbf{u}_x = \mathbf{i} + \frac{\partial f}{\partial x}(a, b)\mathbf{k}.$$

Similarly, $\frac{\partial f}{\partial y}(a, b)$ is the slope of the tangent line T_2 to the trace of the graph of f in the plane $x = a$. The direction of this tangent line is given by

$$\mathbf{u}_y = \mathbf{j} + \frac{\partial f}{\partial y}(a, b)\mathbf{k}.$$

Now, a normal vector to the tangent plane to $z = f(x, y)$ has to be orthogonal to both $\mathbf{u}_x$ and $\mathbf{u}_y$. Thus, one such a vector is

$$\mathbf{n} = \mathbf{u}_y \times \mathbf{u}_x = \begin{vmatrix} \mathbf{i} & \mathbf{j} & \mathbf{k} \\ 0 & 1 & \frac{\partial f}{\partial y}(a, b) \\ 1 & 0 & \frac{\partial f}{\partial x}(a, b) \end{vmatrix} = \frac{\partial f}{\partial x}(a, b)\mathbf{i} + \frac{\partial f}{\partial y}(a, b)\mathbf{j} - \mathbf{k}.$$

(ii) The normal line passes through (a, b, c) and has the direction given by the vector $\mathbf{n}$. Its equation follows from Theorem 1.5.

(iii) The tangent plane to the surface $z = f(x, y)$ that passes through $P(a, b, c)$ consists of all points $Q(x, y, z)$ such that $\mathbf{n} \perp \overrightarrow{PQ}$. This yields

$$\mathbf{n} \bullet \overrightarrow{PQ} = 0 \Longrightarrow \left(\frac{\partial f}{\partial x}(a, b), \frac{\partial f}{\partial y}(a, b), -1 \right) \bullet (x - a, y - z, z - c) = 0,$$

that is,

$$z - c = \frac{\partial f}{\partial x}(a, b)(x - a) + \frac{\partial f}{\partial y}(a, b)(y - b).$$

Using $c = f(a, b)$ we obtain the desired formula. □

Example 4.11 *Let* $f(x, y) = \ln(xy^2)$.

(i) *Write the equation of the tangent plane to the surface* $z = f(x, y)$ *at* $(1, 1, 0)$;

(ii) *Write the equation of the normal line to the above surface;*

(iii) *Are there points on the graph of* f *such that the tangent plane at these points is parallel to the plane* $2x + 3y + 6z = 10$?

Solution. The domain of definition of f is $D = \{(x, y) \in \mathbb{R}^2 : x > 0, y \neq 0\}$.
(i) The equation of the tangent plane is

$$z - 0 = \frac{\partial f}{\partial x}(1, 1)(x - 1) + \frac{\partial f}{\partial y}(1, 1)(y - 1).$$

We find

$$\frac{\partial f}{\partial x}(x, y) = \frac{1}{x} \quad \text{and} \quad \frac{\partial f}{\partial y}(x, y) = \frac{2}{y}.$$

The equation of the tangent plane at $(1, 1, 0)$ is

$$z - 0 = 1(x - 1) + 2(y - 1) \quad \Longrightarrow x + 2y - z = 3.$$

(ii) The equation is

$$(x, y, z) = (1, 1, f(1, 1)) + t\left(\frac{\partial f}{\partial x}(1, 1), \frac{\partial f}{\partial y}(1, 1), -1 \right), \ t \in \mathbb{R},$$

that is, $(x, y, z) = (1 + t, 1 + 2t, -t)$, $t \in \mathbb{R}$.
(iii) Suppose there is $(a, b, f(a, b))$ a point on the graph of f at which the tangent plane is parallel to $2x + 3y + 6z = 10$. The equation of the tangent plane is

$$z - f(a, b) = \frac{\partial f}{\partial x}(a, b)(x - a) + \frac{\partial f}{\partial y}(a, b)(y - b).$$

Since $\frac{\partial f}{\partial x}(a, b) = \frac{1}{a}$ and $\frac{\partial f}{\partial y}(a, b) = \frac{2}{b}$, the equation of the tangent plane is

$$\frac{1}{a}x + \frac{2}{b}y - z = 3 - f(a, b).$$

Thus, this tangent plane is parallel to $2x + 3y + 6z = 10$ if

$$\frac{2}{\frac{1}{a}} = \frac{3}{\frac{2}{b}} = \frac{6}{-1} \implies 2a = \frac{3b}{2} = -6 \implies a = -3, b = -4.$$

However, $(a, b) = (-3, -4)$ is not in the domain of definition of f, so, there are no points on the graph that satisfy this condition.

Exercises

Exercise 9. Write the equation of the tangent plane and the normal line to the graph of the functions below at the indicated points:

(i) $f(x, y) = x^2 \ln \dfrac{x}{y}$ at $(1, 1)$;

(ii) $f(x, y) = x \cos(x + 2y)$ at $(\pi, 0)$;

(iii) $f(x, y) = x^2 e^{-\frac{1}{y}}$ at $(2, -1)$;

(iv) $f(x, y) = \dfrac{x}{\sqrt{x^2 + y^2 + 1}}$ at $(-2, 2)$.

Exercise 10. Find the equation of the planes that pass through $M(2, -1, 9)$ and $N(1, 0, 7)$ and are tangent to the graph of $f(x, y) = 3x^2 - y^2$.

Exercise 11. Find the points on the surface $z = x^3 - 2xy$ at which the tangent plane is orthogonal to the planes $2x + y = 0$ and $x + 2z = 0$.

Exercise 12. The function $f : \mathbb{R}^2 \to \mathbb{R}$ has continuous partial derivatives at $(-1, 3)$. The equation of the tangent plane to $z = f(x, y)$ is $4x - 3y + 2z = 7$. Find $f(-1, 3)$, $\dfrac{\partial f}{\partial x}(-1, 3)$ and $\dfrac{\partial f}{\partial x}(-1, 3)$.

Exercise 13. Verify that all tangent planes to the graph of $f(x, y) = xe^{y^2}$ make with the plane $z = 0$ an angle larger than or equal to $\pi/4$.

Exercise 14. Consider the elliptic paraboloid $z = x^2 + 4y^2$ and $P(1, -2, 8)$.

(i) Check that P is an exterior point to the surface of the paraboloid;

(ii) Verify that there are infinitely many tangent planes to the paraboloid which pass through P;

(iii) How many of these tangent planes also contain $Q(-2, 1, 14)$?

Exercise 15. Find all points on the surface $z = ye^{xy}$ at which the tangent plane is parallel to $x - z = 5$.

Exercise 16. The functions $f, g : \mathbb{R}^2 \to \mathbb{R}$ have continuous partial derivatives.

(i) If the surfaces $z = f(x, y)$ and $z = g(x, y)$ intersect at (a, b, c) and their tangent planes at (a, b, c) are orthogonal, check that

$$\nabla f(a, b) \bullet \nabla g(a, b) = -1;$$

(ii) If the surfaces $z = f(x, y)$ and $z = g(x, y)$ intersect at (a, b, c) and have the same tangent plane at (a, b, c), check that

$$\nabla f(a, b) = \nabla g(a, b).$$

4.3 Differentiable Functions

So far, in this chapter, we introduced the partial derivatives of a function of several variables and discussed their geometrical interpretation. We are now ready to lead further the study of differentiability. To this aim, let us place ourselves again in the case of a single variable function $f : I \subset \mathbb{R} \to \mathbb{R}$. Recall that f is differentiable at $a \in I$ is equivalent to the existence of the limit in (4.1), which we can write

$$\lim_{t \to 0} \frac{f(a + t) - \{f(a) + f'(a)t\}}{t} = 0. \tag{4.7}$$

Taking sided limits, it follows from (4.7) that

$$\lim_{t \to 0} \frac{f(a + t) - \{f(a) + f'(a)t\}}{|t|} = 0. \tag{4.8}$$

Let now assume $f : D \subset \mathbb{R}^2 \to \mathbb{R}$ and $(a, b) \in D$. To translate (4.8) in terms of functions of two variables, we have to replace:

- t by the pair (h, k);

- $|t|$ by $\|(h, k)\|$;

- $f'(a)$ by $\nabla f(a, b) = \left(\frac{\partial f}{\partial x}(a, b), \frac{\partial f}{\partial y}(a, b) \right)$.

Thus, the analogue of (4.8) reads

$$\lim_{(h,k) \to (0,0)} \frac{f(a + h, b + k) - \{f(a, b) + \frac{\partial f}{\partial x}(a, b)h + \frac{\partial f}{\partial y}(a, b)k\}}{\|(h, k)\|} = 0,$$

or in an equivalent notation

$$\lim_{(h,k) \to (0,0)} \frac{f(a + h, b + k) - \{f(a, b) + \nabla f(a, b) \bullet (h, k)\}}{\|(h, k)\|} = 0.$$

We are now ready to state the definition of a differentiable function of m variables.

Definition 4.12 *A function $f : D \subset \mathbb{R}^m \to \mathbb{R}$ is said to be differentiable at* $\mathbf{a} \in D$ *if:*

(i) $\dfrac{\partial f}{\partial x_j}(\mathbf{a})$ *exists for all $1 \leq j \leq m$;*

(ii) *we have*

$$\frac{f(\mathbf{a} + \mathbf{h}) - \{f(\mathbf{a}) + \nabla f(\mathbf{a}) \bullet \mathbf{h}\}}{\|\mathbf{h}\|} \to 0 \quad \text{as } \|\mathbf{h}\| \to 0. \qquad (4.9)$$

Remark 4.13 *Letting* $\mathbf{x} = \mathbf{a} + \mathbf{h}$, *(4.9) reads*

$$\frac{f(\mathbf{x}) - \{f(\mathbf{a}) + \nabla f(\mathbf{a}) \bullet (\mathbf{x} - \mathbf{a})\}}{\|\mathbf{x} - \mathbf{a}\|} \to 0 \quad \text{as } \mathbf{x} \to \mathbf{a}. \qquad (4.10)$$

Example 4.14 *The function*

$$f : \mathbb{R}^2 \to \mathbb{R}, \quad f(x,y) = f(x,y) = \begin{cases} \dfrac{x^4}{x^2 + y^2} & \text{if } (x,y) \neq (0,0) \\ 0 & \text{if } (x,y) = (0,0), \end{cases}$$

is differentiable at $(0,0)$.

Solution. With Definition 4.1 we compute the partial derivatives of f at $(0,0)$ as follows:

$$\frac{\partial f}{\partial x}(0,0) = \lim_{t \to 0} \frac{f(t,0) - f(0,0)}{t} = \lim_{t \to 0} \frac{t^4}{t^3} = 0$$

$$\frac{\partial f}{\partial y}(0,0) = \lim_{t \to 0} \frac{f(0,t) - f(0,0)}{t} = \lim_{t \to 0} \frac{0}{t^3} = 0.$$

Hence, $\nabla f(0,0) = (0,0)$. Now, f is differentiable at $(0,0)$ if and only if it satisfies (4.10). We have

$$\frac{f(x,y) - \{f(0,0) + \nabla f(0,0) \bullet (x,y)\}}{\|(x,y)\|} = \frac{f(x,y)}{\sqrt{x^2 + y^2}} \quad \text{for all } (x,y) \neq (0,0).$$

Denote by $g(x,y)$ the above quotient, so that

$$g(x,y) = \frac{x^4}{(x^2 + y^2)\sqrt{x^2 + y^2}} \quad \text{for all } (x,y) \neq (0,0).$$

Observe that

$$|g(x,y)| = \underbrace{\frac{x^2}{x^2 + y^2}}_{\leq 1} \cdot \underbrace{\sqrt{\frac{x^2}{x^2 + y^2}}}_{\leq 1} \cdot |x| \leq |x| \to 0 \text{ as } (x,y) \to (0,0).$$

Hence, by Corollary 3.12 we deduce $\lim\limits_{(x,y) \to (0,0)} g(x,y) = 0$ which yields f is differentiable at $(0,0)$.

Example 4.15 *The function*

$$f : \mathbb{R}^2 \to \mathbb{R}, \quad f(x,y) = f(x,y) = \begin{cases} \dfrac{x^3 + y^3}{x^2 + y^2} & \text{if } (x,y) \neq (0,0) \\ 0 & \text{if } (x,y) = (0,0), \end{cases}$$

is not differentiable at $(0,0)$.

Solution. We first compute the partial derivatives of f at $(0,0)$ as follows:

$$\frac{\partial f}{\partial x}(0,0) = \lim_{t \to 0} \frac{f(t,0) - f(0,0)}{t} = \lim_{t \to 0} \frac{t^3}{t^3} = 1,$$

$$\frac{\partial f}{\partial y}(0,0) = \lim_{t \to 0} \frac{f(0,t) - f(0,0)}{t} = \lim_{t \to 0} \frac{t^3}{t^3} = 1.$$

Hence $\nabla f(0,0) = (1,1)$. Having (4.10) in mind, we compute

$$\frac{f(x,y) - \{f(0,0) + \nabla f(0,0) \bullet (x,y)\}}{\|(x,y)\|} = \frac{f(x,y) - x - y}{\sqrt{x^2 + y^2}} \quad \text{for all } (x,y) \neq (0,0).$$

Denote by $g(x,y)$ the above quotient, so that

$$g(x,y) = -\frac{x^2 y + x y^2}{(x^2 + y^2)\sqrt{x^2 + y^2}} \quad \text{for all } (x,y) \neq (0,0).$$

Observe that $g(x,y)$ approaches different values along the x axis and along the semiline $y = x$, $x > 0$. This shows that g has no limit as $(x,y) \to (0,0)$, that is, f is not differentiable at $(0,0)$.

The following result is a reformulation of Definition 4.12.

Theorem 4.16 *A function* $f : D \subset \mathbb{R}^m \to \mathbb{R}$ *is differentiable at* $\mathbf{a} \in D$ *if and only if there exist:*

$$r > 0 \quad \text{with} \quad B_r(\mathbf{a}) \subset D, \quad \text{a vector } \mathbf{u} \in \mathbb{R}^m \quad \text{and} \quad \eta : B_r(\mathbf{0}) \to \mathbb{R}$$

such that:

(i) $\displaystyle\lim_{\mathbf{h} \to 0} \frac{\eta(\mathbf{h})}{\|\mathbf{h}\|} = 0$;

(ii) *we have*

$$f(\mathbf{a} + \mathbf{h}) = f(\mathbf{a}) + \mathbf{u} \bullet \mathbf{h} + \eta(\mathbf{h}) \quad \text{for all } \|\mathbf{h}\| < r. \tag{4.11}$$

Moreover, if (4.11) *holds, then* $\nabla f(\mathbf{a}) = \mathbf{u}$.

Proof Since D is open, there exists $r > 0$ such that $B_r(\mathbf{a}) \subset D$.
Assume first that f is differentiable at $\mathbf{a}$ and let

$$\eta(\mathbf{h}) = f(\mathbf{a} + \mathbf{h}) - \{f(\mathbf{a}) + \nabla f(\mathbf{a}) \bullet \mathbf{h}\} \quad \text{for all } \mathbf{h} \in B_r(\mathbf{0}).$$

Then (4.11) holds with $\mathbf{u} = \nabla f(\mathbf{a})$ and by condition (4.9) in Definition 4.12 we also have that $\lim_{\mathbf{h} \to \mathbf{0}} \dfrac{\eta(\mathbf{h})}{\|\mathbf{h}\|} = 0$.

Conversely, assume now that (i) and (ii) in Theorem 4.16 hold and let us show that f is differentiable at $\mathbf{a}$. First, we have to show that the partial derivatives $\dfrac{\partial f}{\partial x_j}(\mathbf{a})$ exist for all $1 \leq j \leq m$. Indeed, letting $\mathbf{h} = t\mathbf{e}_j$, $|t| < r$ in (4.11) we obtain

$$\frac{f(\mathbf{a} + t\mathbf{e}_j) - f(\mathbf{a})}{t} = u_j + \frac{\eta(t\mathbf{e}_j)}{t} \quad \text{for all } |t| < r. \tag{4.12}$$

Since by our hypothesis (i), we have

$$\left| \frac{\eta(t\mathbf{e}_j)}{t} \right| = \frac{|\eta(t\mathbf{e}_j)|}{\|t\mathbf{e}_j\|} \to 0 \quad \text{as } t \to 0,$$

we may use Corollary 3.12 to derive

$$\lim_{t \to 0} \frac{\eta(t\mathbf{e}_j)}{t} = 0$$

and now from (4.12) we find

$$\lim_{t \to 0} \frac{f(\mathbf{a} + t\mathbf{e}_j) - f(\mathbf{a})}{t} = u_j.$$

This shows that $\dfrac{\partial f}{\partial x_j}(\mathbf{a}) = u_j$ for all $1 \leq j \leq m$. Thus, (4.11) reads

$$f(\mathbf{a} + \mathbf{h}) = f(\mathbf{a}) + \nabla f(\mathbf{a}) \bullet \mathbf{h} + \eta(\mathbf{h}) \quad \text{for all } \|\mathbf{h}\| < r.$$

Hence, using again (i) we obtain

$$\frac{f(\mathbf{a} + \mathbf{h}) - \{f(\mathbf{a}) + \nabla f(\mathbf{a}) \bullet \mathbf{h}\}}{\|\mathbf{h}\|} = \frac{\eta(\mathbf{h})}{\|\mathbf{h}\|} \to 0 \quad \text{as } \mathbf{h} \to \mathbf{0}.$$

In light of Definition 4.12, we have now that f is differentiable at $\mathbf{a}$. ☐

Corollary 4.17 *Let* $f : D \subset \mathbb{R}^m \to \mathbb{R}$. *If* f *is differentiable at* $\mathbf{a} \in D$, *then* f *is continuous at* $\mathbf{a}$.

Proof Since f is differentiable at $\mathbf{a}$, (4.11) holds. Thus,

$$\lim_{\mathbf{x} \to \mathbf{a}} f(\mathbf{x}) = \lim_{\mathbf{h} = \mathbf{x} - \mathbf{a} \to \mathbf{0}} f(\mathbf{a} + \mathbf{h}) = \lim_{\mathbf{h} \to \mathbf{0}} \left(f(\mathbf{a}) + \mathbf{u} \bullet \mathbf{h} + \frac{\eta(\mathbf{h})}{\|\mathbf{h}\|} \cdot \|\mathbf{h}\| \right) = f(\mathbf{a}).$$

Thus, f is continuous at $\mathbf{a} \in D$. ☐

Exercises

Exercise 17. Let $f : \mathbb{R}^2 \to \mathbb{R}$, $f(x, y) = x\sqrt{x^2 + y^2}$.

(i) Show that f is differentiable at $(0, 0)$;

(ii) Are there points on the graph of f at which the tangent plane is orthogonal to the plane $x - 2y - mz = 6$, $m > 0$?

Exercise 18. Study the differentiability at $(0, 0)$ for each of the following functions:

(i) $f : \mathbb{R}^2 \to \mathbb{R}$, $f(x) = \begin{cases} \dfrac{(x+y)^3}{x^2 + y^2} & \text{if } (x, y) \neq (0, 0) \\ 0 & \text{if } (x, y) = (0, 0); \end{cases}$

(ii) $f : \mathbb{R}^2 \to \mathbb{R}$, $f(x) = \begin{cases} \dfrac{x^2 y^2}{x^4 + y^4} \sin x \sin y & \text{if } (x, y) \neq (0, 0) \\ 0 & \text{if } (x, y) = (0, 0). \end{cases}$

Exercise 19. Let $f : \mathbb{R}^2 \to \mathbb{R}$, $f(x, y) = \begin{cases} xy & \text{if } |x| \neq |y| \\ 0 & \text{if } |x| = |y|. \end{cases}$

(i) Find $\dfrac{\partial f}{\partial x}(x, y)$ and $\dfrac{\partial f}{\partial y}(x, y)$ wherever they exist;

(ii) Is f differentiable at $(0, 0)$?

(iii) Is f differentiable at $(-1, 1)$?

Exercise 20. Let

$$f : \mathbb{R}^2 \to \mathbb{R}, \quad f(x, y) = \begin{cases} x^{4/3} \sin \dfrac{y}{x} & \text{if } x \neq 0 \\ 0 & \text{if } x = 0. \end{cases}$$

(i) Check that f is a continuous function on $\mathbb{R}^2$;

(ii) Find $\dfrac{\partial f}{\partial x}(x, y)$ and $\dfrac{\partial f}{\partial y}(x, y)$;

(iii) Use Definition 3.1 to deduce that $\dfrac{\partial f}{\partial y}(x, y)$ has no limit at $(0, 1)$;

(iv) Is f differentiable on $\mathbb{R}^2$?

Exercise 21. Assume $f : B_1(0, 0) \subset \mathbb{R}^2 \to \mathbb{R}$ is a function of two variables which satisfies

$$|f(x, y) - \sin(xy)| \leq x^2 + y^2 \quad \text{for all } (x, y) \in B_1(0, 0).$$

(i) Find $f(0, 0)$;

(ii) Find $\dfrac{\partial f}{\partial x}(0,0)$ and $\dfrac{\partial f}{\partial y}(0,0)$;

(iii) Show that f is differentiable at $(0,0)$.

Exercise 22. Suppose $f : \mathbb{R}^m \to \mathbb{R}$ is a function of m variables which is differentiable at $\mathbf{a} \in \mathbb{R}^m$. Fix $\mathbf{h} \in \mathbb{R}^m \setminus \{\mathbf{0}\}$ and define

$$g : \mathbb{R} \to \mathbb{R}, \text{ by } g(t) = f(\mathbf{a} + t\mathbf{h}).$$

Verify that g is differentiable at 0 and $g'(0) = \nabla f(\mathbf{a}) \bullet \mathbf{h}$.

Exercise 23. Let $f : D \subset \mathbb{R}^m \to \mathbb{R}$ be a function of m variables and $\mathbf{a} \in D$. Assume that

$$\lim_{\mathbf{x} \to \mathbf{a}} \frac{f(\mathbf{x}) - f(\mathbf{a})}{\|\mathbf{x} - \mathbf{a}\|} = 0.$$

Using Theorem 4.16, check that f is differentiable at $\mathbf{a}$ and $\nabla f(\mathbf{a}) = \mathbf{0}$.

Exercise 24. Assume $f : \mathbb{R}^m \to \mathbb{R}$ is a function of $m \geq 1$ variables which satisfies

$$|f(\mathbf{x} + \mathbf{y}) - f(\mathbf{x}) - \mathbf{x} \bullet \mathbf{y}| \leq \|\mathbf{y}\|^2 \quad \text{for all } \mathbf{x}, \mathbf{y} \in \mathbb{R}^m.$$

(i) Using Theorem 4.16, check that f is differentiable at each $\mathbf{x} \in \mathbb{R}^m$ and $\nabla f(\mathbf{x}) = \mathbf{x}$;

(ii) Letting $g(\mathbf{x}) = f(\mathbf{x}) - \frac{\|\mathbf{x}\|^2}{2}$, deduce that $\nabla g(\mathbf{x}) = 0$ for all $\mathbf{x} \in \mathbb{R}^m$;

(iii) Use Exercise 6 to derive that there exists a constant $c \in \mathbb{R}$ such that

$$f(x) = \frac{\|\mathbf{x}\|^2}{2} + c \quad \text{for all } \mathbf{x} \in \mathbb{R}^m.$$

Exercise 25. Let $f : \mathbb{R}^2 \to \mathbb{R}$ be a continuous function of two variables such that:

(i) f admits partial derivatives at all points $(x,y) \neq (a,b)$;

(ii) There exists $\displaystyle\lim_{(x,y) \to (a,b)} \frac{\partial f}{\partial x}(x,y)$ and $\displaystyle\lim_{(x,y) \to (a,b)} \frac{\partial f}{\partial y}(x,y)$.

Use Theorem 4.16 to check that f is differentiable at (a,b).

4.4   A Criterion for Differentiability

The main result of this section is stated below and provides a sufficient condition for the differentiability of a function at a point.

Theorem 4.18 *Let $f : D \subset \mathbb{R}^m \to \mathbb{R}$ be a function of m variables and $\mathbf{a} \in D$. Assume that:*

(i) f *has partial derivatives on a ball* $B_r(\mathbf{a}) \subset D$;

(ii) *For all* $1 \leq j \leq m$ *the partial derivatives* $\dfrac{\partial f}{\partial x_j}$ *are continuous at* **a**.

Then f is differentiable at **a**.

A function $f : D \subset \mathbb{R}^m \to \mathbb{R}$ which has continuous partial derivatives at all points $\mathbf{a} \in D$ is called *continuously differentiable* or a *function of class C^1* on D.

Proof The approach relies on a similar argument to that employed in the proof of Theorem 4.7, namely the use of the one-variable Mean Value Theorem. This time, we take the bold path and prove Theorem 4.18 in its full generality, without resorting to the case of two-variable functions.

Let $\mathbf{x} \in B_r(\mathbf{a})$. For $1 \leq j \leq m$ we denote

$$\mathbf{x}_j = (x_1, \ldots, x_j, a_{j+1}, \ldots, a_m) \quad \text{and} \quad \mathbf{x}_0 = \mathbf{a}.$$

We estimate

$$|f(\mathbf{x}) - f(\mathbf{a}) - \nabla f(\mathbf{a}) \bullet (\mathbf{x} - \mathbf{a})|$$

$$= \left| \sum_{j=1}^{m} \left\{ f(\mathbf{x}_j) - f(\mathbf{x}_{j-1}) - \frac{\partial f}{\partial x_j}(\mathbf{a})(x_j - a_j) \right\} \right|$$

$$\leq \sum_{j=1}^{m} \left| f(\mathbf{x}_j) - f(\mathbf{x}_{j-1}) - \frac{\partial f}{\partial x_j}(\mathbf{a})(x_j - a_j) \right| \qquad (4.13)$$

$$= \sum_{j=1}^{m} \left| g_j(x_j) - g_j(a_j) \right|,$$

where

$$g_j(t) = f(x_1, \ldots, x_{j-1}, t, a_{j+1}, \ldots, a_m) - \frac{\partial f}{\partial x_j}(\mathbf{a})t.$$

By the Mean Value Theorem applied to g_j on the interval with endpoints at x_j and a_j, there exists $c_j = c_j(\mathbf{x}) \in \mathbb{R}$ between x_j and a_j such that

$$\left| g_j(x_j) - g_j(a_j) \right| = \left| \frac{\partial f}{\partial x_j}(\mathbf{z}_j) - \frac{\partial f}{\partial x_j}(\mathbf{a}) \right| \cdot |x_j - a_j|$$

$$\leq \left| \frac{\partial f}{\partial x_j}(\mathbf{z}_j) - \frac{\partial f}{\partial x_j}(\mathbf{a}) \right| \cdot \|\mathbf{x} - \mathbf{a}\|, \qquad (4.14)$$

where $\mathbf{z}_j = (x_1, \ldots, x_{j-1}, c_j, a_{j+1}, \ldots, a_m)$. Next we define

$$g : B_r(\mathbf{a}) \to \mathbb{R}, \quad g(\mathbf{x}) = \sum_{j=1}^{m} \left| \frac{\partial f}{\partial x_j}(\mathbf{z}_j) - \frac{\partial f}{\partial x_j}(\mathbf{a}) \right|.$$

Because $c_j(\mathbf{x}) \to a_j$ and $\mathbf{z}_j \to \mathbf{a}$ as $\mathbf{x} \to \mathbf{a}$, it follows that $g(\mathbf{x}) \to 0$ as $\mathbf{x} \to \mathbf{a}$. Using the estimates (4.13) and (4.14) we find

$$\left| \frac{f(\mathbf{x}) - f(\mathbf{a}) - \nabla f(\mathbf{a}) \bullet (\mathbf{x} - \mathbf{a})}{\|\mathbf{x} - \mathbf{a}\|} \right| \leq g(\mathbf{x}).$$

Thus,

$$\frac{f(\mathbf{x}) - f(\mathbf{a}) - \nabla f(\mathbf{a}) \bullet (\mathbf{x} - \mathbf{a})}{\|\mathbf{x} - \mathbf{a}\|} \to 0 \quad \text{as } \mathbf{x} \to \mathbf{a}.$$

In light of Definition 4.12, this shows that f is differentiable at $\mathbf{a}$. □

Remark 4.19 As an immediate consequence of Theorem 4.18 we deduce that all elementary functions (polynomial, rational, trigonometric, exponential and logarithmic functions) of m variables are continuously differentiable. Indeed, the partial derivatives of such functions are also elementary, thus continuous by Remark 3.23. By virtue of Theorem 4.18 above, the partial derivatives being continuous imply that the original functions are continuously differentiable on their domain of definition.

To summarise the main points of this chapter, there are two ways of studying the differentiability of a function at a point. For functions of two variables, these are presented in the note below.

Important

To check the differentiability of a function $f(x, y)$ at a point (a, b), we have:

(A) Use the Definition 4.12 and prove that

$$\lim_{(x,y) \to (a,b)} \frac{f(x,y) - \{f(a,b) + \nabla f(a,b) \bullet (x - a, y - b)\}}{\|(x - a, y - b)\|} = 0;$$

(B) Or, alternatively, prove that the partial derivatives $\dfrac{\partial f}{\partial x}$ and $\dfrac{\partial f}{\partial y}$ exists on a disc centred at (a, b) and are continuous at (a, b).

Note that the method (B) described above applies only when the partial derivatives exist in a neighbourhood of (a, b) and are continuous at (a, b); this is a sufficient condition for a function to be differentiable at a point. In general, if f is differentiable at a point, it is not necessary true that the partial derivatives are continuous at that point. We can see this fact from the example below.

Example 4.20 *Let*

$$f : \mathbb{R}^2 \to \mathbb{R}, \quad f(x,y) = \begin{cases} \dfrac{x^2 y^2}{x^4 + y^2} & \text{if } (x,y) \neq (0,0) \\ 0 & \text{if } (x,y) = (0,0). \end{cases}$$

Then f is differentiable at $(0,0)$ although $\dfrac{\partial f}{\partial y}$ is not continuous at $(0,0)$.

Solution. We have

$$\frac{\partial f}{\partial x}(x,y) = \begin{cases} \dfrac{2xy^2(y^2 - x^4)}{(x^4 + y^2)^2} & \text{if } (x,y) \neq (0,0) \\ 0 & \text{if } (x,y) = (0,0) \end{cases}$$

and

$$\frac{\partial f}{\partial y}(x,y) = \begin{cases} \dfrac{2x^6 y}{(x^4 + y^2)^2} & \text{if } (x,y) \neq (0,0) \\ 0 & \text{if } (x,y) = (0,0). \end{cases}$$

Along the line $y = 0$, we have $\dfrac{\partial f}{\partial y}(x,y) \to 0$ as $(x,y) \to (0,0)$. Also, along the curve $y = x^2$, we have

$$\frac{\partial f}{\partial y}(x,y) = \frac{2x^8}{(2x^4)^2} = \frac{1}{2} \to \frac{1}{2} \quad \text{as } (x,y) \to (0,0).$$

Hence, $\dfrac{\partial f}{\partial y}(x,y)$ approaches different values along two different paths as (x,y) approaches $(0,0)$. This shows that $\dfrac{\partial f}{\partial y}(x,y)$ has no limit at the origin. Due to this fact, we can only verify the differentiability of f at $(0,0)$ by using Definition 4.12. Thus, f is differentiable at $(0,0)$ if and only if the function

$$g(x,y) = \frac{f(x,y) - \{f(0,0) + \nabla f(0,0) \bullet (x,y)\}}{\|(x,y)\|}$$

tends to zero as $(x,y) \to (0,0)$. Note that

$$g(x,y) = \frac{x^2 y^2}{(x^4 + y^2)\sqrt{x^2 + y^2}} \quad \text{for all } (x,y) \neq (0,0).$$

We have

$$|g(x,y)| = \underbrace{\frac{y^2}{x^4 + y^2}}_{\leq 1} \cdot \underbrace{\frac{|x|}{\sqrt{x^2 + y^2}}}_{\leq 1} \cdot |x| \leq |x| \to 0 \text{ as } (x,y) \to (0,0).$$

Hence, g has limit 0 at $(0,0)$ which means that f is differentiable at this point.

Exercises

Exercise 26. Let $k > 0$ be a real number and

$$f : \mathbb{R}^2 \to \mathbb{R}, \quad f(x) = \begin{cases} \dfrac{x^k}{x^2 + y^2} & \text{if } (x, y) \neq (0, 0) \\ 0 & \text{if } (x, y) = (0, 0). \end{cases}$$

(i) Check that if $k \leq 3$, then f is not differentiable at $(0, 0)$;

(ii) Assume $k > 3$. Verify that $\dfrac{\partial f}{\partial x}$ and $\dfrac{\partial f}{\partial y}$ are continuous on $\mathbb{R}^2$. Is f differentiable at the origin in this case?

4.5 Differentiability of Vector-Valued Functions

Let $m, p \geq 1$ and $F = (f_1, f_2, \ldots, f_p) : D \subset \mathbb{R}^m \to \mathbb{R}^p$ be a vector valued function. If each component f_i, $1 \leq i \leq p$, has partial derivatives at $\mathbf{a} \in D$ we denote

$$J_F(\mathbf{a}) = \begin{pmatrix} \dfrac{\partial f_1}{\partial x_1}(\mathbf{a}) & \dfrac{\partial f_1}{\partial x_2}(\mathbf{a}) & \cdots & \dfrac{\partial f_1}{\partial x_m}(\mathbf{a}) \\[2mm] \dfrac{\partial f_2}{\partial x_1}(\mathbf{a}) & \dfrac{\partial f_2}{\partial x_2}(\mathbf{a}) & \cdots & \dfrac{\partial f_2}{\partial x_m}(\mathbf{a}) \\[2mm] \cdots & \cdots & \cdots & \cdots \\[2mm] \dfrac{\partial f_p}{\partial x_1}(\mathbf{a}) & \dfrac{\partial f_p}{\partial x_2}(\mathbf{a}) & \cdots & \dfrac{\partial f_p}{\partial x_m}(\mathbf{a}) \end{pmatrix}$$

and we call $J_F(\mathbf{a})$ *the Jacobian matrix of F at* $\mathbf{a}$.

Similar to the Definition 4.12, we have:

Definition 4.21 *A vector valued function* $F = (f_1, f_2, \ldots, f_p) : D \subset \mathbb{R}^m \to \mathbb{R}^p$ *is said to be differentiable at* $\mathbf{a} \in D$ *if each component f_i, $1 \leq i \leq p$, is differentiable at* $\mathbf{a}$*, that is:*

(i) $\dfrac{\partial f_i}{\partial x_j}(\mathbf{a})$ *exists for all* $1 \leq i \leq p$, $1 \leq j \leq m$;

(ii) *for all* $1 \leq i \leq p$, *we have*

$$\frac{f_i(\mathbf{a} + \mathbf{h}) - \{f_i(\mathbf{a}) + \nabla f_i(\mathbf{a}) \bullet \mathbf{h}\}}{\|\mathbf{h}\|} \to 0 \quad \text{as } \|\mathbf{h}\| \to 0. \quad (4.15)$$

Differentiable Functions

Condition (4.15) is equivalent to

$$\frac{\|F(\mathbf{a}+\mathbf{h}) - \{F(\mathbf{a}) + J_F(\mathbf{a}) \cdot \mathbf{h}\}\|}{\|\mathbf{h}\|} \to 0 \quad \text{as } \|\mathbf{h}\| \to 0,$$

or, by letting $\mathbf{x} = \mathbf{a} + \mathbf{h}$ one has

$$\frac{\|F(\mathbf{x}) - \{F(\mathbf{a}) + J_F(\mathbf{a}) \cdot (\mathbf{x} - \mathbf{a})\}\|}{\|\mathbf{x} - \mathbf{a}\|} \to 0 \quad \text{as } \mathbf{x} \to \mathbf{a}.$$

Example 4.22 *Let $\alpha > 0$. Then $F : \mathbb{R}^m \to \mathbb{R}^m$, $F(\mathbf{x}) = \mathbf{x}\|\mathbf{x}\|^\alpha$ is differentiable at $\mathbf{0}$.*

Solution. We have $F = (f_1, \ldots, f_m)$ where $f_i(\mathbf{x}) = x_i\|\mathbf{x}\|^\alpha$, $1 \le i \le m$. Observe that

$$\frac{\partial f_i}{\partial x_j}(\mathbf{0}) = \lim_{t \to 0} \frac{f_i(te_j) - f_i(\mathbf{0})}{t} = \lim_{t \to 0} \delta_{ij}|t|^\alpha = 0,$$

where δ_{ij} denotes the Kronecker symbol, that is, $\delta_{ij} = 1$ if $i = j$ and $\delta_{ij} = 0$ otherwise. For $\mathbf{x} \ne \mathbf{0}$ we also have $\dfrac{\partial f_i}{\partial x_j}(\mathbf{x}) = \delta_{ij}\|\mathbf{x}\|^\alpha + \alpha x_i x_j\|\mathbf{x}\|^{\alpha-2}$. This shows that the partial derivatives $\dfrac{\partial f_i}{\partial x_j}$ are continuous on $\mathbb{R}^m \setminus \{\mathbf{0}\}$. To check their continuity at $\mathbf{0}$, we have

$$\left|\frac{\partial f_i}{\partial x_j}(\mathbf{x})\right| \le \|\mathbf{x}\|^\alpha + \alpha|x_i||x_j|\|\mathbf{x}\|^{\alpha-2} \le (1 + \alpha)\|\mathbf{x}\|^\alpha \to 0,$$

as $\mathbf{x} \to \mathbf{0}$. This implies that the partial derivatives are continuous at $\mathbf{0}$ and by Theorem 4.18 we deduce f_i, and thus F, is differentiable at $\mathbf{0}$.

Similar to Theorem 4.16, we have:

Theorem 4.23 *A function $F : D \subset \mathbb{R}^m \to \mathbb{R}^p$ is differentiable at $\mathbf{a} \in D$ if and only if there exist:*

$$r > 0 \quad \text{with} \quad B_r(\mathbf{a}) \subset D, \quad a \; p \times m \; matrix \; U \quad and \quad E : B_r(\mathbf{0}) \to \mathbb{R}^p$$

such that:

(i) $\displaystyle\lim_{\mathbf{h} \to 0} \frac{\|E(\mathbf{h})\|}{\|\mathbf{h}\|} = 0;$

(ii) *we have*

$$F(\mathbf{a}+\mathbf{h})^T = F(\mathbf{a})^T + U \cdot \mathbf{h}^T + E(\mathbf{h}) \quad \text{for all } \|\mathbf{h}\| < r. \quad (4.16)$$

Moreover, if (4.16) holds, then $J_F(\mathbf{a}) = U$.

Exercises

Exercise 27. Compute the Jacobian matrix for each of the functions below:

(i) $F(x, y) = (x^2 \cos y, ye^{-x}, x^y)$, where $x, y > 0$;

(ii) $F(x, y, z) = \left(\sqrt{y} - \sqrt{z}, \dfrac{x}{yz} \right)$, where $y, z > 0$;

(iii) $F(x, y, z) = \left(\dfrac{y}{\sqrt{x^2 + z^2}}, \ x^{y^z}, \ \ln\left(x \ln(y \ln z) \right) \right)$, where $x, y > 1$ and $z > e$.

Exercise 28. Find the differentiable function $F : \mathbb{R}^2 \to \mathbb{R}^3$ knowing that $F(1, 0) = (1, 1, -1)$ and its Jacobian matrix is given by

$$J_F(x, y) = \begin{pmatrix} 2xy^3 - 2 & 3x^2y^2 - 6y \\ \sin y & x \cos y - \sin y \\ -e^{x-y-1} & e^{x-y-1} \end{pmatrix}.$$

Exercise 29. Find the differentiable function $F : \mathbb{R}^3 \to \mathbb{R}^2$ knowing that $F(0, 0, 1) = (-2, 0)$ and its Jacobian matrix is given by

$$J_F(x, y) = \begin{pmatrix} y & x + 2yz & y^2 - 6z^2 \\ z \cos(xz) & 1 & x \cos(xz) \end{pmatrix}.$$

Exercise 30. Let $\phi : \mathbb{R}^2 \to \mathbb{R}$ be a differentiable function. Define $F : \mathbb{R}^2 \to \mathbb{R}^3$ and $G : \mathbb{R}^3 \to \mathbb{R}^3$ by

$$F(x, y) = (x, y, \phi(x, y)) \quad \text{and} \quad G(x, y, z) = (x, y, z\phi(x, y)).$$

Find the Jacobian matrix of F and G in terms of ϕ, $\dfrac{\partial \phi}{\partial x}$ and $\dfrac{\partial \phi}{\partial y}$.

Exercise 31. Verify that for any $\alpha > 0$ the function

$$F : \mathbb{R}^m \to \mathbb{R}^m, \quad F(\mathbf{x}) = \begin{cases} \mathbf{x} \|\mathbf{x}\|^\alpha \ln\|\mathbf{x}\| & \text{if } \mathbf{x} \neq \mathbf{0} \\ 0 & \text{if } \mathbf{x} = \mathbf{0}, \end{cases}$$

is differentiable at $\mathbf{0}$. Find the Jacobian matrix of F.

Exercise 32. Let $F, G : \mathbb{R}^m \to \mathbb{R}^p$ be two differentiable functions. Define

$$h : \mathbb{R}^m \times \mathbb{R}^m \to \mathbb{R} \quad \text{by} \quad h(\mathbf{x}, \mathbf{y}) = F(\mathbf{x}) \bullet G(\mathbf{y}).$$

Show that h is differentiable and

$$\nabla h(\mathbf{x}, \mathbf{y}) = \left(G(\mathbf{y}) \cdot J_F(\mathbf{x}), F(\mathbf{x}) \cdot J_G(\mathbf{y}) \right) \quad \text{for all } \mathbf{x}, \mathbf{y} \in \mathbb{R}^m.$$

Exercise 33. Let $F : \mathbb{R}^m \to \mathbb{R}^p$, $m, p \geq 1$ and $C, \alpha > 0$ be such that

$$\|F(\mathbf{x}) - F(\mathbf{y})\| \leq C \|\mathbf{x} - \mathbf{y}\|^{\alpha} \quad \text{for all } \mathbf{x}, \mathbf{y} \in \mathbb{R}^m.$$

(i)   Check that F is continuous on $\mathbb{R}^m$;

(ii) If $\alpha > 1$ show that F is constant;

(iii) A function which satisfies the above condition with $\alpha = 1$ is called a *Lipschitz function*. Construct a nonconstant Lipschitz function $F : \mathbb{R}^m \to \mathbb{R}^p$.

(iv) Verify that $F : \mathbb{R}^m \to \mathbb{R}$, $F(\mathbf{x}) = \|\mathbf{x}\|$ is a Lipschitz function which is not differentiable at $\mathbf{0}$.

5

Chain Rule

This chapter discusses the Chain Rule for functions of several variables. As a consequence of this result, differentiation of functions defined implicitly is presented. The Mean Value Theorem in several variables setting is also included.

5.1 Chain Rule for Several Variable Functions

Recall from the one-variable Calculus that if $I \subset \mathbb{R}$ is an interval on the real line and $f : \mathbb{R} \to \mathbb{R}$, $g : I \to \mathbb{R}$ are such that

- g is differentiable at $a \in I$;

- f is differentiable at $g(a) \in \mathbb{R}$;

then $f \circ g : I \to \mathbb{R}$ is differentiable at $a \in I$ and

$$(f \circ g)(a) = f'(g(a)) \cdot g'(a).$$

A similar result holds for functions of several variables with appropriate changes. We have to replace the derivatives in the above equality by the corresponding gradient and Jacobian matrix. This is stated in the theorem below.

Theorem 5.1 (Chain Rule)
  Let $D \subset \mathbb{R}^m$ be an open set and $f : \mathbb{R}^p \to \mathbb{R}$, $g : D \to \mathbb{R}^p$. If g is differentiable at $\mathbf{a} \in D$ and f is differentiable at $g(\mathbf{a}) \in \mathbb{R}^p$, then $f \circ g$ is differentiable at $\mathbf{a}$ and

$$\nabla(f \circ g)(\mathbf{a}) = \nabla f(g(\mathbf{a})) \cdot J_g(\mathbf{a}). \tag{5.1}$$

In the above equality the dot $\cdot$ symbol denotes the product between the $1 \times p$ vector $\nabla f(g(\mathbf{a}))$ and the $p \times m$ matrix $J_g(\mathbf{a})$.
Proof We shall make use of Theorem 4.16 and Theorem 4.23. Since f is differentiable at $g(\mathbf{a})$, there exist $\rho > 0$ and

$$\eta : B_\rho(\mathbf{0}) \subset \mathbb{R}^p \to \mathbb{R} \quad \text{such that} \quad \lim_{\mathbf{k} \to 0} \frac{\eta(\mathbf{k})}{\|\mathbf{k}\|} = 0$$

DOI: 10.1201/9781003449652-5

and

$$f(g(\mathbf{a}) + \mathbf{k}) = f(g(\mathbf{a})) + \nabla f(g(\mathbf{a})) \bullet \mathbf{k} + \eta(\mathbf{k}) \quad \text{for all } \|\mathbf{k}\| < \rho. \tag{5.2}$$

Similarly, by Theorem 4.23, there exist:

(i) $r > 0$ with $B_r(\mathbf{a}) \subset D$;

(ii) $E : B_r(\mathbf{0}) \to \mathbb{R}^p$ such that $\displaystyle\lim_{\mathbf{h}\to 0} \frac{\|E(\mathbf{h})\|}{\|\mathbf{h}\|} = 0$ and

$$g(\mathbf{a} + \mathbf{h}) = g(\mathbf{a}) + J_g(\mathbf{a}) \cdot \mathbf{h}^T + E(\mathbf{h}) \quad \text{for all } \|\mathbf{h}\| < r. \tag{5.3}$$

Note that $\displaystyle\lim_{\mathbf{h}\to 0} E(\mathbf{h}) = 0$, so letting

$$\mathbf{k} := J_g(\mathbf{a}) \cdot \mathbf{h}^T + E(\mathbf{h}) \tag{5.4}$$

and by taking $r > 0$ sufficiently small, one has $\|\mathbf{k}\| < \rho$. Thus, letting $\mathbf{k}$ defined by (5.4) in (5.2), we have

$$\begin{aligned}
(f \circ g)(\mathbf{a} + \mathbf{h}) = f(g(\mathbf{a} + \mathbf{h})) &= f(g(\mathbf{a}) + \mathbf{k}) \quad \text{(by (5.3))} \\
&= f(g(\mathbf{a})) + \nabla f(g(\mathbf{a})) \bullet \mathbf{k} + \eta(\mathbf{k}) \quad \text{(by (5.2))}.
\end{aligned}$$

Now from (5.4) we derive

$$(f \circ g)(\mathbf{a} + \mathbf{h}) = f(g(\mathbf{a})) + \nabla f(g(\mathbf{a})) \bullet \{J_g(\mathbf{a}) \cdot \mathbf{h}^T + E(\mathbf{h})\} + \eta(J_g(\mathbf{a}) \cdot \mathbf{h}^T + E(\mathbf{h})).$$

Hence,

$$(f \circ g)(\mathbf{a} + \mathbf{h}) = f(g(\mathbf{a})) + \left[\nabla f(g(\mathbf{a})) \cdot J_g(\mathbf{a}) \right] \bullet \mathbf{h} + \delta(\mathbf{h}), \tag{5.5}$$

where $\delta : B_r(\mathbf{0}) \to \mathbb{R}$ is defined by

$$\delta(\mathbf{h}) = \nabla f(g(\mathbf{a})) \bullet E(\mathbf{h}) + \eta(J_g(\mathbf{a}) \cdot \mathbf{h}^T + E(\mathbf{h})).$$

We claim that $\displaystyle\lim_{\mathbf{h}\to 0} \frac{\delta(\mathbf{h})}{\|\mathbf{h}\|} = 0$. Indeed, we have

$$\frac{\delta(\mathbf{h})}{\|\mathbf{h}\|} = \nabla f(g(\mathbf{a})) \cdot \frac{E(\mathbf{h})}{\|\mathbf{h}\|} + \frac{\eta(J_g(\mathbf{a}) \cdot \mathbf{h}^T + E(\mathbf{h}))}{\|J_g(\mathbf{a}) \cdot \mathbf{h}^T + E(\mathbf{h})\|} \cdot \frac{\|J_g(\mathbf{a}) \cdot \mathbf{h}^T + E(\mathbf{h})\|}{\|\mathbf{h}\|}. \tag{5.6}$$

By the properties of η and E one has

$$\lim_{\mathbf{h}\to 0} \frac{\|E(\mathbf{h})\|}{\|\mathbf{h}\|} = 0 \quad \text{and} \quad \lim_{\mathbf{h}\to 0} \frac{\eta(J_g(\mathbf{a}) \cdot \mathbf{h}^T + E(\mathbf{h}))}{\|J_g(\mathbf{a}) \cdot \mathbf{h}^T + E(\mathbf{h})\|} = 0.$$

Further,

$$\begin{aligned}
\frac{\|J_g(\mathbf{a}) \cdot \mathbf{h}^T + E(\mathbf{h})\|}{\|\mathbf{h}\|} &\leq \frac{|J_g(\mathbf{a})|\|\mathbf{h}\| + \|E(\mathbf{h})\|}{\|\mathbf{h}\|} \\
&\leq |J_g(\mathbf{a})| + \frac{\|E(\mathbf{h})\|}{\|\mathbf{h}\|} \\
&\leq M \text{ (bounded)}.
\end{aligned}$$

In the above estimates $|J_g(\mathbf{a})|$ denotes the sum of the absolute value of all partial derivatives of g. Thus, from (5.6) we deduce $\lim\limits_{\mathbf{h}\to 0}\dfrac{\delta(\mathbf{h})}{\|\mathbf{h}\|} = 0$ and from (5.5) and Theorem 4.16 it follows that $f \circ g$ is differentiable at $\mathbf{a}$ and that (5.1) holds. $\qquad\square$

Example 5.2 *Let $f(x,y) = x^2 e^{-y}$. Assume $x = 2t - 1$ and $y = t + 2$. In this way, $f(x,y)$ becomes a function of t variable only. Find $\dfrac{df}{dt}$.*

Solution. We can replace x and y by their expressions of t in the definition of the function f and thus the calculation is reduced to finding the derivative of a one variable function. However, we prefer to use the Chain Rule as given in Theorem 5.1. Let $g(t) = (2t - 1, t + 2)$. Observe that $f : \mathbb{R}^2 \to \mathbb{R}$ and $g : \mathbb{R} \to \mathbb{R}^2$ are differentiable so we may apply Theorem 5.1 with $m = 1$ and $p = 2$. Thus, $f \circ g : \mathbb{R} \to \mathbb{R}$ is differentiable and

$$\frac{df}{dt} = (\nabla f)(g(t)) \bullet g'(t).$$

We have

$$\nabla f(x,y) = (2xe^{-y}, -x^2 e^{-y})$$
$$(\nabla f)(g(t)) = \left(2(2t - 1)e^{-t-2}, -(2t - 1)^2 e^{-t-2}\right)$$

and finally

$$\frac{df}{dt} = \left(2(2t - 1)e^{-t-2}, -(2t - 1)^2 e^{-t-2}\right) \bullet (2, 1) = -(4t^2 - 12t + 5)e^{-t-2}.$$

Example 5.3 *Suppose f is a differentiable function such that*

$$\frac{\partial f}{\partial x}(-1, 2) = 0 \quad and \quad \frac{\partial f}{\partial y}(-1, 2) = 3.$$

Find the rate of change of f at $(-1, 2)$ along the parabola $y = x^2 - 3x - 2$.

Solution. Let $\phi = f \circ g$, where $g(t) = (t, t^2 - 3t - 2)$. The rate of change of f at $(-1, 2)$ along the parabola $y = x^2 - 3x - 2$ is $\frac{d\phi}{dt}(-1)$. By Chain Rule we compute

$$\frac{d\phi}{dt}(t) = \nabla f(g(t)) \bullet J_g(t) = \frac{\partial f}{\partial x}(t, t^2 - 3t - 2) + (2t - 3)\frac{\partial f}{\partial y}(t, t^2 - 3t - 2).$$

For $t = -1$ we find

$$\frac{d\phi}{dt}(-1) = \frac{\partial f}{\partial x}(-1, 2) - 5\frac{\partial f}{\partial y}(-1, 2) = -15.$$

Example 5.4 *Let $f : \mathbb{R}^2 \to \mathbb{R}$ be a differentiable function and $x = r\cos\theta$, $y = r\sin\theta$ be the polar coordinates of $(x,y) \in \mathbb{R}^2$. Then*

$$\frac{\partial f}{\partial r} = \frac{\partial f}{\partial x}\cos\theta + \frac{\partial f}{\partial y}\sin\theta$$

$$\frac{\partial f}{\partial \theta} = -\frac{\partial f}{\partial x}r\sin\theta + \frac{\partial f}{\partial y}r\cos\theta.$$

Solution. Let $g(r,\theta) = (r\cos\theta, r\sin\theta)$. The above identities are obtained by using the Chain Rule for the composite function $\phi = f \circ g$ whose variables are r and θ. We have

$$\nabla\phi = \nabla f(g(r,\theta)) \cdot J_g(r,\theta).$$

Since

$$J_g(r,\theta) = \begin{pmatrix} \cos\theta & \sin\theta \\ -r\sin\theta & r\cos\theta \end{pmatrix},$$

the conclusion follows.

Remark 5.5 It is more convenient to use the following approach: $(x,y) = (r\cos\theta, r\sin\theta)$ so x and y become themselves functions of r and θ. We next write

$$\frac{\partial f}{\partial r} = \frac{\partial f}{\partial x}\frac{\partial x}{\partial r} + \frac{\partial f}{\partial y}\frac{\partial y}{\partial r} = \frac{\partial f}{\partial x}\cos\theta + \frac{\partial f}{\partial y}\sin\theta$$

$$\frac{\partial f}{\partial \theta} = \frac{\partial f}{\partial x}\frac{\partial x}{\partial \theta} + \frac{\partial f}{\partial y}\frac{\partial y}{\partial \theta} = -\frac{\partial f}{\partial x}r\sin\theta + \frac{\partial f}{\partial y}r\cos\theta.$$

Example 5.6 *The length a of a cuboid increases at a rate of 2 cm/min, the width b decreases at a rate of 3 cm/min while the height c increases at a rate of 4 cm/min. Use the Chain Rule to find the rate of change of the volume of the cuboid at $a = 5$ cm, $b = 6$ cm and $c = 7$ cm.*

Solution. We proceed as explained in Remark 5.5 using

$$V = abc \quad \text{and} \quad \frac{da}{dt} = 2, \quad \frac{db}{dt} = -3, \quad \frac{dc}{dt} = 4.$$

By the Chain Rule, we have

$$\frac{dV}{dt} = \frac{\partial V}{\partial a}\frac{da}{dt} + \frac{\partial V}{\partial b}\frac{db}{dt} + \frac{\partial V}{\partial c}\frac{dc}{dt}$$

so that

$$\frac{dV}{dt} = bc\frac{da}{dt} + ac\frac{db}{dt} + ab\frac{dc}{dt} = 2bc - 3ac + 4ab.$$

At $(a,b,c) = (5,6,7)$ we find $\dfrac{dV}{dt} = 99$ cm^3/min.

Example 5.7 $f(x, y, z)$ *is a differentiable function where*

$$x = st, \quad y = t^3 \quad and \quad z = e^{s-t}.$$

Use the Chain Rule to express the partial derivatives $\dfrac{\partial f}{\partial s}$ *and* $\dfrac{\partial f}{\partial t}$ *in terms of* $\dfrac{\partial f}{\partial x}, \dfrac{\partial f}{\partial y}, \dfrac{\partial f}{\partial z}$ *and s, t.*

Solution. By Chain Rule, we have

$$\frac{\partial f}{\partial s} = \frac{\partial f}{\partial x}\frac{\partial x}{\partial s} + \frac{\partial f}{\partial y}\frac{\partial y}{\partial s} + \frac{\partial f}{\partial z}\frac{\partial z}{\partial s}$$

$$\frac{\partial f}{\partial t} = \frac{\partial f}{\partial x}\frac{\partial x}{\partial t} + \frac{\partial f}{\partial y}\frac{\partial y}{\partial t} + \frac{\partial f}{\partial z}\frac{\partial z}{\partial t}$$

and

$$\begin{cases} \dfrac{\partial x}{\partial s} = t, & \dfrac{\partial y}{\partial s} = 0, & \dfrac{\partial z}{\partial s} = e^{s-t} \\[2mm] \dfrac{\partial x}{\partial t} = s, & \dfrac{\partial y}{\partial t} = 3t^2, & \dfrac{\partial z}{\partial t} = -e^{s-t}. \end{cases}$$

Hence

$$\begin{cases} \dfrac{\partial f}{\partial s} = t\dfrac{\partial f}{\partial x} + e^{s-t}\dfrac{\partial f}{\partial z} \\[3mm] \dfrac{\partial f}{\partial t} = s\dfrac{\partial f}{\partial x} + 3t^2\dfrac{\partial f}{\partial y} - e^{s-t}\dfrac{\partial f}{\partial z}. \end{cases}$$

Exercises

Exercise 1. The radius r of a cone increases at the rate of 2 cm/s and the height h of the cone decreases at the rate of -3 cm/s. Use the Chain Rule to find the rate of change of the volume of the cone where $r = 5$ cm and $h = 12$ cm.

Exercise 2. The temperature at the surface of a sea is given by

$$T(x, y) = e^{-x}y - \frac{x^2}{1 + y^2}.$$

If a person swims along the circles $x(t) = \cos t$, $y(t) = \sin t$, find $\dfrac{dT}{dt}$.

Exercise 3. The differentiable function $f : \mathbb{R}^2 \to \mathbb{R}$ satisfies

$$\frac{\partial f}{\partial x}(3, 2) = 1 \quad and \quad \frac{\partial f}{\partial y}(3, 2) = -3.$$

Find the rate of change of f at $(3,2)$ along the curve $xy = 6$.

Exercise 4. Let $f : \mathbb{R}^3 \to \mathbb{R}$ be a differentiable function such that $\nabla f(2,-2,-2) = (1,0,-4)$. What is the rate of change of f at $(2,-2,-2)$ along the surface $xyz = 8$?

Exercise 5. The differentiable function $f(x,y,z)$ satisfies

$$\frac{\partial f}{\partial x}(1,-3,0) = 2, \quad \frac{\partial f}{\partial y}(1,-3,0) = 3 \quad \text{and} \quad \frac{\partial f}{\partial y}(1,-3,0) = -4.$$

Find the rate of change of f at $(1,-3,0)$ along the line ℓ which is the intersection of the planes $3x + y + z = 0$ and $x + z = 1$.
Hint: First, find a parametric equation of the line ℓ.

Exercise 6. Let $g : \mathbb{R}^2 \to \mathbb{R}$ be a differentiable function and

$$x = 4te^{2s}, \quad y = 2s^2 - t^2.$$

Compute $\dfrac{\partial g}{\partial s} - 2t\dfrac{\partial g}{\partial t}$ in terms of $\dfrac{\partial g}{\partial x}, \dfrac{\partial g}{\partial y}$ and s, t.

Exercise 7. Let $f : \mathbb{R} \to \mathbb{R}$ be a differentiable function and $h : \mathbb{R}^2 \to \mathbb{R}$ be given by $h(x,y) = f(x^2 y)$. Prove that

$$\frac{\partial h}{\partial x}(-1,1) + 2\frac{\partial h}{\partial y}(-1,1) = 0.$$

Exercise 8. Let $f(x,y)$ be a function which has continuous partial derivatives and let

$$x = r^3 - 3rs^2 \quad \text{and} \quad y = 3r^2 s - s^3,$$

where r and s are independent variables.

(i) Use the Chain Rule to express the partial derivatives $\dfrac{\partial f}{\partial r}$ and $\dfrac{\partial f}{\partial s}$ in terms of $\dfrac{\partial f}{\partial x}, \dfrac{\partial f}{\partial y}$, r and s;

(ii) Hence, show that

$$\left(\frac{\partial f}{\partial r}\right)^2 + \left(\frac{\partial f}{\partial s}\right)^2 = 9(r^2 + s^2)^2 \left[\left(\frac{\partial f}{\partial x}\right)^2 + \left(\frac{\partial f}{\partial y}\right)^2\right].$$

Exercise 9. $f(x,y,z)$ is a differentiable function where

$$x = e^r \cos t, \quad y = e^r \sin t \quad \text{and} \quad z = r^2.$$

Find $\dfrac{\partial f}{\partial r}$ and $\dfrac{\partial f}{\partial t}$ in terms of $\dfrac{\partial f}{\partial x}, \dfrac{\partial f}{\partial y}, \dfrac{\partial f}{\partial z}$ and r, t.

Exercise 10. $f(x, y, z)$ is a differentiable function where

$$x = r \cos \varphi \sin \theta, \quad y = r \sin \varphi \sin \theta \quad \text{and} \quad z = r \cos \theta$$

are the spherical coordinates of (x, y, z). Use the Chain Rule to express $\dfrac{\partial f}{\partial r}$, $\dfrac{\partial f}{\partial \varphi}$ and $\dfrac{\partial f}{\partial \theta}$ in terms of $\dfrac{\partial f}{\partial x}$, $\dfrac{\partial f}{\partial y}$, $\dfrac{\partial f}{\partial z}$ and r, φ, θ.

Exercise 11. Let $f : \mathbb{R} \to \mathbb{R}$ be a differentiable function and $\phi(x, y) = f\left(\dfrac{xy}{x + y}\right)$.

(i) What is the domain D of definition of ϕ?

(ii) Show that

$$x^2 \frac{\partial \phi}{\partial x} = y^2 \frac{\partial \phi}{\partial x} \quad \text{on } D.$$

Exercise 12. Let $f : \mathbb{R} \to \mathbb{R}$ be a differentiable function and $\phi : \mathbb{R}^2 \to \mathbb{R}$ be defined as $\phi(x, y) = xf(x^3 - y^2)$. Show that

$$2y\phi(x, y) = 2xy\frac{\partial \phi}{\partial x}(x, y) + 3x^4\frac{\partial \phi}{\partial y}(x, y).$$

Exercise 13. Let $f : \mathbb{R}^2 \to \mathbb{R}$ be a differentiable function and $\phi : \mathbb{R}^3 \to \mathbb{R}$ be given by $\phi(x, y, t) = f(tx, ty)$.

(i) Show that $\nabla \phi(0, 0, 0) = (0, 0)$;

(ii) Show that

$$\frac{\partial \phi}{\partial x}(1, 1, 1) = \frac{\partial f}{\partial x}(1, 1), \quad \frac{\partial \phi}{\partial y}(1, 1, 1) = \frac{\partial f}{\partial y}(1, 1),$$

$$\frac{\partial \phi}{\partial t}(1, 1, 1) = \frac{\partial f}{\partial x}(1, 1) + \frac{\partial f}{\partial y}(1, 1).$$

Exercise 14. A differentiable function $f : \mathbb{R}^2 \to \mathbb{R}$ is called homogeneous of degree $n \geq 1$ if

$$f(tx, ty) = t^n f(x, y) \quad \text{for all } t, x, y \in \mathbb{R}.$$

(i) Check that $f(x, y) = x^4 - 3x^3y + 5x^2y^2$ is homogeneous of degree 4;

(ii) Show that if f is homogeneous of degree $n \geq 2$, then $\dfrac{\partial f}{\partial x}$ and $\dfrac{\partial f}{\partial y}$ are homogeneous of degree $n - 1$;

(iii)   Show that if f is homogeneous of degree $n \geq 1$, then

$$x\frac{\partial f}{\partial x} + y\frac{\partial f}{\partial y} = nf.$$

Exercise 15. A function $f : \mathbb{R}^m \to \mathbb{R}$, $m \geq 2$ is said to be *radially symmetric* if there exists $\phi : \mathbb{R} \to \mathbb{R}$ such that $f(x) = \phi(\|\mathbf{x}\|)$ for all $\mathbf{x} \in \mathbb{R}^m$.

Check that if ϕ is differentiable on $\mathbb{R}$, then f is differentiable on $\mathbb{R}^m \setminus \{\mathbf{0}\}$ and

$$\nabla f(\mathbf{x}) = \phi'(\|\mathbf{x}\|)\frac{\mathbf{x}}{\|\mathbf{x}\|} \qquad \text{for all} \quad \mathbf{x} \in \mathbb{R}^m \setminus \{\mathbf{0}\}.$$

5.2   Implicit Differentiation

Let $y = y(x)$ be a differentiable function that is defined on a open interval $I \subset \mathbb{R}$. Assume that, instead of knowing the analytical expression of $y(x)$, we know that $(x, y(x))$ satisfies some relation of the type

$$f(x, y(x)) = c \quad \text{for all } x \in I, \tag{5.7}$$

where f is a function of two variables. This means that y is defined *implicitely* by the equation (5.7). To better grasp the concept, think of the equation

$$x^3 + xy + y^3 = 5. \tag{5.8}$$

Then, equation (5.8) defines implicitly y as a function of x. Turning back to (5.7), we have that for all $x \in I$, the pair $(x, y(x))$ belongs to the level curve $L(f, c)$ of f. Assuming that f is differentiable in $\mathbb{R}^2$, we can use the Chain Rule to differentiate in (5.7) and obtain:

$$\frac{\partial f}{\partial x}(x, y(x)) + \frac{\partial f}{\partial x}(x, y(x))\frac{dy}{dx} = 0 \quad \text{for all } x \in I. \tag{5.9}$$

If $x_0 \in I$, $y_0 = y(x_0)$ and $\dfrac{\partial f}{\partial x}(x_0, y_0) \neq 0$, then (5.9) yields

$$\frac{dy}{dx}(x_0) = -\frac{\dfrac{\partial f}{\partial x}(x_0, y_0)}{\dfrac{\partial f}{\partial y}(x_0, y_0)}.$$

To summarize, we have:

- We do not know the analytic expression of $y(x)$, but we do know the equation that defines $y(x)$ implicitly;

- We do not know the analytic expression of $y(x)$, but we can compute the derivative $\dfrac{dy}{dx}$ at a specific point x_0 under some conditions.

Let us come back to our example (5.8) and take $(x_0, y_0) = (-1, 2)$. Differentiating with respect to x one obtains

$$3x^2 + y + x\frac{dy}{dx} + 3y^2\frac{dy}{dx} = 0.$$

At $(x_0, y_0) = (-1, 2)$ one has

$$3x_0^2 + y_0 + (x_0 + 3y_0^2)\frac{dy}{dx}(x_0) = 0 \implies \frac{dy}{dx}(-1) = -\frac{5}{11}.$$

Although we do not have an analytic expression of y as a function of x, the above method allows us to find the derivative $\dfrac{dy}{dx}$ at a specific point x_0. The curve $y(x)$ that satisfies (5.8) is depicted in Figure 5.1. Inspecting the graphical representation, we see that $x \mapsto y(x)$ is a differentiable function (as there are no sharp points on the graph); this fact will be rigorously justified in Chapter 9.

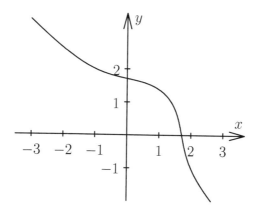

Figure 5.1
The implicit curve $y = y(x)$ given by (5.8).

In the following, we shall apply the above implicit differentiation method to several variable functions.

Example 5.8 *Suppose $z = z(x, y)$ is a differentiable function such that*

$$2x^2y - 6yz + e^{x+z} = 1 \quad and \quad z(-3, 1) = 3.$$

Find the derivatives $\dfrac{\partial z}{\partial x}(-3, 1)$ *and* $\dfrac{\partial z}{\partial y}(-3, 1).$

We shall justify in Example 9.5 the existence of a such a differentiable function $z = z(x, y)$ which lives in a disc centred at $(-3, 1)$. For the time being, we are only concerned with the way one computes its derivatives at $(-3, 1)$.

Solution. Differentiating the implicit equation with respect to x and y respectively, we obtain

$$\begin{cases} 4xy - 6y\dfrac{\partial z}{\partial x} + e^{x+z} + e^{x+z}\dfrac{\partial z}{\partial x} = 0 \\[2mm] 2x^2 - 6z - 6y\dfrac{\partial z}{\partial y} + e^{x+z}\dfrac{\partial z}{\partial y} = 0 \end{cases} \implies \begin{cases} (e^{x+z} - 6y)\dfrac{\partial z}{\partial x} = -4xy - e^{x+z} \\[2mm] (e^{x+z} - 6y)\dfrac{\partial z}{\partial y} = -2x^2 + 6z. \end{cases}$$

At $(x, y, z) = (-3, 1, 3)$ we deduce $\dfrac{\partial z}{\partial x}(-3, 1) = -\dfrac{11}{5}$ and $\dfrac{\partial z}{\partial y}(-3, 1) = 0$.

Example 5.9 *The equation*

$$z^3 - xy + yz + y^3 = 2$$

defines the differentiable function z in terms of the independent variables x and y.

    (i) *Find $z(1, 1)$;*

    (ii) *Use implicit differentiation to find the values of the partial derivatives $\dfrac{\partial z}{\partial x}$ and $\dfrac{\partial z}{\partial y}$ at the point $(1, 1)$;*

    (iii) *Find the points (x, y) where $\nabla z(x, y) = (0, 0)$.*

Solution. (i) Letting $(x, y) = (1, 1)$ in the implicit equation satisfied by z one gets $z^3 + z - 2 = 0$, that is, $(z - 1)(z^2 + z + 2) = 0$. From here we find $z(1, 1) = 1$.

    (ii) Given the implicit equation satisfied by $z = z(x, y)$, we differentiate it with respect to x and y respectively to obtain

$$3z^2\frac{\partial z}{\partial x} - y + y\frac{\partial z}{\partial x} = 0$$

$$3z^2\frac{\partial z}{\partial y} - x + z + y\frac{\partial z}{\partial y} + 3y^2 = 0.$$

Hence, solving for $\dfrac{\partial z}{\partial x}$ and $\dfrac{\partial z}{\partial y}$ we find

$$\frac{\partial z}{\partial x} = \frac{y}{3z^2 + y} \quad \text{and} \quad \frac{\partial z}{\partial y} = \frac{x - z - 3y^2}{3z^2 + y}. \tag{5.10}$$

Observe that $\dfrac{\partial z}{\partial x}$ and $\dfrac{\partial z}{\partial y}$ are defined by the above equalities as long as $3z^2 + y \neq 0$. At $(x, y, z) = (1, 1, 1)$ we find

$$\frac{\partial z}{\partial x}(1, 1) = \frac{1}{4}, \quad \frac{\partial z}{\partial y}(1, 1) = -\frac{3}{4}.$$

(iii) If $\dfrac{\partial z}{\partial x} = \dfrac{\partial z}{\partial y} = 0$, then our equalities (5.10) give

$$y = 0 \quad \text{and} \quad x - z - 3y^2 = 0,$$

and hence $x = z$. We substitute $y = 0$ into the implicit equation

$$z^3 - xy + yz + y^3 - 2 = 0,$$

and derive $z^3 = 2$. Hence the required point is $(x, y) = (\sqrt[3]{2}, 0)$.

_____

Exercises

Exercise 16. The differentiable function $z = z(r, \theta)$ is defined implicitly by

$$z \cos \theta + z^2 = \frac{r \sin \theta}{z}.$$

Find $z\left(1, \dfrac{\pi}{2}\right)$, $\dfrac{\partial z}{\partial r}\left(1, \dfrac{\pi}{2}\right)$ and $\dfrac{\partial z}{\partial \theta}\left(1, \dfrac{\pi}{2}\right)$.

Exercise 17. A particle moves on the surface of equation $xy - yz + 2zx = 6$ in $\mathbb{R}^3$. If the particle is now located at $(-1, 2, -2)$, find how fast is z changing with respect to x and y.

Exercise 18. The function $z = z(x, y)$ satisfies the equation

$$e^{x+y-z} - x^2 z + y^3 = 8 \quad \text{for all } (x, y) \in (-\infty, 0) \times \mathbb{R}.$$

(i) Check that for all $(x, y) \in (-\infty, 0) \times \mathbb{R}$ there exists a unique $z(x, y)$ that satisfies the above equation;

(ii) Assume the function z is differentiable (which is true, as we shall see in Chapter 9). Find $z(-1, 2)$, $\dfrac{\partial z}{\partial x}(-1, 2)$ and $\dfrac{\partial z}{\partial y}(-1, 2)$.

Exercise 19. The differentiable function $u(x, y)$ and $v(x, y)$ satisfy

$$\begin{cases} e^u = 3(v + x) + 1 \\ e^v = 2(u + y) + 1 \\ u(0, 0) = v(0, 0) = 0. \end{cases}$$

Find $\nabla u(0, 0)$ and $\nabla v(0, 0)$.

5.3   Mean Value Theorem in Several Variables

The Mean Value Theorem for functions of several variables is an application
of the Chain Rule which was introduced in Section 5.1. More precisely, we
have:

Theorem 5.10 (Mean Value Theorem in Several Variables)
*Let $f : D \subset \mathbb{R}^m \to \mathbb{R}$ be a differentiable function and $\mathbf{a}, \mathbf{b} \in D$ be such that
$[\mathbf{a}, \mathbf{b}] \subset D$. Then, there exists $\mathbf{c} \in (\mathbf{a}, \mathbf{b})$ such that*

$$f(\mathbf{b}) - f(\mathbf{a}) = \nabla f(\mathbf{c}) \bullet (\mathbf{b} - \mathbf{a}).$$

Proof Let $g : [0, 1] \to \mathbb{R}$, $g(t) = \mathbf{a} + t(\mathbf{b} - \mathbf{a})$ and $\phi(t) = (f \circ g)(t)$. Note that
ϕ is well defined since $[\mathbf{a}, \mathbf{b}] \subset D$. Also, by the Chain Rule one has

$$\phi'(t) = \nabla f(g(t)) \bullet J_g(t) = \nabla f\big(\mathbf{a} + t(\mathbf{b} - \mathbf{a})\big) \bullet (\mathbf{b} - \mathbf{a}) \quad \text{for all } t \in (0, 1).$$

Now, by the Mean Value Theorem for single variable there exists $\tau \in (0, 1)$
such that
$$\begin{aligned}
f(\mathbf{b}) - f(\mathbf{a}) = \phi(1) - \phi(0) &= \phi'(\tau) \\
&= \nabla f\big(\mathbf{a} + \tau(\mathbf{b} - \mathbf{a})\big) \bullet (\mathbf{b} - \mathbf{a}) \\
&= \nabla f(\mathbf{c}) \bullet (\mathbf{b} - \mathbf{a}),
\end{aligned}$$
where $\mathbf{c} = \mathbf{a} + \tau(\mathbf{b} - \mathbf{a}) \in (\mathbf{a}, \mathbf{b})$. $\square$

Example 5.11 *Let $f : \mathbb{R}^2 \to \mathbb{R}$ be given by $f(x, y) = xy(x - y)$ and $\mathbf{a} =
(1, -1)$, $\mathbf{b} = (-1, 1)$. Apply the Mean Value Theorem for several variables to
f on the interval $[\mathbf{a}, \mathbf{b}]$ and find the corresponding point(s) $\mathbf{c}$ in this result.*

Solution. f is a differentiable function so we are well entitled to apply the
Mean Value Theorem. Thus, there exists at least one $\mathbf{c} \in (\mathbf{a}, \mathbf{b})$ such that
$f(\mathbf{b}) - f(\mathbf{a}) = \nabla f(\mathbf{c}) \bullet (\mathbf{b} - \mathbf{a})$. Observe that

$$f(\mathbf{a}) = f(1, -1) = -2, \quad f(\mathbf{b}) = f(-1, 1) = 2$$

and if $\mathbf{c} = (c_1, c_2)$ one has $\nabla f(\mathbf{c}) = (2c_1c_2 - c_2^2, c_1^2 - 2c_1c_2)$. Thus, the Mean
Value Theorem yields

$$4 = (2c_1c_2 - c_2^2, c_1^2 - 2c_1c_2) \bullet (-2, 2).$$

It follows that $c_1^2 + c_2^2 - 4c_1c_2 = 2$. Note that $\mathbf{c} \in (\mathbf{a}, \mathbf{b})$ so $c_2 = -c_1$. Thus,
the previous equality implies $6c_1^2 = 2$ and from here $c_2 = \pm\frac{1}{\sqrt{3}}$. We find

$$\mathbf{c} = \left(\frac{1}{\sqrt{3}}, -\frac{1}{\sqrt{3}}\right) \quad \text{and} \quad \mathbf{c} = \left(-\frac{1}{\sqrt{3}}, \frac{1}{\sqrt{3}}\right).$$

Remark 5.12 Sometimes we may find a unique point $\mathbf{c}$ or infinitely many
points $\mathbf{c}$ that satisfy the Mean Value Theorem 5.10 as illustrated in Exercises
21 and 22.

Exercises

Exercise 20. Assume $f : B_r(\mathbf{a}) \subset \mathbb{R}^m \to \mathbb{R}$, $m \geq 1$, is differentiable and that $\nabla f(\mathbf{x}) = 0$ for all $\mathbf{x} \in B_r(\mathbf{a})$. Using the Mean Value Theorem 5.10 show that f is constant.

Note: We have seen in Chapter 4 Exercise 6 that the above result holds under a weaker assumption, namely f has partial derivatives at each point in $B_r(\mathbf{a})$ (instead of being differentiable). However, the solution of the above exercise is more elegant when one uses Theorem 5.10.

Exercise 21. Let $f : \mathbb{R}^2 \to \mathbb{R}$ be given by $f(x, y) = x^2 - xy$. Show that there are infinitely many points $\mathbf{c}$ on the open interval that joins $\mathbf{a} = (-1, 0)$ and $\mathbf{b} = (1, 2)$ which satisfy the Mean Value Theorem 5.10 on $[\mathbf{a}, \mathbf{b}]$.

Exercise 22. Let $f : \mathbb{R}^2 \to \mathbb{R}$ be given by $f(x, y) = x^2 e^y$. Show that there is a unique point $\mathbf{c}$ on the open interval that joins $\mathbf{a} = (0, -1)$ and $\mathbf{b} = (-1, 1)$ which satisfies the Mean Value Theorem 5.10 on $[\mathbf{a}, \mathbf{b}]$.

Exercise 23. Let $f : [a, b] \subset \mathbb{R} \to \mathbb{R}^p$, $p \geq 2$, be a function with the following properties:

(i) f is continuous on $[a, b]$;

(ii) f is differentiable on (a, b);

(iii) $f(a)$ and $f(b)$ are orthogonal to a vector $\mathbf{u} \in \mathbb{R}^p$.

Prove that there exists $c \in (a, b)$ such that $f'(c)$ is orthogonal to $\mathbf{u}$.

Hint: Apply the Mean Value Theorem to $g : [a, b] \to \mathbb{R}$, $g(t) = f(t) \bullet \mathbf{u}$.

Exercise 24. Let $f : \mathbb{R}^2 \to \mathbb{R}$ be a differentiable function. Show that there exists $c \in (0, 1)$ such that

$$f(1, 1) - f(0, 0) = 2c\frac{\partial f}{\partial x}(c^2, c^2) + 2c\frac{\partial f}{\partial y}(c^2, c^2).$$

Hint: Apply the Mean Value Theorem to $g(t) = f(t^2, t^2)$.

Exercise 25. Let $\mathbf{u} \in \mathbb{R}^p$ be a unit vector and $F : D \subset \mathbb{R}^m \to \mathbb{R}^p$ be a differentiable function. Let also $\mathbf{a}, \mathbf{b} \in D$ be such that the line segment joining $\mathbf{a}$ and $\mathbf{b}$ is contained in D. Prove that there exists $\mathbf{c} \in (\mathbf{a}, \mathbf{b})$ such that

$$\mathbf{u} \bullet (F(\mathbf{b}) - F(\mathbf{a})) = (\mathbf{u}\, J_F(\mathbf{c})) \bullet (\mathbf{b} - \mathbf{a}).$$

Hint: Apply the Mean Value Theorem to

$$g : [0, 1] \to \mathbb{R}, \ g(t) = \mathbf{u} \bullet F(\mathbf{a} + t(\mathbf{b} - \mathbf{a})).$$

Exercise 26. Prove or otherwise disprove by constructing a counterexample the following vesion of Rolle's Theorem: If $f : D \subset \mathbb{R}^m \to \mathbb{R}$ is differentiable and $\mathbf{a}, \mathbf{b} \subset D$ are such that $[\mathbf{a}, \mathbf{b}] \subset D$ and $f(\mathbf{b}) = f(\mathbf{a})$, then there exists $\mathbf{c} \in (\mathbf{a}, \mathbf{b})$ such that $\nabla f(\mathbf{c}) = \mathbf{0}$.

Exercise 27. Let $f : \mathbb{R}^2 \to \mathbb{R}$ be a differentiable function such that $\dfrac{\partial f}{\partial x} = \dfrac{\partial f}{\partial y}$.

(i) Give example of a nonconstant function with the above property;

(ii) Using the Mean Value Theorem, show that $f(x, y) = f(x + y, 0)$ for all $x, y \in \mathbb{R}^2$;

(iii) Deduce that there exists a differentiable function $g : \mathbb{R} \to \mathbb{R}$ such that $f(x, y) = g(x + y)$;

(iv) Formulate a similar result if $\dfrac{\partial f}{\partial x} = -\dfrac{\partial f}{\partial y}$.

6

Directional Derivative

6.1 Directional Derivative

Let $f : D \subset \mathbb{R}^2 \to \mathbb{R}$ be a function defined on an open set D containing (a, b). Then $\dfrac{\partial f}{\partial x}(a, b)$ and $\dfrac{\partial f}{\partial y}(a, b)$ measure the rate of change of f in the direction of x and y axis respectively.

Consider now a line ℓ in the plane $z = 0$ that passes through (a, b), whose direction is given by the unit vector $\mathbf{u} = (u_1, u_2)$ (see Figure 6.1).

The parametric equation of the line ℓ is thus given by

$$\ell : \begin{cases} x(t) = a + tu_1 \\ y(t) = b + tu_2 \end{cases} \quad t \in \mathbb{R}.$$

Denote by σ the plane that contains the line ℓ and is orthogonal to the plane $z = 0$; see Figure 6.1. The trace of the surface $z = f(x, y)$ in the plane σ is

$$g(t) = f(x(t), y(t)) = f(a + tu_1, b + tu_2).$$

Note that $g(0) = f(a, b)$. Thus, the slope of the tangent T to the graph of $g(t)$ at $t = 0$ is given by

$$g'(0) = \lim_{t \to 0} \frac{g(t) - g(0)}{t - 0} = \lim_{t \to 0} \frac{f(a + tu_1, b + tu_2) - f(a, b)}{t}.$$

This is exactly the directional derivative of f at (a, b) in the direction of $\mathbf{u}$. More precisely, we have:

Definition 6.1 *Let $f : D \subset \mathbb{R}^m \to \mathbb{R}$, $\mathbf{a} \in D$ and let $\mathbf{u}$ be a unit vector. The directional derivative of f at $\mathbf{a}$ in the direction $\mathbf{u}$ is given by*

$$D_{\mathbf{u}} f(\mathbf{a}) = \lim_{t \to 0} \frac{f(\mathbf{a} + t\mathbf{u}) - f(\mathbf{a})}{t}.$$

DOI: 10.1201/9781003449652-6

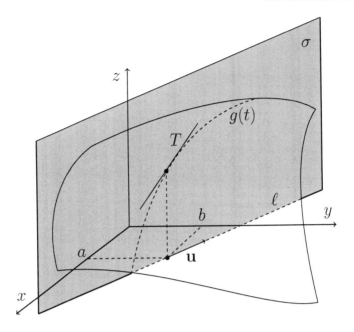

Figure 6.1
The directional derivative as the slope of tangent line T.

Remark 6.2 If $\mathbf{u} = \mathbf{e}_j$ for some $1 \leq j \leq m$, then the above limit is exactly the partial derivative $\dfrac{\partial f}{\partial x_j}(\mathbf{a})$ (see Definition 4.1). More precisely,

$$D_{\mathbf{e}_j} f(\mathbf{a}) = \frac{\partial f}{\partial x_j}(\mathbf{a}).$$

Example 6.3 *Let $f : \mathbb{R}^2 \to \mathbb{R}$ be defined by*

$$f(x,y) = \begin{cases} \dfrac{x^2 y}{x^4 + 3y^2} & \text{if } (x,y) \neq (0,0) \\ 0 & \text{if } (x,y) = (0,0). \end{cases}$$

(i) *Compute the directional derivative of f at $(0,0)$ in the direction $(1,1)$;*

(ii) *Show that f is not differentiable at $(0,0)$ but has directional derivative at $(0,0)$ in any direction.*

Solution. (i) A unit vector in the direction $(1,1)$ is

$$\mathbf{u} = \frac{1}{\|(1,1)\|}(1,1) = \left(\frac{1}{\sqrt{2}}, \frac{1}{\sqrt{2}} \right).$$

Hence

$$D_{\mathbf{u}}f(0,0) = \lim_{t \to 0} \frac{f((0,0)+t\mathbf{u}) - f(0,0)}{t} = \lim_{t \to 0} \frac{f(tu_1, tu_2)}{t}$$

$$= \lim_{t \to 0} \frac{f(\frac{t}{\sqrt{2}}, \frac{t}{\sqrt{2}})}{t} = \lim_{t \to 0} \frac{1}{t} \cdot \frac{\frac{t^3}{2\sqrt{2}}}{\frac{t^4}{4} + \frac{3t^2}{2}}$$

$$= \lim_{t \to 0} \frac{\frac{1}{2\sqrt{2}}}{\frac{t^2}{4} + \frac{3}{2}} = \frac{1}{3\sqrt{2}}.$$

(ii) Note first that f is not continuous at $(0,0)$ (see Example 3.8) hence, f is not differentiable at $(0,0)$. Similar to the above calculations, if $\mathbf{u} = (u_1, u_2)$ then

$$D_{\mathbf{u}}f(0,0) = \lim_{t \to 0} \frac{f(tu_1, tu_2)}{t} = \lim_{t \to 0} \begin{cases} \dfrac{u_1^2 u_2}{t^2 u_1^4 + 3u_2^2} & \text{if } u_2 \neq 0 \\ 0 & \text{if } u_2 = 0 \end{cases}$$

$$= \begin{cases} \dfrac{u_1^2}{3u_2} & \text{if } u_2 \neq 0 \\ 0 & \text{if } u_2 = 0. \end{cases}$$

Example 6.3 provides a pathological function of two variables which is not continuous (hence, not differentiable) at $(0,0)$ but has directional derivative at $(0,0)$ in any direction. The next example shows that there are continuous functions without directional derivative at a point in any direction.

Example 6.4 *Let $f : \mathbb{R}^m \to \mathbb{R}$, $m \geq 2$ be defined by $f(\mathbf{x}) = \|\mathbf{x}\|$.*
Then for any unit vector $\mathbf{u} \in \mathbb{R}^m$, f has no directional derivative at $\mathbf{0}$ in the direction $\mathbf{u}$.

Solution. According to Definition 6.1, we have to compute

$$\lim_{t \to 0} \frac{f(t\mathbf{u}) - f(\mathbf{0})}{t} = \lim_{t \to 0} \frac{\|t\mathbf{u}\|}{t} = \lim_{t \to 0} \frac{|t|}{t}.$$

The sided limits of $\frac{|t|}{t}$ as $t \to 0^-$ and as $t \to 0^+$ are -1 and $+1$ respectively. This shows that the above limit does not exist, so f has no directional derivative at $\mathbf{0}$ in the direction $\mathbf{u}$.

If a function is differentiable at a point, we can link the directional derivative with its gradient. This is illustrated in the result below.

Theorem 6.5 *If $f : D \subset \mathbb{R}^m \to \mathbb{R}$ is a differentiable function at $\mathbf{a} \in D$ and $\mathbf{u} = (u_1, u_2, \ldots, u_m)$ is a unit vector, then*

$$D_{\mathbf{u}}f(\mathbf{a}) = \nabla f(\mathbf{a}) \bullet \mathbf{u},$$

that is,

$$D_{\mathbf{u}}f(\mathbf{a}) = \frac{\partial f}{\partial x_1}(\mathbf{a})u_1 + \frac{\partial f}{\partial x_2}(\mathbf{a})u_2 + \cdots + \frac{\partial f}{\partial x_m}(\mathbf{a})u_m.$$

Proof Let $\mathbf{x}(t) = \mathbf{a} + t\mathbf{u}$. Then $\dfrac{d\mathbf{x}}{dt} = \mathbf{u}$ and

$$D_\mathbf{u}f(\mathbf{a}) = \lim_{t \to 0} \frac{f(\mathbf{a} + t\mathbf{u}) - f(\mathbf{a})}{t}$$

$$= \lim_{t \to 0} \frac{f(\mathbf{x}(t)) - f(\mathbf{x}(0))}{t}$$

$$= \lim_{t \to 0} \frac{w(t) - w(0)}{t} \qquad \text{where } w(t) = f(\mathbf{x}(t))$$

$$= \frac{dw}{dt}\Big|_{t=0}$$

$$= \left(\nabla f(\mathbf{x}(t)) \bullet \frac{d\mathbf{x}}{dt}\right)\Big|_{t=0} \qquad \text{(by Chain Rule)}$$

$$= \nabla f(\mathbf{a}) \bullet \mathbf{u}.$$

□

Example 6.6 *Find the directional derivative of* $f(x, y, z) = e^x \cos y + xz$ *at* $(1, \pi, -1)$ *in the direction* $\mathbf{w} = \mathbf{i} - 3\mathbf{j} + 4\mathbf{k}$.

Solution. The unit vector in the direction $\mathbf{w}$ is

$$\mathbf{u} = \frac{1}{\|w\|}w = \frac{1}{\sqrt{26}}(1, -3, 4).$$

Since f is differentiable, by Theorem 6.5, one has

$$D_\mathbf{u}f(1, \pi, -1) = \left(\frac{\partial f}{\partial x}\frac{1}{\sqrt{26}} + \frac{\partial f}{\partial y}\frac{-3}{\sqrt{26}} + \frac{\partial f}{\partial z}\frac{4}{\sqrt{26}}\right)\Big|_{(1,\pi,-1)}$$

$$= \frac{1}{\sqrt{26}}\left(e^x \cos y + z + 3e^x \sin y + 4x\right)\Big|_{(1,\pi,-1)}$$

$$= \frac{3 - e}{\sqrt{26}}.$$

Exercises

Exercise 1. Compute the directional derivative of the following functions at the indicated point.

(i) $f(x, y) = x^{2y}$ at $(e, 1)$ in the direction $\mathbf{i} - \mathbf{j}$;

(ii) $f(x, y) = \cos(xy)$ at $\left(1, \dfrac{\pi}{2}\right)$ in the direction $-3\mathbf{i} + 4\mathbf{j}$;

(iii) $f(x, y, z) = x^2 z e^{-x+y+z}$ at $(1, 0, 1)$ in the direction $\mathbf{i} + 2\mathbf{j} - 2\mathbf{k}$;

(iv) $f(x, y) = x \cos y + y \sin z$ at $\left(\dfrac{\pi}{2}, \dfrac{\pi}{2}, \dfrac{\pi}{2}\right)$ in the direction $-2\mathbf{i} - 6\mathbf{j} + 3\mathbf{k}$.

Exercise 2. Let $f : \mathbb{R}^2 \to \mathbb{R}$, $f(x,y) = \begin{cases} \dfrac{xy^2}{x^2 + 3y^2} & \text{if } (x,y) \neq (0,0) \\ 0 & \text{if } (x,y) = (0,0). \end{cases}$

(i) Check that f is continuous at $(0,0)$;

(ii) Find $\dfrac{\partial f}{\partial x}(0,0)$ and $\dfrac{\partial f}{\partial y}(0,0)$;

(iii) Show that f is not differentiable at $(0,0)$;

(iv) Show that f has directional derivative at $(0,0)$ in any direction.

Exercise 3. Let $f(x,y) = \begin{cases} \dfrac{(x^2 + y^2)(x^3 + y^3)}{x^4 + y^4} & \text{if } (x,y) \neq (0,0) \\ 0 & \text{if } (x,y) = (0,0). \end{cases}$

(i) Check that f is continuous at $(0,0)$.

(ii) Compute $\dfrac{\partial f}{\partial x}(0,0)$, $\dfrac{\partial f}{\partial y}(0,0)$ and show that f is not differentiable at $(0,0)$.

(iii) If $\mathbf{u} = (u_1, u_2)$ is a unit vector, show that $D_{\mathbf{u}}f(0,0) = \dfrac{u_1^3 + u_2^3}{u_1^4 + u_2^4}$.

Exercise 4. Let $f : \mathbb{R}^2 \to \mathbb{R}$, $f(x,y) = \begin{cases} 1 & \text{if } xy \neq 0 \\ 0 & \text{if } xy = 0. \end{cases}$

For which unit vectors $\mathbf{u} \in \mathbb{R}^2$ does $D_{\mathbf{u}}f(1,0)$ exist?

Exercise 5. So far, we have seen derivatives in many shapes and forms. Let $D \subset \mathbb{R}^m$, $m \geq 2$ be an open set and $f : D \to \mathbb{R}$. Consider the following statements:

(A) f is continuous at $\mathbf{a} \in D$;

(B) f has partial derivatives at $\mathbf{a} \in D$;

(C) f has directional derivatives at $\mathbf{a} \in D$ in any direction $\mathbf{u} \in \mathbb{R}^m$;

(D) f is differentiable at $\mathbf{a} \in D$.

(i) Check that the following implications hold:

$$(A) \Longleftarrow (D) \Longrightarrow (C) \Longrightarrow (B)$$

(ii) Construct counterexamples to show that the implications

$$(A) \Longrightarrow (B), \quad (C) \Longrightarrow (D) \quad \text{and} \quad (B) \Longrightarrow (C)$$

are not always true.

Hint: For an example related to the implication $(B) \Longrightarrow (C)$ you may use the function in Exercise 4 above.

Exercise 6. $f : \mathbb{R}^2 \to \mathbb{R}$ is a differentiable function at $(a, b) \in \mathbb{R}^2$. It is known that:

   (i)   The directional derivative of f at (a, b) in the direction $\mathbf{i} - \mathbf{j}$ is $\sqrt{5}$.

   (ii) The directional derivative of f at (a, b) in the direction $2\mathbf{i} + \mathbf{j}$ is $2\sqrt{2}$.

Find the directional derivative of f at (a, b) in the direction $3\mathbf{i} + \mathbf{j}$.

Exercise 7. Let $f : D \subset \mathbb{R}^m \to \mathbb{R}$ be a differentiable function and $\mathbf{a}, \mathbf{b} \in D$ be two distinct points such that $[\mathbf{a}, \mathbf{b}] \subset D$. If $\mathbf{u}$ is the unit vector in the direction $\mathbf{b} - \mathbf{a}$, show that there exists $\mathbf{c} \in (\mathbf{a}, \mathbf{b})$ such that

$$D_{\mathbf{u}}f(\mathbf{c}) = \frac{f(\mathbf{b}) - f(\mathbf{a})}{\|\mathbf{b} - \mathbf{a}\|}.$$

Exercise 8. Let $f, g : D \subset \mathbb{R}^m \to \mathbb{R}$ and $\mathbf{u} \in \mathbb{R}^m$ be a unit vector such that:

- f or g are continuous at $\mathbf{a} \in D$;

- f and g have directional derivative at $\mathbf{a}$ in the direction $\mathbf{u}$.

   (i)   Check that fg has directional derivatives at $\mathbf{a}$ in direction $\mathbf{u}$ and

$$D_{\mathbf{u}}(fg)(\mathbf{a}) = f(\mathbf{a})D_{\mathbf{u}}g(\mathbf{a}) + g(\mathbf{a})D_{\mathbf{u}}f(\mathbf{a}).$$

   (ii) By constructing a counterexample, show that the above equality does not hold if neither of f and g is continuous at $\mathbf{a}$.

Exercise 9. Let $f : \mathbb{R}^m \to \mathbb{R}$, $m \geq 2$ be a *radially symmetric* function, that is,
$$f(\mathbf{x}) = \phi(\|\mathbf{x}\|) \quad \text{for all} \quad \mathbf{x} \in \mathbb{R}^m, \quad \text{where} \quad \phi : \mathbb{R} \to \mathbb{R}.$$
Assume that ϕ is differentiable on $\mathbb{R}$ and let $\mathbf{u}, \mathbf{v} \in \mathbb{R}^m$ be two unit vectors.

   (i)   Check that $D_{\mathbf{u}}f(\mathbf{v}) = D_{\mathbf{v}}f(\mathbf{u})$;

   (ii) Check that if $\mathbf{a} \in \mathbb{R}^m \setminus \{\mathbf{0}\}$ is such that $D_{\mathbf{u}}f(\mathbf{a}) = D_{\mathbf{v}}f(\mathbf{a})$, then $\mathbf{u} - \mathbf{v} \perp \phi'(\|\mathbf{a}\|)\mathbf{a}$.

6.2   A Geometric Insight

We start first with the following geometric property that connects the directional derivative with the gradient.

Theorem 6.7 *Let $D \subset \mathbb{R}^m$ be an open set. Suppose $f : D \subset \mathbb{R}^m \to \mathbb{R}$ is a differentiable function at $\mathbf{a} \in D$ and let $\mathbf{u} \in \mathbb{R}^m$ be a unit vector.*
If θ is the angle between $\mathbf{u}$ and $\nabla f(\mathbf{a})$, then

$$D_{\mathbf{u}}f(\mathbf{a}) = \|\nabla f(\mathbf{a})\| \cdot \cos \theta.$$

In particular,

$$-\|\nabla f(\mathbf{a})\| \leq D_{\mathbf{u}}f(\mathbf{a}) \leq \|\nabla f(\mathbf{a})\|.$$

Proof We know from Theorem 6.5 that $D_{\mathbf{u}}f(\mathbf{a}) = \nabla f(\mathbf{a}) \bullet \mathbf{u}$. Thus, by formula (1.2) in Chapter 1, we have

$$D_{\mathbf{u}}f(\mathbf{a}) = \|\nabla f(\mathbf{a})\| \cdot \|\mathbf{u}\| \cdot \cos \theta = \|\nabla f(\mathbf{a})\| \cdot \cos \theta.$$

Hence, $-\|\nabla f(\mathbf{a})\| \leq D_{\mathbf{u}}f(\mathbf{a}) \leq \|\nabla f(\mathbf{a})\|$. $\qquad \square$

Remark 6.8 From Theorem 6.7 we deduce:

- The maximum rate of change of f at $\mathbf{a}$ equals $\|\nabla f(\mathbf{a})\|$; that is, the maximum value of the directional derivative $D_{\mathbf{u}}f(\mathbf{a})$ equals $\|\nabla f(\mathbf{a})\|$ and this occurs when $\cos \theta = 1$; in this case $\mathbf{u} = \dfrac{1}{\|\nabla f(\mathbf{a})\|}\nabla f(\mathbf{a})$.

- The minimum rate of change of f at $\mathbf{a}$, that is, the minimum value of the directional derivative $D_{\mathbf{u}}f(\mathbf{a})$ equals $-\|\nabla f(\mathbf{a})\|$ and this occurs when $\cos \theta = -1$; in this case $\mathbf{u} = -\dfrac{1}{\|\nabla f(\mathbf{a})\|}\nabla f(\mathbf{a})$.

The above conclusion is illustrated in Figure 6.2.

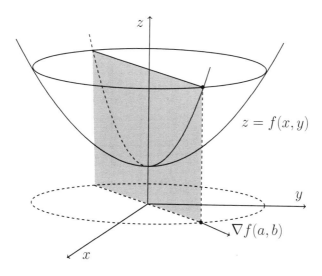

Figure 6.2
The maximum rate of change is achieved along the gradient vector.

Example 6.9 *Let $f(x,y) = xy - y^3$. Find the unit vector $\mathbf{u} \in \mathbb{R}^2$ for which $D_{\mathbf{u}}f(2,1)$ is maximum. Find this maximum value.*

Solution. We have $\nabla f(2,1) = (y, x - 3y^2)\big|_{(2,1)} = (1,-1)$. Hence, $D_{\mathbf{u}}f(2,1)$ is maximum in the direction $\nabla f(2,1) = (1,-1)$ so $\mathbf{u} = \frac{1}{\|(1,-1)\|}(1,-1) = \frac{1}{\sqrt{2}}(1,-1)$. The value of this directional derivative is $\|\nabla f(2,1)\| = \sqrt{2}$.

Example 6.10 *Let $f(x,y) = x^2 - 2xy^2$. Are there unit vectors $\mathbf{u}$ for which:*
(i) $D_{\mathbf{u}}f(1,-2) = 0$? (ii) $D_{\mathbf{u}}f(1,-2) = 11$? (iii) $D_{\mathbf{u}}f(1,-2) = 10$?

Solution. We have $\nabla f(1,2) = (2x - 2y^2, -4xy)\big|_{(1,-2)} = (-6,8)$.

(i) Let $\mathbf{u} = (u_1, u_2) \in \mathbb{R}^2$ be a unit vector such that $D_{\mathbf{u}}f(1,-2) = 0$. Since f is differentiable at $(1,-2)$ it follows from Theorem 6.5 that
$$D_{\mathbf{u}}f(1,-2) = \nabla f(1,-2) \bullet \mathbf{u}.$$

Thus, $\mathbf{u}$ satisfies
$$\begin{cases} -6u_1 + 8u_2 = 0 \\ u_1^2 + u_2^2 = 1 \end{cases} \implies \begin{cases} u_2 = \frac{3}{4}u_1 \\ u_1^2 + u_2^2 = 1 \end{cases} \implies \mathbf{u} = \pm\left(\frac{4}{5}, \frac{3}{5}\right).$$

(ii) We could proceed as above and write first the equations satisfied by the components u_1 and u_2 of $\mathbf{u}$. Let us instead adopt a different approach that relies on Theorem 6.7. By this result, we know that
$$D_{\mathbf{u}}f(1,-2) \le \|\nabla f(1,-2)\| = \sqrt{(-6)^2 + 8^2} = 10,$$
so $D_{\mathbf{u}}f(1,2) \ne 11$.

(iii) If $D_{\mathbf{u}}f(1,2) = 10$, then $D_{\mathbf{u}}f(1,2) = \|\nabla f(1,-2)\|$ and this occurs only when
$$\mathbf{u} = \frac{1}{\|\nabla f(1,-2)\|}\nabla f(1,-2) = \left(-\frac{3}{5}, \frac{4}{5}\right).$$

Exercises

Exercise 10. Let $f(x,y) = \ln(x^2 + 3y^2)$. In which direction is the rate of change of f at $(-1,1)$ equal to $\frac{3}{2}$?

Exercise 11. Let $g(x,y) = e^{-2x^2} + e^{-8y^2}$ and $P(-2,1)$, $Q(-3,4)$. At which point (x,y) in the plane does the rate of change of g in the direction $\overrightarrow{PQ}$ reach its maximum? At which point does it reach its minimum?

Exercise 12. The temperature at a point (x,y,z) in space changes according to the rule
$$T(x,y,z) = \frac{(x+y)^2}{1+z^2}.$$

(i) On which surface that passes through $(-1, 3, 1)$ is the temperature constant?

(ii) In which direction at $(-1, 3, 1)$ should you go in order to cool down as quickly as possible?

Exercise 13. The height of a hill at (x, y) is given by

$$z = 800 - \frac{1}{720}x^4 - \frac{1}{40}(y+1)^2.$$

Suppose the x axis points east and the y axis points north.

(i) If you stand at $(6, -5)$ and move north, will you ascend or descend?

(ii) If you stand at $(-6, 3)$ and move south-west, will you ascend or descend?

Exercise 14. Let $f(x, y) = x^2(1 - y) + y^2 - 3\ln(x - xy)$.

(i) Find the directional derivative $D_{\mathbf{u}}f(1, -2)$ where $\mathbf{u}$ is the unit vector in the direction of $2\mathbf{i} + \mathbf{j}$.

(ii) In which direction does f decrease the most rapidly at $(1, -2)$?

(iii) Two unit vectors $\mathbf{u}$ and $\mathbf{v}$ satisfy

$$D_{\mathbf{u}}f(1, -2) + D_{\mathbf{v}}f(1, -2) = 10.$$

Can we conclude that $\mathbf{u} = \mathbf{v}$? Justify your answer.

Exercise 15. A function $f : \mathbb{R}^2 \to \mathbb{R}$ is said to be *even* if it satisfies

$$f(-x, -y) = f(x, y) \quad \text{for all} \quad (x, y) \in \mathbb{R}^2.$$

Assume f is even and differentiable.

(i) Check that $\nabla f(0, 0) = (0, 0)$;

(ii) Verify that f increases the most rapidly at $(-2, 3)$ in the same direction along which f decreases the most rapidly at $(2, -3)$.

Exercise 16. A function $g : \mathbb{R}^2 \to \mathbb{R}$ is said to be *odd* if it satisfies

$$g(-x, -y) = -g(x, y) \quad \text{for all} \quad (x, y) \in \mathbb{R}^2.$$

Verify that if g is odd and differentiable, then g increases the most rapidly at $(-2, 3)$ in the same direction along which g increases the most rapidly at $(2, -3)$.

Exercise 17. The function $f : \mathbb{R}^2 \to \mathbb{R}$ given by $f(x, y) = x^2 + axy + by^3$, $a, b \in \mathbb{R}$ increases the most rapidly at $(2, -1)$ in the direction $3\mathbf{i} + \mathbf{j}$. Show that

$$a < 4, \quad b > -7/3 \quad \text{and} \quad 7a + 9b = 4.$$

Exercise 18. Let $D \subset \mathbb{R}^2$ be an open set. Assume $f : D \subset \mathbb{R}^2 \to \mathbb{R}$ is a differentiable function at $\mathbf{a} \in D$ and let $\mathbf{u}, \mathbf{v}$ be two unit vectors which are not collinear (that is, $\mathbf{u} \neq \pm\mathbf{v}$). Show that if $D_{\mathbf{u}}f(\mathbf{a}) = D_{\mathbf{v}}f(\mathbf{a}) = 0$, then $\nabla f(\mathbf{a}) = (0,0)$.

Exercise 19. $f, g : \mathbb{R}^2 \to \mathbb{R}$ are differentiable functions at $(a,b) \in \mathbb{R}^2$. It is known that:

(i) f increases the most rapidly at (a,b) in the direction $(2,-1)$ and this increase is 3.

(i) g increases the most rapidly at (a,b) in the direction $(-1,2)$ and this increase is 4.

In which direction does $f - g$ increase the most rapidly at (a,b) and what is the value of this increase?

Exercise 20. Let $D \subset \mathbb{R}^m$ be an open set and let $f : D \to \mathbb{R}$ be a differentiable function at $\mathbf{a} \in D$.

(i) Assume that $\mathbf{u}, \mathbf{v} \in \mathbb{R}^m$ are two unit vectors such that

$$D_{\mathbf{u}}f(\mathbf{a})D_{\mathbf{v}}f(\mathbf{a}) = \|\nabla f(\mathbf{a})\|^2.$$

Check that $\mathbf{u} = \mathbf{v}$;

(ii) Assume that $\mathbf{u}, \mathbf{v}, \mathbf{w} \in \mathbb{R}^m$ are three unit vectors such that

$$D_{\mathbf{u}}f(\mathbf{a})D_{\mathbf{v}}f(\mathbf{a})D_{\mathbf{w}}f(\mathbf{a}) = \|\nabla f(\mathbf{a})\|^3.$$

Check that at least two of the vectors $\mathbf{u}$, $\mathbf{v}$ and $\mathbf{w}$ are equal.

Exercise 21. Let $D \subset \mathbb{R}^2$ be an open set and let $f : D \to \mathbb{R}$ be a differentiable function at $\mathbf{a}$. Show that if $\mathbf{u}$ and $\mathbf{v}$ are two orthogonal unit vectors, then:

(i) $D_{\mathbf{u}}f(\mathbf{a}) + D_{\mathbf{v}}f(\mathbf{a}) \leq \sqrt{2}\,\|\nabla f(\mathbf{a})\|$;

(ii) $D_{\mathbf{u}}f(\mathbf{a}) \cdot D_{\mathbf{v}}f(\mathbf{a}) \leq \dfrac{1}{2}\,\|\nabla f(\mathbf{a})\|^2.$

6.3   The Gradient and the Level Sets

The following result provides a nice geometrical property of the gradient.

Theorem 6.11 *Suppose $f : D \subset \mathbb{R}^m \to \mathbb{R}$ is differentiable at $\mathbf{a} \in D$. If $\nabla f(\mathbf{a}) \neq \mathbf{0}$, then $\nabla f(\mathbf{a})$ is orthogonal to the level set of f containing $\mathbf{a}$.*

In other words, the gradient vector at regular points of f is orthogonal to the corresponding level set.

Proof Let $c = f(\mathbf{a})$ and let $L(f, c)$ be the level set of f corresponding to c. Let also $\mathbf{r}(t)$ be a curve in $L(f, c)$ passing through $\mathbf{a}$. Then $f(\mathbf{r}(t)) = c$ for all t. Differentiating in this equality, by the Chain Rule we find

$$0 = \frac{d}{dt} f(\mathbf{r}(t)) = \nabla f(\mathbf{r}(t)) \bullet \mathbf{r}'(t).$$

Recall that $\mathbf{r}'(t)$ is the tangent vector (see Definition 2.17) to the curve $\mathbf{r}(t)$ and the above equality states that $\nabla f(\mathbf{a})$ is orthogonal to it. Thus, $\nabla f(\mathbf{a})$ is orthogonal to any curve in $L(f, c)$ through $\mathbf{a}$, that is, $\nabla f(\mathbf{a})$ is orthogonal to $L(f, c)$. $\square$

The table below summarizes the conclusions from Theorem 6.7, Remark 6.8 and Theorem 6.11.

Important

Let $f : D \subset \mathbb{R}^m \to \mathbb{R}$ be differentiable at $\mathbf{a} \in D$ and $\mathbf{u} \in \mathbb{R}^m$ be a unit vector. Assume that $\nabla f(\mathbf{a}) \neq \mathbf{0}$.

(A) The maximum value of $D_{\mathbf{u}} f(\mathbf{a})$ over all possible choices of unit vectors $\mathbf{u}$ is $\|\nabla f(\mathbf{a})\|$; it occurs when $\mathbf{u} = \dfrac{1}{\|\nabla f(\mathbf{a})\|} \nabla f(\mathbf{a})$.

(B) The minimum value of $D_{\mathbf{u}} f(\mathbf{a})$ over all possible choices of unit vectors $\mathbf{u}$ is $-\|\nabla f(\mathbf{a})\|$; it occurs when $\mathbf{u} = -\dfrac{1}{\|\nabla f(\mathbf{a})\|} \nabla f(\mathbf{a})$.

(C) The gradient $\nabla f(\mathbf{a})$ points in the direction of the most rapid increase of f at $\mathbf{a}$.

(D) The gradient $\nabla f(\mathbf{a})$ is orthogonal to the level set of f containing $\mathbf{a}$.

Example 6.12 Let $g : \mathbb{R}^2 \to \mathbb{R}$, $g(x, y) = x^2 + 2y^2$. Find the equation of the tangent plane to the surface $z = g(x, y)$ at $(1, -2, 9)$.

Solution. One method we have already encountered relies on the use of Theorem 4.10(iii). We explore a different approach here which relies on Theorem 6.11. The surface $z = g(x, y)$ is the level set $L(f, 0)$ where $f(x, y, z) = x^2 + 2y^2 - z$. By Theorem 6.11, the gradient $\nabla f(1, -2, 9) = (2, -8, -1)$ is orthogonal to the level set $L(f, 0)$. Thus, the tangent plane has the equation

$$(2, -8, -1) \bullet \big[(x, y, z) - (1, -2, 9)\big] = 0 \implies 2x - 8y - z = 9.$$

Remark 6.13 There are many situations where one cannot use Theorem 4.10(iii). For instance, the equation $z^3 + z = x^2 + 2y^2$ cannot be written in the form $z = f(x, y)$ that allows us to use Theorem 4.10(iii). Instead, Theorem 6.11 can be applied directly, the only condition one needs to check is that the gradient is not identically the zero vector. The next example illustrate precisely this fact.

Example 6.14 *Consider the surface $\mathcal{S}_1$ of equation $x^2 - xyz + z^2 = 7$ and the surface $\mathcal{S}_2$ of equation $x^2y + z^3 = 3$.*

(i) *Write the equation of the normal line and tangent plane to $\mathcal{S}_1$ at $(3, -1, -1)$.*

(ii) *The surfaces $\mathcal{S}_1$ and $\mathcal{S}_2$ intersect each other along a curve $\mathcal{C}$ that contains $(2, 1, -1)$.*

Write an equation of the tangent line to $\mathcal{C}$ at $(2, 1, -1)$.

Solution. Let us first note that $\mathcal{S}_1$ and $\mathcal{S}_2$ can be described as level surfaces: $\mathcal{S}_1 = L(f, 7)$ and $\mathcal{S}_2 = L(g, 3)$ where

$$f(x, y, z) = x^2 - xyz + z^2 \quad \text{and} \quad g(x, y, z) = x^2y + z^3.$$

(i) The normal vector to $\mathcal{S}_1$ at $(3, -1, -1)$ is

$$\mathbf{n} = \nabla f(3, -1, -1) = (2x - yz, -xz, -xy + 2z)\Big|_{(3,-1,-1)} = (5, 3, 1).$$

The normal line has the direction $(5, 3, 1)$ and passes through $(3, -1, -1)$. Its equation is

$$\begin{pmatrix} x \\ y \\ z \end{pmatrix} = \begin{pmatrix} 3 \\ -1 \\ -1 \end{pmatrix} + t \begin{pmatrix} 5 \\ 3 \\ 1 \end{pmatrix} \implies \begin{pmatrix} x \\ y \\ z \end{pmatrix} = \begin{pmatrix} 5t + 3 \\ 3t - 1 \\ t - 1 \end{pmatrix}, t \in \mathbb{R},$$

and the equation of the tangent plane is

$$(5, 3, 1) \bullet \left\{(x, y, z) - (3, -1, -1)\right\} = 0 \implies 5x + 3y + z = 11.$$

(ii) A normal vector to the level surface $L(f, 7)$ at $(2, 1, -1)$ is

$$\mathbf{n}_1 = \nabla f(2, 1, -1) = (2x - yz, -xz, -xy + 2z)\Big|_{(2,1,-1)} = (5, 2, -4).$$

Similarly, a normal vector to the level surface $L(g, 3)$ at $(2, 1, -1)$ is

$$\mathbf{n}_2 = \nabla g(2, 1, -1) = (2xy, x^2, 3z^2)\Big|_{(2,1,-1)} = (4, 4, 3).$$

The tangent line must be perpendicular to $\mathbf{n}_1$ and $\mathbf{n}_2$ so has the direction of the cross product $\mathbf{n}_1 \times \mathbf{n}_2$ given by

$$\mathbf{n}_1 \times \mathbf{n}_2 = \det \begin{pmatrix} \mathbf{i} & \mathbf{j} & \mathbf{k} \\ 5 & 2 & -4 \\ 4 & 4 & 3 \end{pmatrix} = 22\mathbf{i} - 31\mathbf{j} + 12\mathbf{k}.$$

An equation of the tangent line to C at the point $(2, 1, -1)$ is

$$\begin{pmatrix} x \\ y \\ z \end{pmatrix} = \begin{pmatrix} 2 \\ 1 \\ -1 \end{pmatrix} + t \begin{pmatrix} 22 \\ -31 \\ 12 \end{pmatrix} \implies \begin{pmatrix} x \\ y \\ z \end{pmatrix} = \begin{pmatrix} 22t + 2 \\ -31t + 1 \\ 12t - 1 \end{pmatrix}, t \in \mathbb{R}.$$

Exercises

Exercise 22. The function $f : \mathbb{R}^2 \to \mathbb{R}$ is continuously differentiable and satisfies:

(i) The level curve $L(f, 2)$ of f at level $c = 2$ contains the point $(3, -1)$;

(ii) The equation of the tangent line to the level curve $L(f, 2)$ at $(3, -1)$ is $x - 2y = 5$;

(iii) If $\mathbf{u}$ is the unit vector in the direction $\mathbf{i} - 2\mathbf{j}$, then $D_{\mathbf{u}} f(3, -1)$.

Find $f(3, -1)$ and $\nabla f(3, -1)$.

Exercise 23. The function $f : \mathbb{R}^2 \to \mathbb{R}$ is continuously differentiable and satisfies:

(i) The equation of the tangent plane to the surface $z = f(x, y)$ at $(2, -3, f(2, -3))$ is $z = -x + by - 2$;

(ii) For some $c \in \mathbb{R}$, the level curve $L(f, c)$ of f at level c is the parabola $y^2 - 2x = 5$.

Find b and c and hence find $f(2, -3)$ and $\nabla f(2, -3)$.

Hint: Observe that $(2, -3)$ lies on the parabola $y^2 - 2x = 5$, so it belongs to the level curve $L(f, c)$.

Exercise 24. At which points on the ellipsoid $2(x - 1)^2 + (y + 1)^2 + z^2 = 12$ is the tangent plane parallel to the plane $2x - 3y + z = 7$?

Exercise 25. Show that the points on the surface $x^2 + yz + z^3 = 1$ at which the tangent plane is orthogonal to $2x + 5y + 4z = 8$ lie on the quadric

$$\left(x + \frac{z}{2}\right)^2 + (z - y)^2 = 4x^2 + y^2 - 3.$$

Exercise 26. Let $f : \mathbb{R}^3 \to \mathbb{R}$ be a differentiable function and $c \in \mathbb{R}$. It is known that the level surface $L(f, c)$ is the hyperboloid of one sheet $x^2 + y^2 - z^2 = 4$ that passes through $(1, 2, 1)$. Show that

$$\frac{\partial f}{\partial x}(1, 2, 1) + \frac{\partial f}{\partial z}(1, 2, 1) = 0.$$

Exercise 27. Let $f : \mathbb{R}^3 \to \mathbb{R}$ be a continuously differentiable function such that the level surface $L(f, 1)$ of f at level $c = 1$ is the sphere centred at $(0, 1, 2)$ that passes through $(-1, 0, 3)$.

(i) Find the equation of the tangent plane to $L(f, 1)$ at $(-1, 0, 3)$;

(ii) Show that

$$\frac{\partial f}{\partial x}(-1, 0, 3) + \frac{\partial f}{\partial x}(-1, 0, 3) + 2\frac{\partial f}{\partial z}(-1, 0, 3) = 0.$$

Exercise 28. Let $a, b, c > 0$. The plane $x + y + z = 0$ is tangent at the point $P(x_0, y_0, z_0)$ to both the cone $z^2 = ax^2 + by^2$ and the paraboloid $z = bx^2 + ay^2 + c$.

(i) Show that $a = b$;

(ii) Find a, b, c and x_0, y_0, z_0.

Exercise 29. Show that if a tangent plane to the cone $x^2 + (y-1)^2 = z^2$ is parallel to the plane $\alpha x + \beta y + \gamma z = d$, $\alpha, \beta, \gamma, d \in \mathbb{R}$, $(\alpha, \beta, \gamma) \neq (0, 0, 0)$, then $\alpha^2 + \beta^2 = \gamma^2$.

Is the converse of this statement also true?

Exercise 30. Consider the surface S_1 of equation $x^2 + 2xyz + 2y^2 = 1$ and the surface S_2 of equation $y^3 - x^2 z = 6$.

(i) Write the equation of the normal line and tangent plane to S_1 at $(1, 0, 2)$.

(ii) Find all the points on surface S_2 at which the normal line is parallel to the vector $-4\mathbf{i} + 3\mathbf{j} - \mathbf{k}$.

(iii) The surfaces S_1 and S_2 intersect each other along a curve C that contains $(-1, 2, 2)$.

Write an equation of the tangent line to C at $(-1, 2, 2)$.

7

Second-Order Derivatives

The focus of the current chapter is on second order derivatives of functions of several variables. The Hessian matrix and its properties are discussed. As an application, the Laplace operator is introduced which is a key differential operator in mathematical modelling.

7.1   Second-Order Derivatives

Let $f : D \subset \mathbb{R}^2 \to \mathbb{R}$ be a function of two variables such that $\dfrac{\partial f}{\partial x}$ and $\dfrac{\partial f}{\partial y}$ exist on D. If both $\dfrac{\partial f}{\partial x}$ and $\dfrac{\partial f}{\partial y}$ admit partial derivatives with respect to x and y we say that f has second order partial derivatives. We thus write

$$\frac{\partial^2 f}{\partial x^2} = \frac{\partial}{\partial x}\left(\frac{\partial f}{\partial x}\right) \qquad \frac{\partial^2 f}{\partial y \partial x} = \frac{\partial}{\partial y}\left(\frac{\partial f}{\partial x}\right)$$

$$\frac{\partial^2 f}{\partial x \partial y} = \frac{\partial}{\partial x}\left(\frac{\partial f}{\partial y}\right) \qquad \frac{\partial^2 f}{\partial y^2} = \frac{\partial}{\partial y}\left(\frac{\partial f}{\partial y}\right).$$

In subscript notation, we have

$$f_{xx} := (f_x)_x = \frac{\partial}{\partial x}\left(\frac{\partial f}{\partial x}\right) = \frac{\partial^2 f}{\partial x^2} \qquad f_{xy} := (f_x)_y = \frac{\partial}{\partial y}\left(\frac{\partial f}{\partial x}\right) = \frac{\partial^2 f}{\partial y \partial x}$$

$$f_{yx} := (f_y)_x = \frac{\partial}{\partial x}\left(\frac{\partial f}{\partial y}\right) = \frac{\partial^2 f}{\partial x \partial y} \qquad f_{yy} := (f_y)_y = \frac{\partial}{\partial y}\left(\frac{\partial f}{\partial y}\right) = \frac{\partial^2 f}{\partial y^2}.$$

We can extend this definition to functions of several variables as follows.

Definition 7.1 *Let $f : D \subset \mathbb{R}^m \to \mathbb{R}$ be a function of m variables. We say that f has second order partial derivatives, if:*

(i) *f has first order partial derivatives;*

(ii) *for each $1 \leq i \leq m$ the function $\dfrac{\partial f}{\partial x_i} : D \to \mathbb{R}$ has first order partial derivatives.*

DOI: 10.1201/9781003449652-7

We denote

$$\frac{\partial^2 f}{\partial x_i \partial x_j} := \frac{\partial}{\partial x_i}\left(\frac{\partial f}{\partial x_j}\right) = (f_{x_j})_{x_i} = f_{x_j x_i}.$$

For $1 \le i,j \le m$, $i \ne j$, we call $\dfrac{\partial^2 f}{\partial x_i \partial x_j}$ the mixed second order partial derivatives of f.

Example 7.2 *Let $f : \mathbb{R}^2 \to \mathbb{R}$, $f(x,y) = ye^{x^2-2y}$. Compute $\dfrac{\partial^2 f}{\partial x^2}$, $\dfrac{\partial^2 f}{\partial x \partial y}$, $\dfrac{\partial^2 f}{\partial y \partial x}$ and $\dfrac{\partial^2 f}{\partial y^2}$.*

Solution. The first order derivatives of f are

$$\frac{\partial f}{\partial x}(x,y) = 2xye^{x^2-2y} \quad \text{and} \quad \frac{\partial f}{\partial y}(x,y) = (1-2y)e^{x^2-2y}.$$

We further compute

$$\frac{\partial^2 f}{\partial x^2} = \frac{\partial}{\partial x}\left(\frac{\partial f}{\partial x}\right) = 2y(1+2x^2)e^{x^2-2y}$$

$$\frac{\partial^2 f}{\partial x \partial y} = \frac{\partial}{\partial x}\left(\frac{\partial f}{\partial y}\right) = 2x(1-2y)e^{x^2-2y}$$

$$\frac{\partial^2 f}{\partial y \partial x} = \frac{\partial}{\partial y}\left(\frac{\partial f}{\partial x}\right) = 2x(1-2y)e^{x^2-2y}$$

$$\frac{\partial^2 f}{\partial y^2} = \frac{\partial}{\partial y}\left(\frac{\partial f}{\partial y}\right) = 4(y-1)e^{x^2-2y}.$$

In the above calculations we observe that the mixed second order partial derivatives $\dfrac{\partial^2 f}{\partial x \partial y}$ and $\dfrac{\partial^2 f}{\partial y \partial x}$ are equal. This is not a mere coincidence, rather a property that holds more generally as stated in Theorem 7.4.

We should point out that not all functions admit second order derivatives. Revisiting Example 4.3 in Chapter 4, we have:

Example 7.3 *Let*

$$f : \mathbb{R}^2 \to \mathbb{R}, \quad f(x,y) = \begin{cases} \dfrac{xy}{x^2 + y^2} & \text{if } (x,y) \ne (0,0) \\ 0 & \text{if } (x,y) = (0,0). \end{cases}$$

Then, $\frac{\partial^2 f}{\partial x^2}(0,0) = \frac{\partial^2 f}{\partial y^2}(0,0) = 0$ while the mixed second order derivatives $\frac{\partial^2 f}{\partial x \partial y}(0,0)$ and $\frac{\partial^2 f}{\partial y \partial x}(0,0)$ do not exist (as real numbers).

Solution. We have seen in Examples 3.4 and 4.3 that although f is not continuous, it has first order derivative at every single point in $\mathbb{R}^2$ and

$$\frac{\partial f}{\partial x}(x,y) = \begin{cases} \dfrac{y(y^2 - x^2)}{(x^2 + y^2)^2} & \text{if } (x,y) \neq (0,0) \\ 0 & \text{if } (x,y) = (0,0), \end{cases}$$

and

$$\frac{\partial f}{\partial y}(x,y) = \begin{cases} \dfrac{x(x^2 - y^2)}{(x^2 + y^2)^2} & \text{if } (x,y) \neq (0,0) \\ 0 & \text{if } (x,y) = (0,0). \end{cases}$$

We compute the second order derivatives by using Definition 7.1 as follows:

$$\frac{\partial^2 f}{\partial x^2}(0,0) = \frac{\partial}{\partial x}\left(\frac{\partial f}{\partial x}\right)(0,0) = \lim_{t \to 0} \frac{\frac{\partial f}{\partial x}(t,0) - \frac{\partial f}{\partial x}(0,0)}{t} = 0.$$

In the same way one has

$$\frac{\partial^2 f}{\partial y^2}(0,0) = \frac{\partial}{\partial y}\left(\frac{\partial f}{\partial y}\right)(0,0) = \lim_{t \to 0} \frac{\frac{\partial f}{\partial y}(0,t) - \frac{\partial f}{\partial y}(0,0)}{t} = 0.$$

Finally, for the mixed second order derivatives, we have

$$\frac{\partial^2 f}{\partial x \partial y}(0,0) = \frac{\partial}{\partial x}\left(\frac{\partial f}{\partial y}\right)(0,0)$$

$$= \lim_{t \to 0} \frac{\frac{\partial f}{\partial y}(t,0) - \frac{\partial f}{\partial y}(0,0)}{t}$$

$$= \lim_{t \to 0} \frac{1}{t^2} = +\infty.$$

In the same way one has $\dfrac{\partial^2 f}{\partial xy}(0,0) = +\infty$.

Throughout this chapter, $D \subset \mathbb{R}^m$, $m \geq 2$ denotes an open set.

Theorem 7.4 (Equality of the Mixed Second Order Derivatives)
Let $f : D \subset \mathbb{R}^m \to \mathbb{R}$ be a function of m variables. Suppose that for some $1 \leq i, j \leq m$, $i \neq j$, we have:

(i) $\dfrac{\partial f}{\partial x_i}$, $\dfrac{\partial f}{\partial x_j}$, $\dfrac{\partial^2 f}{\partial x_i \partial x_j}$, $\dfrac{\partial^2 f}{\partial x_j \partial x_i}$ *exist on D;*

(ii) $\dfrac{\partial^2 f}{\partial x_i \partial x_j}$ *and* $\dfrac{\partial^2 f}{\partial x_j \partial x_i}$ *are continuous at $\mathbf{a} \in D$.*

Then $\dfrac{\partial^2 f}{\partial x_i \partial x_j}(\mathbf{a}) = \dfrac{\partial^2 f}{\partial x_j \partial x_i}(\mathbf{a}).$

Proof It is enough to prove the above result for functions of two variables. Indeed, only the x_i and x_j variables are involved in the process of the computation of the above mixed second partial derivatives. Assume $f : D \subset \mathbb{R}^2 \to \mathbb{R}$ is such that

- $\dfrac{\partial f}{\partial x}, \dfrac{\partial f}{\partial y}, \dfrac{\partial^2 f}{\partial x \partial y}, \dfrac{\partial^2 f}{\partial y \partial x}$ exist on D;

- $\dfrac{\partial^2 f}{\partial x \partial y}$ and $\dfrac{\partial^2 f}{\partial y \partial x}$ are continuous at $(a, b) \in D$.

Since D is open, there exists $r > 0$ such that $(a + h, b + k) \in D$ for all $|h| < r$ and $|k| < r$.

The proof relies on successive use of the one-variable Mean Value Theorem to the quotient $Q(h, k)$ defined for $0 < |h| < r$, $0 < |k| < r$ by

$$Q(h, k) := \frac{f(a + h, b + k) - f(a + h, b) - f(a, b + k) + f(a, b)}{hk}.$$

We first write $Q(h, k)$ as

$$Q(h, k) = \frac{\phi(h) - \phi(0)}{hk}, \quad \text{where} \quad \phi(h) = f(a + h, b + k) - f(a + h, b).$$

By the Mean Value Theorem there exists h_1 between 0 and h such that

$$\phi(h) - \phi(0) = h\phi'(h_1) = h\left(\frac{\partial f}{\partial x}(a + h_1, b + k) - \frac{\partial f}{\partial x}(a + h_1, b)\right).$$

Thus,

$$Q(h, k) = \frac{\overline{\phi}(k) - \overline{\phi}(0)}{k} \quad \text{where} \quad \overline{\phi}(k) = \frac{\partial f}{\partial x}(a + h_1, b + k).$$

By the Mean Value Theorem there exists k_1 between 0 and k such that $Q(h, k) = \overline{\phi}'(k_1)$, that is,

$$Q(h, k) = \frac{\partial^2 f}{\partial y \partial x}(a + h_1, b + k_1) \quad \text{where} \quad 0 < |h_1| < |h|, 0 < |k_1| < |k|. \quad (7.1)$$

We now start again the calculation of $Q(h, k)$ by writing

$$Q(h, k) = \frac{\psi(k) - \psi(0)}{hk} \quad \text{where} \quad \psi(k) = f(a + h, b + k) - f(a, b + k).$$

By the Mean Value Theorem there exists k_2 between 0 and k such that

$$\psi(k) - \psi(0) = k\psi'(k_2) = k\left(\frac{\partial f}{\partial y}(a + h, b + k_2) - \frac{\partial f}{\partial y}(a, b + k_2)\right)$$

and then

$$Q(h, k) = \frac{\overline{\psi}(h) - \overline{\psi}(0)}{h} \quad \text{where} \quad \overline{\psi}(h) = \frac{\partial f}{\partial y}(a + h, b + k_2).$$

By the Mean Value Theorem again, there exists h_2 between 0 and h such that $Q(h, k) = \bar{\psi}'(h_2)$, that is,

$$Q(h, k) = \frac{\partial^2 f}{\partial x \partial y}(a + h_2, b + k_2) \quad \text{where} \quad 0 < |h_2| < |h|, 0 < |k_2| < |k|. \quad (7.2)$$

Now, from (7.1) and (7.2) we deduce

$$\frac{\partial^2 f}{\partial y \partial x}(a + h_1, b + k_1) = \frac{\partial^2 f}{\partial x \partial y}(a + h_2, b + k_2),$$

where $0 < |h_1|, |h_2| < |h|$ and $0 < |k_1|, |k_2| < |k|$. Letting $(h, k) \to (0, 0)$ in the above equality, we have $(h_i, k_i) \to (0, 0)$, $i = 1, 2$ and using the fact that $\frac{\partial^2 f}{\partial x \partial y}, \frac{\partial^2 f}{\partial y \partial x}$ are continuous at (a, b) we deduce

$$\frac{\partial^2 f}{\partial y \partial x}(a, b) = \frac{\partial^2 f}{\partial x \partial y}(a, b).$$

This concludes our proof. □

The continuity assumption on the mixed second order partial derivatives of f in Theorem 7.4(ii) is necessary. This follows from the example below.

Example 7.5 *Let*

$$f : \mathbb{R}^2 \to \mathbb{R}, \quad f(x, y) = \begin{cases} \dfrac{xy(x^2 - y^2)}{x^2 + y^2} & \text{if } (x, y) \neq (0, 0) \\ 0 & \text{if } (x, y) = (0, 0). \end{cases}$$

Then:

(i) $\dfrac{\partial f}{\partial x}$ *and* $\dfrac{\partial f}{\partial x}$ *exists and are continuous on* $\mathbb{R}^2$;

(ii) $\dfrac{\partial^2 f}{\partial y \partial x}(0, 0) \neq \dfrac{\partial^2 f}{\partial x \partial y}(0, 0).$

Solution. (i) If $(x, y) \neq (0, 0)$, then f is a rational function in a neighbourhood of (x, y), thus (by Remark 4.19) it is differentiable. We compute

$$\frac{\partial f}{\partial x}(x, y) = \frac{y(x^4 + 4x^2 y^2 - y^4)}{(x^2 + y^2)^2} \quad \text{and} \quad \frac{\partial f}{\partial y}(x, y) = \frac{x(x^4 - 4x^2 y^2 - y^4)}{(x^2 + y^2)^2}.$$

Also

$$\frac{\partial f}{\partial x}(0, 0) = \lim_{t \to 0} \frac{f(t, 0) - f(0, 0)}{t} = 0$$

$$\frac{\partial f}{\partial y}(0, 0) = \lim_{t \to 0} \frac{f(0, t) - f(0, 0)}{t} = 0.$$

Since for all $(x, y) \in \mathbb{R}^2$, we have

$$|x^4 \pm 4x^2y^2 \pm y^4| \leq 2(x^2 + 2x^2y^2 + y^4) = (x^2 + y^2)^2,$$

we estimate

$$\left|\frac{\partial f}{\partial x}(x, y)\right| \leq 2|y| \quad \text{and} \quad \left|\frac{\partial f}{\partial x}(x, y)\right| \leq 2|x| \quad \text{for all } (x, y) \in \mathbb{R}^2 \setminus \{(0, 0)\}.$$

By Corollary 3.12 it follows that

$$\lim_{(x,y) \to (0,0)} \frac{\partial f}{\partial x}(x, y) = 0 = \frac{\partial f}{\partial x}(0, 0)$$

$$\lim_{(x,y) \to (0,0)} \frac{\partial f}{\partial y}(x, y) = 0 = \frac{\partial f}{\partial y}(0, 0).$$

Thus, the first order derivatives $\dfrac{\partial f}{\partial x}$ and $\dfrac{\partial f}{\partial y}$ exists and are continuous on $\mathbb{R}^2$.
In particular, it follows by Theorem 4.18 that f is differentiable on $\mathbb{R}^2$.

(ii) By the definition of the partial derivative we compute

$$\frac{\partial^2 f}{\partial x \partial y}(0, 0) = \frac{\partial}{\partial x}\left(\frac{\partial f}{\partial y}\right)(0, 0) = \lim_{t \to 0} \frac{\frac{\partial f}{\partial y}(t, 0) - \frac{\partial f}{\partial y}(0, 0)}{t} = 1,$$

$$\frac{\partial^2 f}{\partial y \partial x}(0, 0) = \frac{\partial}{\partial y}\left(\frac{\partial f}{\partial x}\right)(0, 0) = \lim_{t \to 0} \frac{\frac{\partial f}{\partial x}(0, t) - \frac{\partial f}{\partial x}(0, 0)}{t} = -1.$$

Definition 7.6 *Let $f : D \subset \mathbb{R}^m \to \mathbb{R}$ be a function of m variables which has second order partial derivatives at $\mathbf{a} \in D$. The matrix*

$$H_f(\mathbf{a}) = \left(\frac{\partial^2 f}{\partial x_j \partial x_i}\right)_{1 \leq i,j \leq m} = \begin{pmatrix} \dfrac{\partial^2 f}{\partial x_1^2} & \dfrac{\partial^2 f}{\partial x_2 \partial x_1} & \cdots & \dfrac{\partial^2 f}{\partial x_m \partial x_1} \\ \dfrac{\partial^2 f}{\partial x_1 \partial x_2} & \dfrac{\partial^2 f}{\partial x_2^2} & \cdots & \dfrac{\partial^2 f}{\partial x_m \partial x_2} \\ \cdots & \cdots & \cdots & \cdots \\ \dfrac{\partial^2 f}{\partial x_1 \partial x_m} & \dfrac{\partial^2 f}{\partial x_2 \partial x_m} & \cdots & \dfrac{\partial^2 f}{\partial x_m^2} \end{pmatrix}$$

is called the Hessian matrix of f at $\mathbf{a}$.

Observe that the i-th row of the Hessian matrix contains the gradient of $\dfrac{\partial f}{\partial x_i}$.
If f has continuous second order partial derivatives at $\mathbf{a} \in D$, by Theorem 7.4, it follows that the Hessian matrix $H_f(\mathbf{a})$ is symmetric. In such a case, only the entries on the main diagonal and above it can be distinct and thus, f can have at most

$$m + (m - 1) + (m - 2) + \ldots + 1 = \frac{m(m + 1)}{2}$$

distinct second order partial derivatives.

Example 7.7 *Let* $f : \mathbb{R}^2 \to \mathbb{R}$ *be a function which has second order derivatives and such that its Hessian matrix is* $H_f(x, y) = \begin{pmatrix} 1 & 0 \\ 0 & 1 \end{pmatrix}$ *for all* $(x, y) \in \mathbb{R}^2$.

Show that there exist $a, b, c \in \mathbb{R}$ *such that*

$$f(x, y) = \frac{x^2 + y^2}{2} + ax + by + c \quad \text{for all} \quad (x, y) \in \mathbb{R}^2.$$

Solution. Let $g(x, y) = f(x, y) - \frac{x^2 + y^2}{2}$. Then, its Hessian matrix is the zero matrix. We also have

$$\nabla\left(\frac{\partial g}{\partial x}\right) = (0, 0) \quad \text{and} \quad \nabla\left(\frac{\partial g}{\partial y}\right) = (0, 0).$$

By Exercise 6 in Chapter 4 it follows that $\dfrac{\partial g}{\partial x} = a$ and $\dfrac{\partial g}{\partial y} = b$ are constant functions. Thus, $\nabla(g - ax - by) = (0, 0)$ which implies $g(x, y) = ax + by + c$ for some constants $a, b, c \in \mathbb{R}$. Hence, $f(x, y) = \frac{x^2 + y^2}{2} + ax + by + c$ for all $(x, y) \in \mathbb{R}^2$.

Definition 7.8 *A function* $f : D \subset \mathbb{R}^m \to \mathbb{R}$ *is called of class* C^2 *if it has partial derivatives of order one and two and these derivatives are continuous on* D.

Exercises

Exercise 1. Compute the second order derivatives for each of the functions below.

(i) $f(x, y) = ye^{xy-1}$;

(ii) $f(x, y) = \dfrac{x^3 y}{\ln^2 y}$;

(iii) $f(x, y, z) = x^{y+z}$;

(iv) $f(\mathbf{x}) = \|\mathbf{x}\|^4$, $\mathbf{x} \in \mathbb{R}^m$, $m \geq 2$.

Exercise 2. Let $f : \mathbb{R}^2 \to \mathbb{R}$, $f(x, y) = \begin{cases} \dfrac{x^3 + y^3}{x^2 + y^2} & \text{if } (x, y) \neq (0, 0) \\ 0 & \text{if } (x, y) = (0, 0). \end{cases}$

Check that:

(i) $\dfrac{\partial^2 f}{\partial x^2}(0, 0) = \dfrac{\partial^2 f}{\partial y^2}(0, 0) = 0$;

(ii) $\dfrac{\partial^2 f}{\partial x \partial y}(0,0)$ and $\dfrac{\partial^2 f}{\partial y \partial x}(0,0)$ do not exist.

Exercise 3. Let $f : \mathbb{R}^2 \to \mathbb{R}$, $f(x,y) = \begin{cases} \dfrac{(x^2 - y^2)^2}{x^2 + y^2} & \text{if } (x,y) \neq (0,0) \\ 0 & \text{if } (x,y) = (0,0). \end{cases}$

Find the Hessian matrix $H_f(0,0)$ of f at $(x,y) = (0,0)$.

Exercise 4. Let $f : \mathbb{R}^2 \to \mathbb{R}$, $f(x,y) = \begin{cases} x^2 & \text{if } x \neq 0 \\ y^2 & \text{if } x = 0. \end{cases}$

(i) Show that $\dfrac{\partial f}{\partial y}(x,y)$ exists for all $(x,y) \in \mathbb{R}^2$;

(ii) Compute $\dfrac{\partial^2 f}{\partial x \partial y}(0,0)$ and $\dfrac{\partial^2 f}{\partial y^2}(0,0)$.

Exercise 5. Find the smallest positive integer $a > 0$ such that the function

$$f : \mathbb{R}^2 \to \mathbb{R}, \quad f(x) = \begin{cases} \dfrac{y^a}{x^2 + y^4} & \text{if } (x,y) \neq (0,0) \\ 0 & \text{if } (x,y) = (0,0) \end{cases} \qquad \text{is twice differentiable.}$$

Exercise 6. Find $f : \mathbb{R}^2 \to \mathbb{R}$ which has second order partial derivatives and whose Hessian matrix is $H_f(x,y) = \begin{pmatrix} 0 & 1 \\ 1 & 2 \end{pmatrix}$ for all $(x,y) \in \mathbb{R}^2$.

Hint: As in Example 7.7 consider $g(x,y) = f(x,y) - y^2 - xy$.

Exercise 7. Find $f : \mathbb{R}^2 \to \mathbb{R}$ which has second order partial derivatives and whose Hessian matrix is $H_f(x,y) = \begin{pmatrix} y^2 & 2xy \\ 2xy & x^2 \end{pmatrix}$ for all $(x,y) \in \mathbb{R}^2$.

7.2   Chain Rule for Second-Order Derivatives

The Chain Rule we discussed in Section 5 can be extended to second order derivatives as well. Instead of providing a counterpart of Theorem 5.1, we prefer to illustrated the situation through a couple of specific examples. The first one is the extension of Example 5.4.

Example 7.9 *Let $f : \mathbb{R}^2 \to \mathbb{R}$ be a function of class C^2 and*

$$x = r \cos \theta, \quad y = r \sin \theta$$

be the polar coordinates of $(x,y) \in \mathbb{R}^2$. Find $\dfrac{\partial^2 f}{\partial r^2}$, $\dfrac{\partial^2 f}{\partial r \partial \theta}$ and $\dfrac{\partial^2 f}{\partial \theta^2}$.

Solution. By the Chain Rule, we have

$$\frac{\partial f}{\partial r} = \frac{\partial f}{\partial x}\cos\theta + \frac{\partial f}{\partial y}\sin\theta, \tag{7.3a}$$

$$\frac{\partial f}{\partial \theta} = -\frac{\partial f}{\partial x}r\sin\theta + \frac{\partial f}{\partial y}r\cos\theta. \tag{7.3b}$$

Then, using (7.3a) we find

$$\frac{\partial^2 f}{\partial r^2} = \frac{\partial}{\partial r}\left(\frac{\partial f}{\partial r}\right) = \frac{\partial}{\partial r}\left(\frac{\partial f}{\partial x}\cos\theta + \frac{\partial f}{\partial y}\sin\theta\right)$$

$$= \frac{\partial}{\partial r}\left(\frac{\partial f}{\partial x}\right)\cos\theta + \frac{\partial}{\partial r}\left(\frac{\partial f}{\partial y}\right)\sin\theta. \tag{7.4}$$

We next compute the second order derivatives $\dfrac{\partial}{\partial r}\left(\dfrac{\partial f}{\partial x}\right)$ and $\dfrac{\partial}{\partial r}\left(\dfrac{\partial f}{\partial y}\right)$. Again by the Chain Rule one has

$$\frac{\partial}{\partial r}\left(\frac{\partial f}{\partial x}\right) = \frac{\partial}{\partial x}\left(\frac{\partial f}{\partial x}\right)\cdot\frac{\partial x}{\partial r} + \frac{\partial}{\partial y}\left(\frac{\partial f}{\partial x}\right)\cdot\frac{\partial y}{\partial r} = \frac{\partial^2 f}{\partial x^2}\cos\theta + \frac{\partial^2 f}{\partial y\partial x}\sin\theta$$

$$\frac{\partial}{\partial r}\left(\frac{\partial f}{\partial y}\right) = \frac{\partial}{\partial x}\left(\frac{\partial f}{\partial y}\right)\cdot\frac{\partial x}{\partial r} + \frac{\partial}{\partial y}\left(\frac{\partial f}{\partial y}\right)\cdot\frac{\partial y}{\partial r} = \frac{\partial^2 f}{\partial x\partial y}\cos\theta + \frac{\partial^2 f}{\partial y^2}\sin\theta. \tag{7.5}$$

Since f is of class C^2, by Theorem 7.4, we have

$$\frac{\partial^2 f}{\partial x\partial y} = \frac{\partial^2 f}{\partial y\partial x}.$$

Using the above equalities in (7.4) we deduce

$$\frac{\partial^2 f}{\partial r^2} = \frac{\partial^2 f}{\partial x^2}\cos^2\theta + 2\frac{\partial^2 f}{\partial x\partial y}\sin\theta\cos\theta + \frac{\partial^2 f}{\partial y^2}\sin^2\theta.$$

Further, from (7.3b), we have

$$\frac{\partial^2 f}{\partial r\partial \theta} = \frac{\partial}{\partial r}\left(\frac{\partial f}{\partial \theta}\right) = \frac{\partial}{\partial r}\left(-\frac{\partial f}{\partial x}r\sin\theta + \frac{\partial f}{\partial y}r\cos\theta\right).$$

Next, by the product rule one finds

$$\frac{\partial^2 f}{\partial r\partial \theta} = -\frac{\partial f}{\partial x}\sin\theta - \frac{\partial}{\partial r}\left(\frac{\partial f}{\partial x}\right)r\sin\theta + \frac{\partial f}{\partial y}\cos\theta + \frac{\partial}{\partial r}\left(\frac{\partial f}{\partial y}\right)r\cos\theta.$$

Now, by (7.5) we find

$$\frac{\partial^2 f}{\partial r\partial \theta} = -\frac{\partial f}{\partial x}\sin\theta + \frac{\partial f}{\partial y}\cos\theta + \left\{-\frac{\partial^2 f}{\partial x^2} + \frac{\partial^2 f}{\partial y^2}\right\}r\sin\theta\cos\theta$$

$$+ \frac{\partial^2 f}{\partial x\partial y}r(\cos^2\theta - \sin^2\theta).$$

In a similar manner, by (7.3b) we compute

$$\frac{\partial^2 f}{\partial \theta^2} = \frac{\partial}{\partial \theta}\left(\frac{\partial f}{\partial \theta}\right) = \frac{\partial}{\partial \theta}\left(-\frac{\partial f}{\partial x}r\sin\theta + \frac{\partial f}{\partial y}r\cos\theta\right)$$

$$= -\frac{\partial f}{\partial x}r\cos\theta - \frac{\partial f}{\partial y}r\sin\theta - \frac{\partial}{\partial \theta}\left(\frac{\partial f}{\partial x}\right)r\sin\theta + \frac{\partial}{\partial \theta}\left(\frac{\partial f}{\partial y}\right)r\cos\theta. \tag{7.6}$$

It remains to compute $\dfrac{\partial}{\partial \theta}\left(\dfrac{\partial f}{\partial x}\right)$ and $\dfrac{\partial}{\partial \theta}\left(\dfrac{\partial f}{\partial y}\right)$. To do so, we use a similar computation to (7.5), namely

$$\frac{\partial}{\partial \theta}\left(\frac{\partial f}{\partial x}\right) = \frac{\partial}{\partial x}\left(\frac{\partial f}{\partial x}\right)\cdot\frac{\partial x}{\partial \theta} + \frac{\partial}{\partial y}\left(\frac{\partial f}{\partial x}\right)\cdot\frac{\partial y}{\partial \theta} = -\frac{\partial^2 f}{\partial x^2}r\sin\theta + \frac{\partial^2 f}{\partial y\partial x}r\cos\theta$$

$$\frac{\partial}{\partial \theta}\left(\frac{\partial f}{\partial y}\right) = \frac{\partial}{\partial x}\left(\frac{\partial f}{\partial y}\right)\cdot\frac{\partial x}{\partial \theta} + \frac{\partial}{\partial y}\left(\frac{\partial f}{\partial y}\right)\cdot\frac{\partial y}{\partial \theta} = -\frac{\partial^2 f}{\partial x\partial y}r\sin\theta + \frac{\partial^2 f}{\partial y^2}r\cos\theta.$$

Using these equalities in (7.6) we finally obtain

$$\frac{\partial^2 f}{\partial \theta^2} = -\frac{\partial f}{\partial x}r\cos\theta - \frac{\partial f}{\partial y}r\sin\theta + \frac{\partial^2 f}{\partial x^2}r^2\sin^2\theta$$

$$- 2\frac{\partial^2 f}{\partial x\partial y}r^2\sin\theta\cos\theta + \frac{\partial^2 f}{\partial y^2}r^2\cos^2\theta.$$

Example 7.10 *Let $f(x,y)$ be a function which has continuous second order partial derivatives and let $x = 2st$, $y = t^2 - s^2$, where t and s are independent variables. Prove that*

$$\frac{\partial^2 f}{\partial t^2} + \frac{\partial^2 f}{\partial s^2} = 4(s^2 + t^2)\left(\frac{\partial^2 f}{\partial x^2} + \frac{\partial^2 f}{\partial y^2}\right).$$

Solution. By the Chain Rule, we have

$$\frac{\partial f}{\partial t} = \frac{\partial f}{\partial x}\cdot\frac{\partial x}{\partial t} + \frac{\partial f}{\partial y}\cdot\frac{\partial y}{\partial t} \implies \frac{\partial f}{\partial t} = 2s\frac{\partial f}{\partial x} + 2t\frac{\partial f}{\partial y}. \tag{7.7}$$

We differentiate once more with respect to t in the last equality. By product rule one finds

$$\frac{\partial^2 f}{\partial t^2} = \frac{\partial}{\partial t}\left(\frac{\partial f}{\partial t}\right) = \frac{\partial}{\partial t}\left(2s\frac{\partial f}{\partial x} + 2t\frac{\partial f}{\partial y}\right) \qquad (\text{ from (7.7)})$$

$$= 2s\frac{\partial}{\partial t}\left(\frac{\partial f}{\partial x}\right) + 2\frac{\partial f}{\partial y} + 2t\frac{\partial}{\partial t}\left(\frac{\partial f}{\partial y}\right) \qquad (\text{ by product rule}).$$

Using the Chain Rule, one has

$$\frac{\partial}{\partial t}\left(\frac{\partial f}{\partial x}\right) = \frac{\partial}{\partial x}\left(\frac{\partial f}{\partial x}\right)\cdot\frac{\partial x}{\partial t} + \frac{\partial}{\partial y}\left(\frac{\partial f}{\partial x}\right)\cdot\frac{\partial y}{\partial t} = 2s\frac{\partial^2 f}{\partial x^2} + 2t\frac{\partial^2 f}{\partial y\partial x}$$

$$\frac{\partial}{\partial t}\left(\frac{\partial f}{\partial y}\right) = \frac{\partial}{\partial x}\left(\frac{\partial f}{\partial y}\right)\cdot\frac{\partial x}{\partial t} + \frac{\partial}{\partial y}\left(\frac{\partial f}{\partial y}\right)\cdot\frac{\partial y}{\partial t} = 2s\frac{\partial^2 f}{\partial x\partial y} + 2t\frac{\partial^2 f}{\partial y^2}.$$

Thus, we may continue the above computation as follows:

$$
\frac{\partial^2 f}{\partial t^2} = 2s \left\{ 2s \frac{\partial^2 f}{\partial x^2} + 2t \frac{\partial^2 f}{\partial y \partial x} \right\} + 2 \frac{\partial f}{\partial y} + 2t \left\{ 2s \frac{\partial^2 f}{\partial x \partial y} + 2t \frac{\partial^2 f}{\partial y^2} \right\}
$$

$$
= 2 \frac{\partial f}{\partial y} + 4s^2 \frac{\partial^2 f}{\partial x^2} + 8st \frac{\partial^2 f}{\partial x \partial y} + 4t^2 \frac{\partial^2 f}{\partial y^2}.
$$

(7.8)

Similarly,

$$
\frac{\partial^2 f}{\partial s^2} = -2 \frac{\partial f}{\partial y} + 4t^2 \frac{\partial^2 f}{\partial x^2} - 8st \frac{\partial^2 f}{\partial x \partial y} + 4s^2 \frac{\partial^2 f}{\partial y^2}.
$$

(7.9)

Now, adding (7.8) and (7.9) we obtain

$$
\frac{\partial^2 f}{\partial t^2} + \frac{\partial^2 f}{\partial s^2} = 4(s^2 + t^2) \left(\frac{\partial^2 f}{\partial x^2} + \frac{\partial^2 f}{\partial y^2} \right).
$$

Exercises

Exercise 8. Let $g : \mathbb{R}^2 \to \mathbb{R}$ be a function of class C^2 and $x = s^2 - 2t^2$, $y = st^2$. Using the Chain Rule, compute $\dfrac{\partial^2 g}{\partial s \partial t}$.

Exercise 9. Let $h : \mathbb{R}^2 \to \mathbb{R}$ be a function of class C^2 and

$$
x = t(s + r) , \quad y = t(s - r).
$$

Show that

$$
\frac{\partial^2 h}{\partial r \partial t} = \frac{\partial h}{\partial x} - \frac{\partial h}{\partial y} + x \frac{\partial^2 h}{\partial x^2} + (y - x) \frac{\partial^2 h}{\partial x \partial y} - y \frac{\partial^2 h}{\partial y^2}.
$$

Exercise 10. Let $f : \mathbb{R}^2 \to \mathbb{R}$ be a function of class C^2 and $x = e^r \cos t$, $y = e^r \sin t$.

(i) Compute $\dfrac{\partial^2 f}{\partial r^2}, \dfrac{\partial^2 f}{\partial r \partial t}$ and $\dfrac{\partial^2}{\partial t^2}$;

(ii) Prove that $\dfrac{\partial^2 f}{\partial r^2} + \dfrac{\partial^2 f}{\partial t^2} = e^{2r} \left(\dfrac{\partial^2 f}{\partial x^2} + \dfrac{\partial^2 f}{\partial y^2} \right)$.

Exercise 11. The differentiable function $z = z(x, y)$ satisfies the equation

$$
z^3 + xy(z + y) = 0.
$$

(i) Find $z(-1, 2)$, $\dfrac{\partial z}{\partial x}(-1, 2)$ and $\dfrac{\partial z}{\partial y}(-1, 2)$;

(ii) Show that z is twice differentiable at $(-1, 2)$ and compute $\dfrac{\partial^2 z}{\partial x \partial y}(-1, 2)$.

7.3 The Laplace Operator

One important quantity that involves the second order derivatives is the Laplace operator (or simply Laplacian) as defined below.

Definition 7.11 *Let $f : D \subset \mathbb{R}^m \to \mathbb{R}$ be a function of m variables which has second order partial derivatives. The Laplace operator (or Laplacian) of f is given by*

$$\Delta f(\mathbf{x}) = \frac{\partial^2 f}{\partial x_1^2}(\mathbf{x}) + \frac{\partial^2 f}{\partial x_2^2}(\mathbf{x}) + \cdots + \frac{\partial^2 f}{\partial x_m^2}(\mathbf{x}) \quad \text{for all } \mathbf{x} \in D.$$

The term Laplace operator was coined after the French mathematician Pierre-Simon de Laplace (1749–1827) and has since been used in the mathematical modelling of various real life phenomena such as:

- heat and fluid flow through diffusion equation;

- electric and gravitational potentials through Poisson's equation;

- the wave function in quantum mechanics through Schrödinger's equation.

It is straightforward to see that Δf is the trace of the Hessian matrix H_f.

Definition 7.12 *Let $f : D \subset \mathbb{R}^m \to \mathbb{R}$ be a function which has second order partial derivatives. We say that f is harmonic on D if $\Delta f = 0$ in D.*

If $D \subset \mathbb{R}$, then harmonic functions $f(x)$ on D are nothing but affine functions $f(x) = ax + b$ where $a, b \in \mathbb{R}$. However, if $D \subset \mathbb{R}^m$, $m \geq 2$, then the class of harmonic functions is much broader. For instance $e^x \cos y$, $e^x \sin y$, $x^2 - y^2$ are harmonic functions in $\mathbb{R}^2$. The reader having some familiarity with Complex Analysis will have certainly noticed that if $f(x, y) = u(x, y) + iv(x, y)$ is holomorphic (that is, complex differentiable) in an open set $D \subset \mathbb{R}^2$, then both u and v are harmonic functions. This fact follows immediately from the *Cauchy-Riemann equations*

$$\frac{\partial u}{\partial x} = \frac{\partial v}{\partial y} \quad \text{and} \quad \frac{\partial u}{\partial y} = -\frac{\partial v}{\partial x}.$$

For instance, taking $f(x, y) = e^{x+iy}$ one has that

$$\mathrm{Re} f(x, y) = e^x \cos y \quad \text{and} \quad \mathrm{Im} f(x, y) = e^x \sin y$$

are harmonic functions. Conversely, under mild assumptions on the open set $D \subset \mathbb{R}^2$, if $u(x, y)$ is harmonic on D, then it is the real part of a holomorphic function defined on D. There is no counterpart of this result in higher dimensions. The field of mathematics that studies the harmonic functions is called Potential Theory. This is because harmonic functions are used to model the gravity and the electrostatic forces through what we nowadays call gravitational potential and electrostatic potential. Turning back to the Laplace operator, we state and prove below one property of it.

Theorem 7.13 *Let* $f : D \subset \mathbb{R}^m \to \mathbb{R}$ *be a function of class* C^2. *Assume* $\mathbf{a} \in D$ *satisfies* $B_r(\mathbf{a}) \subset D$ *for some* $r > 0$ *and*

$$f(\mathbf{x}) \le f(\mathbf{a}) \quad \text{for all } \mathbf{x} \in B_r(\mathbf{a}). \tag{7.10}$$

Then $\Delta f(\mathbf{a}) \le 0$.

A point $\mathbf{a} \in D$ which satisfies (7.10) is called a *local maximum point of* f and we shall be concerned with the study of these points in Chapter 10.
Proof Let $\mathbf{a} = (a_1, a_2, \ldots, a_m)$. The main point of the proof is to reduce it to the setting of one-variable functions with which the reader is more familiar. Since $B_r(\mathbf{a}) \subset D$, we can define

$$g : (-r, r) \to \mathbb{R} \quad \text{by} \quad g(t) = f(t + a_1, a_2, \ldots, a_m).$$

Then, g is of class C^2 and $t = 0$ is a maximum point of g. By L'Hôpital's rule, we have

$$\lim_{t \to 0} \frac{g(t) + g(-t) - 2g(0)}{t^2} = \lim_{t \to 0} \frac{g'(t) - g'(-t)}{2t} \quad \text{(by L'Hôpital's rule)}$$

$$= \lim_{t \to 0} \frac{g''(t) + g''(-t)}{2} \quad \text{(again, by L'Hôpital's rule)}$$

$$= g''(0).$$

Since $t = 0$ is a maximum point of g, there exists $\eta \in (0, r)$ such that $g(t) \le g(0)$ for all $t \in (-\eta, \eta)$. This shows that

$$\frac{g(t) + g(-t) - 2g(0)}{t^2} \le 0 \quad \text{for all } t \in (-\eta, \eta).$$

From the above calculations we deduce $g''(0) \le 0$ which means $\frac{\partial^2 f}{\partial x_1^2}(\mathbf{a}) \le 0$. With the same idea, we derive $\frac{\partial^2 f}{\partial x_i^2}(\mathbf{a}) \le 0$ for all $1 \le i \le m$ and this yields $\Delta f(\mathbf{a}) \le 0$. $\qquad \square$

Example 7.14 *Let* $f : \mathbb{R}^m \to \mathbb{R}$, $m \ge 2$ *be a radially symmetric function, that is,*

$$f(\mathbf{x}) = \phi(\|\mathbf{x}\|) \quad \text{for all} \quad \mathbf{x} \in \mathbb{R}^m, \quad \text{where} \quad \phi : \mathbb{R} \to \mathbb{R}.$$

If ϕ *is twice differentiable on* $\mathbb{R}$, *then for all* $\mathbf{x} \in \mathbb{R}^m \setminus \{\mathbf{0}\}$, *we have*

$$\Delta f(\mathbf{x}) = \phi''(r) + \frac{m-1}{r}\phi'(r) \quad \text{where } r = \|\mathbf{x}\|.$$

The above equality is called the Laplace operator for radially symmetric functions.

Solution. Since ϕ is twice differentiable, for all $1 \le i \le m$, we have (see also Exercise 15 in Chapter 5):

$$\frac{\partial f}{\partial x_i}(\mathbf{x}) = \phi'(\|\mathbf{x}\|)\frac{x_i}{\|\mathbf{x}\|},$$

$$\frac{\partial^2 f}{\partial x_i^2}(\mathbf{x}) = \phi''(\|\mathbf{x}\|)\frac{x_i^2}{\|\mathbf{x}\|^2} + \phi'(\|\mathbf{x}\|)\frac{\|\mathbf{x}\|^2 - x_i^2}{\|\mathbf{x}\|^3}.$$

Thus,

$$\Delta f(\mathbf{x}) = \sum_{i=1}^{m} \frac{\partial^2 f}{\partial x_i^2}(\mathbf{x}) = \phi''(r) + \frac{m-1}{r}\phi'(r).$$

Exercises

Exercise 12. Compute the Laplace operator of f in each of the following cases.

    (i) $f(x, y, z) = \ln(x^2 + y^2 + z^2)$, $(x, y, z) \in \mathbb{R}^3 \setminus \{(0, 0, 0)\}$;

    (ii) $f(\mathbf{x}) = \|\mathbf{x}\|^k$, $\mathbf{x} \in \mathbb{R}^m \setminus \{\mathbf{0}\}$, $k \in \mathbb{R}$.

Exercise 13. Check that the following functions are harmonic on their domain of definition.

    (i) $f(x, y) = \ln(x^2 + y^2)$, $(x, y) \in \mathbb{R}^2 \setminus \{(0, 0)\}$;

    (ii) $f(\mathbf{x}) = \|\mathbf{x}\|^{2-m}$, $\mathbf{x} \in \mathbb{R}^m \setminus \{\mathbf{0}\}$;

    (iii) $f(\mathbf{x}) = \dfrac{x_m}{\|\mathbf{x}\|^m}$, $\mathbf{x} \in \mathbb{R}^m \setminus \{\mathbf{0}\}$.

Exercise 14. Let $f, g : D \subset \mathbb{R}^m \to \mathbb{R}$ be two functions with second order partial derivatives. Check that

$$\Delta(fg) = f\Delta g + 2\nabla f \bullet \nabla g + 2g\Delta f.$$

Exercise 15. Let $f : \mathbb{R}^m \to \mathbb{R}$ be a function with second order partial derivatives. Prove that if f and f^3 are harmonic functions, then f is constant.

Exercise 16. Let $f : \mathbb{R}^3 \to \mathbb{R}$ be a twice differentiable function and

$$x = r\cos\theta, \quad y = r\sin\theta.$$

Check that

$$\Delta f(x, y, z) = \frac{1}{r}\frac{\partial}{\partial r}\left(r\frac{\partial f}{\partial r}\right) + \frac{1}{r^2}\frac{\partial^2 f}{\partial \theta^2} + \frac{\partial^2 f}{\partial z^2}.$$

The above formula is called the Laplace operator in cylindrical coordinates.

Exercise 17. Let $f : \mathbb{R}^m \to \mathbb{R}$ be defined as

$$f(\mathbf{x}) = \begin{cases} \dfrac{e^{\|\mathbf{x}\|} - e^{-\|\mathbf{x}\|}}{2\|\mathbf{x}\|} & \text{if } \mathbf{x} \neq \mathbf{0} \\[2mm] 1 & \text{if } \mathbf{x} = \mathbf{0}. \end{cases}$$

(i) Check that f is a function of class C^2 on $\mathbb{R}^m$;

(ii) Check that

$$\Delta f(x) = \begin{cases} \left(1 - \dfrac{m-3}{\|\mathbf{x}\|^2}\right) f(\mathbf{x}) + \dfrac{m-3}{2} \dfrac{e^{\|\mathbf{x}\|} + e^{-\|\mathbf{x}\|}}{2\|\mathbf{x}\|} & \text{if } \mathbf{x} \neq \mathbf{0} \\[2mm] \dfrac{m}{3} & \text{if } \mathbf{x} = \mathbf{0} \end{cases}$$

for all $\mathbf{x} \in \mathbb{R}^m$. Hint: Use Example 7.14.

8

Taylor's Theorem

This chapter introduces higher order derivatives of functions of several variables. These are used to state the Taylor's Approximation Theorem. Further, linear and quadratic approximations are derived which approximate a regular function around a given point as a linear or quadratic polynomial.

8.1 Higher Order Derivatives

We can iterate the computation of second order partial derivatives defined in the previous chapter and obtain quantities which are naturally written as

$$\frac{\partial^k f}{\partial x_1^{\alpha_1} \partial x_2^{\alpha_2} \cdots \partial x_m^{\alpha_m}},$$

where $\alpha_i \geq 0$ are integers and $\alpha_1 + \alpha_2 + \cdots + \alpha_m = k$. To denote these quantities in a simpler way, we introduce the multi-index notation as follows.

Definition 8.1 *We say that* $\alpha = (\alpha_1, \alpha_2, \ldots, \alpha_m)$ *is a* m-*multi-index if* $\alpha_i \geq 0$ *are integers for all* $1 \leq i \leq m$. *We denote*

$$|\alpha| = \alpha_1 + \alpha_2 + \cdots + \alpha_m \quad and \quad \alpha! = \alpha_1! \alpha_2! \cdots \alpha_m!.$$

For $\mathbf{z} = (z_1, z_2, \ldots, z_m) \in \mathbb{R}^m$ *we denote* $\mathbf{z}^\alpha = z_1^{\alpha_1} z_2^{\alpha_2} \cdots z_m^{\alpha_m}$. *For a function* $f : D \subset \mathbb{R}^m \to \mathbb{R}$ *we also denote*

$$D^\alpha f = \frac{\partial^{|\alpha|} f}{\partial x_1^{\alpha_1} \partial x_2^{\alpha_2} \cdots \partial x_m^{\alpha_m}}.$$

For instance, if $\alpha = (2, 3, 1)$, then $|\alpha| = 6$, $\alpha! = 2!3!1! = 12$,

$$\mathbf{z}^\alpha = z_1^2 z_2^3 z_3 \quad and \quad D^\alpha f = \frac{\partial^6 f}{\partial x_1^2 \partial x_2^3 \partial x_3}.$$

By direct computation we derive the following result.

DOI: 10.1201/9781003449652-8

Example 8.2 *Let* $\mathbf{a} \in \mathbb{R}^m$ *and* $f, g : \mathbb{R}^m \to \mathbb{R}$, $f(x) = e^{\mathbf{a} \bullet \mathbf{x}}$, $g(\mathbf{x}) = \sin(\mathbf{a} \bullet \mathbf{x})$. *Then*

$$D^\alpha f(\mathbf{x}) = \mathbf{a}^\alpha f(\mathbf{x}), \quad D^\alpha g(\mathbf{x}) = \begin{cases} \mathbf{a}^\alpha \sin(\mathbf{a} \bullet \mathbf{x}) & \text{if } |\alpha| = 4n \\ \mathbf{a}^\alpha \cos(\mathbf{a} \bullet \mathbf{x}) & \text{if } |\alpha| = 4n + 1 \\ -\mathbf{a}^\alpha \sin(\mathbf{a} \bullet \mathbf{x}) & \text{if } |\alpha| = 4n + 2 \\ -\mathbf{a}^\alpha \cos(\mathbf{a} \bullet \mathbf{x}) & \text{if } |\alpha| = 4n + 3. \end{cases}$$

In analogy to Definition 7.8 we state:

Definition 8.3 *Let* $D \subset \mathbb{R}^m$ *be an open set and* $n \geq 1$ *be an integer. We say that* $f : D \to \mathbb{R}$ *is of class* C^n *on* D *if all partial derivatives of* f *up to order* n *exist and are continuous on* D. *We denote*

$$C^n(D) = \{f : D \to \mathbb{R} : f \text{ is of class } C^n \text{ on } D\}.$$

The notion of a C^1 function coincides with that of a continuously differentiable function as it was introduced in Section 4.4. In this spirit, a C^n function is also called a n times continuously differentiable function.

If $f(x, y)$ is a function of two variables of class C^n, $n \geq 1$, then by Theorem 7.4, we have

$$\frac{\partial^n f}{\partial x^a \partial y^b} = \frac{\partial^n f}{\partial y^b \partial x^a},$$

whenever a, b are nonnegative integers such that $a + b = n$. Thus, there are $n + 1$ possibilities for choosing the multi-index $\alpha = (a, b)$, so, f has at most $n + 1$ distinct partial derivatives of order n.

Assume now $f : D \subset \mathbb{R}^3 \to \mathbb{R}$ is a C^n function of three variables. We want to estimate the number of distinct partial derivatives of f of order n. For instance, if $n = 3$, then by Theorem 7.4, we have

$$\frac{\partial^3 f}{\partial x \partial y \partial z} = \frac{\partial^3 f}{\partial y \partial x \partial z} = \frac{\partial^3 f}{\partial y \partial z \partial x}.$$

The number of distinct partial derivatives of order n is at most equal to that of triples (a, b, c) of non-negative integers such that $a + b + c = n$. For $0 \leq c \leq n$ there are $n - c + 1$ pairs (a, b) of non-negative integers such that $a + b = n - c$. Hence, there are at most

$$\sum_{c=0}^{n} (n - c + 1) = (n + 1) + n + (n - 1) + \ldots + 2 + 1 = \frac{(n + 1)(n + 2)}{2}$$

distinct n-th order partial derivatives of f.

Exercises

Exercise 1. Let $f : D \subset \mathbb{R}^m \to \mathbb{R}$ be a three times differentiable function on D and $g(\mathbf{x}) = \mathbf{x} \bullet \nabla f(\mathbf{x})$. Prove that $\Delta g(\mathbf{x}) = 2\Delta f(\mathbf{x}) + \mathbf{x} \bullet \nabla(\Delta f(\mathbf{x}))$.

Exercise 2. (i) Let $f(x, y) = e^{xy}$. Find $D^\alpha f$, where $\alpha = (\alpha_1, \alpha_2)$ is a 2-multi-index;

(ii) Let $f(x, y, z) = e^{xyz}$. Find $D^\alpha f$, where $\alpha = (\alpha_1, \alpha_2, \alpha_3)$ is a 3-multi-index.

Exercise 3. Let $f : \mathbb{R}^2 \to \mathbb{R}$, $f(x, y) = xe^{xy}$.

(i) Show that for any $n \geq 1$ there exists a polynomial $P_n(x, y)$ such that
$$\frac{\partial^n f}{\partial x^n}(x, y) = P_n(x, y)e^{xy} \quad \text{for all } (x, y) \in \mathbb{R}^2;$$

(ii) Deduce the expression of $P_n(x, y)$.

8.2   Taylor's Theorem

The following result is the counterpart of Taylor's Theorem for functions of several variables.

Theorem 8.4 (Taylor's Theorem)
   Let $f : B_r(\mathbf{a}) \subset \mathbb{R}^m \to \mathbb{R}$ be a function of class C^n, $n \geq 1$. Then, there exists a function $g : B_r(\mathbf{a}) \to \mathbb{R}$ such that:

(i) *g is continuous at $\mathbf{a}$ and $g(\mathbf{a}) = 0$;*

(ii) $\displaystyle \lim_{\mathbf{x} \to \mathbf{a}} \frac{g(\mathbf{x})}{\|\mathbf{x} - \mathbf{a}\|^n} = 0$;

(iii) *for any $\mathbf{x} \in B_r(\mathbf{a})$, we have*
$$f(\mathbf{x}) = \sum_{0 \leq |\alpha| \leq n} \frac{D^\alpha f(\mathbf{a})}{\alpha!}(\mathbf{x} - \mathbf{a})^\alpha + g(\mathbf{x}). \tag{8.1}$$

Proof We translate everything to the setting of one-variable functions. Let $\mathbf{x} \in B_r(\mathbf{a})$ and take $\mathbf{z} \in \partial B_r(\mathbf{a})$ such that the points $\mathbf{a}, \mathbf{x}$ and $\mathbf{z}$ are collinear and $\mathbf{x} \in [\mathbf{a}, \mathbf{z})$. Define $h : [0, 1) \to \mathbb{R}$, $h(t) = f(\mathbf{a} + t(\mathbf{z} - \mathbf{a}))$ which is a function of class C^n on $[0, 1)$. Set next

$$\psi : [0, 1) \to \mathbb{R}, \quad \psi(t) = \begin{cases} \dfrac{1}{t^n}\left(h(t) - \sum_{k=0}^{n} \dfrac{h^{(k)}(0)}{k!} t^k \right) & \text{if } 0 < t < 1 \\[2ex] 0 & \text{if } t = 0. \end{cases}$$

We claim that ψ is continuous at $t = 0$. Indeed, applying successively the L'Hôpital's rule, we have

$$\lim_{t \to 0} \psi(t) = \lim_{t \to 0} \frac{h'(t) - \displaystyle\sum_{k=0}^{n-1} \dfrac{h^{(k)}(0)}{k!} t^k}{n t^{n-1}}$$

$$= \lim_{t \to 0} \frac{h''(t) - \displaystyle\sum_{k=0}^{n-2} \dfrac{h^{(k)}(0)}{k!} t^k}{n(n-1) t^{n-2}}$$

$$= \dots$$

$$= \lim_{t \to 0} \frac{h^{(n)}(t) - h^{(n)}(0)}{n!}$$

$$= 0 = \psi(0).$$

From the definition of ψ, we have

$$h(t) = \sum_{k=0}^{n} \frac{h^{(k)}(0)}{k!} t^k + t^n \psi(t) \quad \text{for all } 0 \le t < 1.$$

Letting $t = \|\mathbf{x} - \mathbf{a}\|/r < 1$, we have $\mathbf{x} = \mathbf{a} + t(\mathbf{z} - \mathbf{a})$ and the above equality yields

$$f(\mathbf{x}) = \sum_{k=0}^{n} \frac{h^{(k)}(0)}{k!} \frac{\|\mathbf{x} - \mathbf{a}\|^k}{r^k} + \frac{\|\mathbf{x} - \mathbf{a}\|^n}{r^n} \psi\left(\frac{\|\mathbf{x} - \mathbf{a}\|}{r} \right). \qquad (8.2)$$

Finally, to deduce (8.1) we observe that $h(0) = f(\mathbf{a})$,

$$h'(0) = \left[\sum_{i=1}^{m} (\mathbf{z} - \mathbf{a})_i \frac{\partial}{\partial x_i} \right] f(\mathbf{a}), \quad h''(0) = \left[\sum_{i=1}^{m} (\mathbf{z} - \mathbf{a})_i \frac{\partial}{\partial x_i} \right]^2 f(\mathbf{a}),$$

and in general

$$h^{(k)}(0) = \left[\sum_{i=1}^{m} (\mathbf{z} - \mathbf{a})_i \frac{\partial}{\partial x_i} \right]^k f(\mathbf{a}), \quad 1 \le k \le n.$$

Using the multinomial formula

$$(y_1 + y_2 + \dots + y_m)^k = \sum_{|\alpha|=k} \frac{k!}{\alpha!} \mathbf{y}^\alpha \quad \text{for all } \mathbf{y} = (y_1, y_2, \dots, y_m) \in \mathbb{R}^m,$$

we have

$$\frac{h^{(k)}(0)}{k!} \frac{\|\mathbf{x} - \mathbf{a}\|^k}{r^k} = \sum_{|\alpha|=k} \frac{D^\alpha f(\mathbf{a})}{\alpha!} \left(\frac{\|\mathbf{x} - \mathbf{a}\|}{r} (\mathbf{z} - \mathbf{a}) \right)^\alpha$$

$$= \sum_{|\alpha|=k} \frac{D^\alpha f(\mathbf{a})}{\alpha!} (\mathbf{x} - \mathbf{a})^\alpha. \tag{8.3}$$

Define also

$$g : B_r(\mathbf{a}) \to \mathbb{R} \quad \text{by} \quad g(\mathbf{x}) = \frac{\|\mathbf{x} - \mathbf{a}\|^n}{r^n} \psi \left(\frac{\|\mathbf{x} - \mathbf{a}\|}{r} \right).$$

Using the above equality together with (8.3) in (8.2) we deduce (8.1). □

Definition 8.5 *The quantity*

$$P_{n,\mathbf{a}}(\mathbf{x}) = \sum_{0 \le |\alpha| \le n} \frac{D^\alpha f(\mathbf{a})}{\alpha!} (\mathbf{x} - \mathbf{a})^\alpha$$

is called the Taylor polynomial of degree n associated with f at point $\mathbf{a}$.

The term $g(\mathbf{x})$ in (8.1) is often called the *Taylor remainder* of order n in Taylor's Theorem 8.4. It is frequently denoted $R_{n,\mathbf{a}}(\mathbf{x})$. If f is a more regular function (in particular if f of class C^{n+1}), then we have a better knowledge on the remainder which is given by

$$R_{n,\mathbf{a}}(\mathbf{x}) = \sum_{|\alpha|=n+1} \frac{D^\alpha f(\overline{\mathbf{x}})}{\alpha!} (\mathbf{x} - \mathbf{a})^\alpha, \quad \text{for some } \overline{\mathbf{x}} \in (\mathbf{a}, \mathbf{x}).$$

We will not investigate this avenue in the present book. Instead, we shall introduce the following terminology called *Landau symbols*.

Definition 8.6 *Let $D \subset \mathbb{R}^m$ be an open set, $\mathbf{a} \in \overline{D}$ and $g, h : D \to \mathbb{R}$ be two functions, where $h > 0$ on D.*

(i) (big-O symbol) *We write $g(\mathbf{x}) = O(h(\mathbf{x}))$ as $\mathbf{x} \to \mathbf{a}$ if there exist to constants $C, r > 0$ such that $B_r(\mathbf{a}) \subset D$ and*

$$|f(\mathbf{x})| \le Cg(\mathbf{x}) \quad \text{for all } x \in B_r(\mathbf{a}).$$

(ii) (little-o symbol) *We write $g(\mathbf{x}) = o(h(\mathbf{x}))$ as $\mathbf{x} \to \mathbf{a}$ if*

$$\lim_{\mathbf{x} \to \mathbf{a}} \frac{|f(\mathbf{x})|}{g(\mathbf{x})} = 0.$$

Let us place ourselves in the context of the Taylor's Theorem 8.4. Then, the function $g(\mathbf{x})$ in the statement of this result satisfies

$$\lim_{\mathbf{x} \to \mathbf{a}} \frac{g(\mathbf{x})}{\|\mathbf{x} - \mathbf{a}\|^n} = 0, \quad \text{that is,} \quad g(\mathbf{x}) = o(\|\mathbf{x} - \mathbf{a}\|^n) \text{ as } \mathbf{x} \to \mathbf{a}.$$

We may thus reformulate Theorem 8.4 as follows: If f is a function of class C^n on the ball $B_r(\mathbf{a}) \subset \mathbb{R}^m$, then

$$f(\mathbf{x}) = \sum_{0 \leq |\alpha| \leq n} \frac{D^\alpha f(\mathbf{a})}{\alpha!} (\mathbf{x} - \mathbf{a})^\alpha + o(\|\mathbf{x} - \mathbf{a}\|^n) \quad \text{as } \mathbf{x} \to \mathbf{a}, \tag{8.4}$$

which further yields the Taylor's approximation

$$f(\mathbf{x}) \simeq \sum_{0 \leq |\alpha| \leq n} \frac{D^\alpha f(\mathbf{a})}{\alpha!} (\mathbf{x} - \mathbf{a})^\alpha = P_{n,\mathbf{a}}(\mathbf{x}) \quad \text{as } \mathbf{x} \to \mathbf{a}.$$

8.3 Linear and Quadratic Approximation

Equality (8.4) allows us to derive explicit polynomial approximations of a function f in a neighbourhood of the point $\mathbf{a}$ in its domain. In this section we discuss in more details the particular cases $n = 1$ and $n = 2$ in (8.4).

By taking $n = 1$ in Taylor's Theorem 8.4 we obtain:

Theorem 8.7 (Linear Taylor approximation)
   Assume $f : B_r(\mathbf{a}) \subset \mathbb{R}^m \to \mathbb{R}$ is a function of class C^1. Then, as $\mathbf{x} \to \mathbf{a}$ we have

$$f(\mathbf{x}) = f(\mathbf{a}) + \sum_{i=1}^m \frac{\partial f}{\partial x_i}(\mathbf{a})(x_i - a_i) + o(\|\mathbf{x} - \mathbf{a}\|).$$

The quantity $L_f(\mathbf{x}) = f(\mathbf{a}) + \sum_{i=1}^m \frac{\partial f}{\partial x_i}(\mathbf{a})(x_i - a_i)$ is called the linear Taylor approximation of f at $\mathbf{a}$.

Theorem 8.7 states that

$$f(\mathbf{x}) = f(\mathbf{a}) + \nabla f(\mathbf{a}) \bullet (\mathbf{x} - \mathbf{a}) + o(\|\mathbf{x} - \mathbf{a}\|) \quad \text{as } \mathbf{x} \to \mathbf{a},$$

which yields

$$f(\mathbf{x}) \simeq f(\mathbf{a}) + \nabla f(\mathbf{a}) \bullet (\mathbf{x} - \mathbf{a}) \quad \text{as } \mathbf{x} \to \mathbf{a}.$$

Let us observe that for functions of two variables, the above approximation reads

$$f(x, y) \simeq f(a, b) + \frac{\partial f}{\partial x}(a, b)(x - a) + \frac{\partial f}{\partial y}(a, b)(y - b) \quad \text{as } (x, y) \to (a, b).$$

The above quantity is nothing but the tangent plane to the surface $z = f(x,y)$ (see Theorem 4.10). Thus, the best linear approximation of a continuously differentiable function at a point is the tangent plane to its graph at that point.

Example 8.8 *Find the linear Taylor approximation of $f(x,y) = e^{xy}\sin(2y)$ at $\left(0, \frac{\pi}{3}\right)$.*

Solution. We have $f\left(0, \frac{\pi}{3}\right) = \frac{\sqrt{3}}{2}$. Also,

$$\frac{\partial f}{\partial x} = ye^{xy}\sin(2y) \quad \text{and} \quad \frac{\partial f}{\partial y} = xe^{xy}\sin(2y) + 2e^{xy}\cos(2y).$$

Thus, $\frac{\partial f}{\partial x}\left(0, \frac{\pi}{3}\right) = \frac{\pi}{2\sqrt{3}}$ and $\frac{\partial f}{\partial y}\left(0, \frac{\pi}{3}\right) = -1$. The linear Taylor approximation is

$$L_f(x,y) = f\left(0, \frac{\pi}{3}\right) + \frac{\partial f}{\partial x}\left(0, \frac{\pi}{3}\right)x + \frac{\partial f}{\partial y}\left(0, \frac{\pi}{3}\right)\left(y - \frac{\pi}{3}\right).$$

Hence

$$L_f(x,y) = \frac{\pi}{2\sqrt{3}}x - y + \frac{\sqrt{3}}{2} + \frac{\pi}{3}.$$

Example 8.9 $f : \mathbb{R}^2 \to \mathbb{R}$ *is a continuously differentiable function whose linear Taylor approximation at $(-2,1)$ is $L_f(x,y) = 3x - y + 6$.*

(i) *Find $f(-2,1)$, $\frac{\partial f}{\partial x}(-2,1)$ and $\frac{\partial f}{\partial y}(-2,1)$;*

(ii) *Find the linear Taylor approximation of f^2 and $\frac{1}{f}$ at $(-2,1)$.*

Solution. (i) The linear Taylor approximation of f at $(-2,1)$ is

$$L_f(x,y) = f(-2,1) + \frac{\partial f}{\partial x}(-2,1)(x+2) + \frac{\partial f}{\partial y}(-2,1)(y-1).$$

On the other hand, $L_f(x,y) = 3x - y + 6$. Identifying the coefficients in the above expression, we have

$$\begin{cases} \dfrac{\partial f}{\partial x}(-2,1) = 3 \\[2mm] \dfrac{\partial f}{\partial y}(-2,1) = -1 \\[2mm] 2\dfrac{\partial f}{\partial x}(-2,1) - \dfrac{\partial f}{\partial y}(-2,1) + f(-2,1) = 6 \end{cases} \implies \begin{cases} \dfrac{\partial f}{\partial x}(-2,1) = 3 \\[2mm] \dfrac{\partial f}{\partial y}(-2,1) = -1 \\[2mm] f(-2,1) = -1. \end{cases}$$

(ii) Let $g(x, y) = f(x, y)^2$. Then $g(-2, 1) = 1$ and

$$\frac{\partial g}{\partial x}(-2, 1) = 2f(-2, 1)\frac{\partial f}{\partial x}(-2, 1) = -6$$

$$\frac{\partial g}{\partial y}(-2, 1) = 2f(-2, 1)\frac{\partial f}{\partial y}(-2, 1) = 2.$$

Thus, the linear Taylor approximation of $g = f^2$ at $(-2, 1)$ is

$$L_{f^2}(x, y) = 1 - 6(x + 2) + 2(y - 1) = -6x + 2y - 13.$$

To find the linear Taylor approximation of $h = \dfrac{1}{f}$ at $(-2, 1)$ we compute $h(-2, 1) = -1$ and

$$\frac{\partial h}{\partial x}(-2, 1) = -\frac{\dfrac{\partial f}{\partial x}(-2, 1)}{f(-2, 1)^2} = -3, \quad \frac{\partial h}{\partial y}(-2, 1) = -\frac{\dfrac{\partial f}{\partial y}(-2, 1)}{f(-2, 1)^2} = 1.$$

Hence,

$$L_{1/f}(x, y) = -1 - 3(x + 2) + (y - 1) = -3x + y - 8.$$

Taking $n = 2$ in Taylor's Theorem 8.4 we obtain:

Theorem 8.10 (Quadratic Taylor approximation)
Assume $f : B_r(\mathbf{a}) \subset \mathbb{R}^m \to \mathbb{R}$ is a function of class C^2. Then, as $\mathbf{x} \to \mathbf{a}$ we have

$$f(\mathbf{x}) = f(\mathbf{a}) + \sum_{i=1}^m \frac{\partial f}{\partial x_i}(\mathbf{a})(x_i - a_i)$$

$$+ \frac{1}{2} \sum_{i,j=1}^m \frac{\partial^2 f}{\partial x_i \partial x_j}(\mathbf{a})(x_i - a_i)(x_j - a_j) + o(\|\mathbf{x} - \mathbf{a}\|^2).$$

The quantity

$$Q_f(\mathbf{x}) = f(\mathbf{a}) + \sum_{i=1}^m \frac{\partial f}{\partial x_i}(\mathbf{a})(x_i - a_i) + \frac{1}{2} \sum_{i,j=1}^m \frac{\partial^2 f}{\partial x_i \partial x_j}(\mathbf{a})(x_i - a_i)(x_j - a_j)$$

is called the quadratic Taylor approximation of f at $\mathbf{a}$.

The result in Theorem 8.10 can be written in vectorial form as

$$f(\mathbf{x}) = f(\mathbf{a}) + \nabla f(\mathbf{a}) \bullet (\mathbf{x} - \mathbf{a}) + \frac{1}{2}(\mathbf{x} - \mathbf{a}) \bullet \left[H_f(\mathbf{a}) \cdot (\mathbf{x} - \mathbf{a})^T \right] + o(\|\mathbf{x} - \mathbf{a}\|^2)$$

as $\mathbf{x} \to \mathbf{a}$. This yields the approximation

$$f(\mathbf{x}) \simeq (\mathbf{a}) + \nabla f(\mathbf{a}) \bullet (\mathbf{x} - \mathbf{a}) + \frac{1}{2}(\mathbf{x} - \mathbf{a}) \bullet \left[H_f(\mathbf{a}) \cdot (\mathbf{x} - \mathbf{a})^T \right] \quad \text{as } \mathbf{x} \to \mathbf{a}.$$

Example 8.11 *Find the quadratic Taylor approximation of* $f(x,y) = \sqrt{3 - x^2 + 2y}$ *at* $(1,1)$.

Solution. We compute $f(1,1) = 2$ and

$$\frac{\partial f}{\partial x}(x,y) = -x(3 - x^2 + 2y)^{-1/2}, \quad \frac{\partial f}{\partial y}(x,y) = (3 - x^2 + 2y)^{-1/2}.$$

Thus $\frac{\partial f}{\partial x}(1,1) = -\frac{1}{2}$ and $\frac{\partial f}{\partial y}(1,1) = \frac{1}{2}$. Further, we have

$$\frac{\partial^2 f}{\partial x^2}(x,y) = -(3 + 2y)(3 - x^2 + 2y)^{-3/2},$$

$$\frac{\partial^2 f}{\partial x \partial y}(x,y) = \frac{\partial^2 f}{\partial y \partial x} = x(3 - x^2 + 2y)^{-3/2},$$

$$\frac{\partial^2 f}{\partial y^2}(x,y) = -(3 - x^2 + 2y)^{-3/2}.$$

Hence

$$\frac{\partial^2 f}{\partial x^2}(1,1) = -\frac{5}{8}, \quad \frac{\partial^2 f}{\partial x \partial y}(1,1) = \frac{\partial^2 f}{\partial y \partial x}(1,1) = \frac{1}{8} \quad \text{and} \quad \frac{\partial^2 f}{\partial x^2}(1,1) = -\frac{1}{8}.$$

The quadratic approximation is

$$Q_f(x,y) = 2 - \frac{1}{2}(x-1) + \frac{1}{2}(y-1) + \frac{1}{2}\left[-\frac{5}{8}(x-1)^2 + \frac{1}{4}(x-1)(y-1) - \frac{1}{8}(y-1)^2\right]$$

that is,

$$Q_f(x,y) = 2 - \frac{1}{2}x + \frac{1}{2}y - \frac{5}{16}(x-1)^2 + \frac{1}{8}(x-1)(y-1) - \frac{1}{16}(y-1)^2.$$

The linear and quadratic Taylor approximation can be used to determine more complex limits of functions than those encountered in Chapter 3. This fact is illustrated in the next example.

Example 8.12 *Use the linear or quadratic Taylor approximation to compute the following limits:*

(i) $\displaystyle\lim_{(x,y)\to(0,0)} \frac{\sin(x + y + x^2 y) - x - y}{\sqrt{x^2 + y^2}}$;

(ii) $\displaystyle\lim_{(x,y)\to(0,0)} \frac{\cos(xy - 2x) - 2y^2 - 1}{x^2 + y^2}$;

(iii) $\displaystyle\lim_{(x,y)\to(0,0)} \frac{e^{xy^2} - 1}{x^2 + y^2}$.

Solution. (i) Let $f(x, y) = \sin(x + y + x^2 y)$. Then

$$\frac{\partial f}{\partial x}(x, y) = (1 + 2xy) \cos(x + y + x^2 y), \quad \frac{\partial f}{\partial y}(x, y) = (1 + x^2) \cos(x + y + x^2 y).$$

Thus, by Theorem 8.7, we have $f(x, y) = x + y + g(x, y)$, where g satisfies $g(x, y) = o(\|(x, y)\|)$ as $(x, y) \to (0, 0)$. Hence,

$$\lim_{(x,y) \to (0,0)} \frac{\sin(x + y + x^2 y) - x - y}{\sqrt{x^2 + y^2}} = \lim_{(x,y) \to (0,0)} \frac{f(x, y) - x - y}{\sqrt{x^2 + y^2}}$$

$$= \lim_{(x,y) \to (0,0)} \frac{g(x, y)}{\sqrt{x^2 + y^2}} = 0.$$

(ii) Let $f(x, y) = \cos(xy - 2x)$. Then

$$\frac{\partial f}{\partial x}(x, y) = -(y - 2) \sin(xy - 2x), \quad \frac{\partial f}{\partial y}(x, y) = -x \sin(xy - 2x),$$

and

$$\frac{\partial^2 f}{\partial x^2}(x, y) = -(y - 2)^2 \cos(xy - 2x)$$

$$\frac{\partial^2 f}{\partial x \partial y}(x, y) = -\sin(xy - 2x) - x(y - 2) \cos(xy - 2x)$$

$$\frac{\partial^2 f}{\partial y^2}(x, y) = -x^2 \cos(xy - 2x).$$

By Theorem 8.10, we have $f(x, y) = 1 - 2x^2 + g(x, y)$ where g satisfies $g(x, y) = o(\|(x, y)\|^2)$ as $(x, y) \to (0, 0)$. Now,

$$\lim_{(x,y) \to (0,0)} \frac{\cos(xy - 2x) - 2y^2 - 1}{x^2 + y^2} = \lim_{(x,y) \to (0,0)} \frac{f(x, y) - 2y^2 - 1}{x^2 + y^2}$$

$$= \lim_{(x,y) \to (0,0)} \frac{-2(x^2 + y^2) + g(x, y)}{x^2 + y^2}$$

$$= -2 + \lim_{(x,y) \to (0,0)} \frac{g(x, y)}{x^2 + y^2} = -2.$$

(iii) As above, by Theorem 8.10 applied to $f(x, y) = e^{xy^2}$, we have $f(x, y) = 1 + g(x, y)$ where g satisfies $g(x, y) = o(\|(x, y)\|^2)$ as $(x, y) \to (0, 0)$. Then

$$\lim_{(x,y) \to (0,0)} \frac{e^{xy^2} - 1}{x^2 + y^2} = \lim_{(x,y) \to (0,0)} \frac{f(x, y) - 1}{x^2 + y^2} = \lim_{(x,y) \to (0,0)} \frac{g(x, y)}{x^2 + y^2} = 0.$$

We note in passing that the use of the linear or quadratic Taylor approximation of the function $f(x, y)$ at the top line in the above limits is dictated by the denominator. If for instance the denominator is $\sqrt{x^2 + y^2} = \|(x,y)\|$ as in Example 8.12(i), then one needs to use the linear approximation of f, since the remainder $g(x,y)$ satisfies $g(x,y) = o(\|(x,y)\|)$ as $(x,y) \to (0,0)$. Similarly, if the denominator is $x^2 + y^2$ as in Example 8.12(ii)–(iii), then one has to use the quadratic Taylor approximation as the remainder $g(x,y)$ satisfies in this case the faster decay rate

$$g(x,y) = o(x^2 + y^2) = o(\|(x,y)\|^2) \quad \text{as} \quad (x,y) \to (0,0).$$

More limits of this type are provided in Exercise 10.

Exercises

Exercise 4. Find the quadratic Taylor approximation of the following functions at the indicated points.

(i) $f(x,y) = xe^{-xy}$ at $(1,0)$;

(ii) $f(x,y) = x^y$ at $(2,2)$;

(iii) $f(x,y) = x\ln(y^2 + 1)$ at $(2,1)$;

(iv) $f(x,y) = \sin x \cos y$ at $\left(\frac{\pi}{2}, 0\right)$.

Exercise 5. The continuously differentiable function $f : \mathbb{R}^2 \to \mathbb{R}$ satisfies

$$f(2,3) = 6, \quad f(2.1,3) = 7 \quad \text{and} \quad f(2,2.9) = 5.$$

What would the best approximation for $f(2.1, 3.1)$ be?

Exercise 6. Let $f : \mathbb{R}^2 \to \mathbb{R}$ be a twice continuously differentiable function and

$$g : \mathbb{R}^2 \to \mathbb{R}, \quad g(x,y) = f(x^2, y^2).$$

Find the quadratic Taylor approximation of g at $(0,0)$.

Exercise 7. Let $g : \mathbb{R} \to \mathbb{R}$ be a C^2 function such that $g(0) = 2$, $g'(0) = 1$ and $g''(0) = 4$. Define

$$f : \mathbb{R}^2 \to \mathbb{R}, \quad f(x,y) = g(x^2 - y^3).$$

Find the quadratic Taylor approximation of f at $(1,1)$.

Exercise 8. Let $g : \mathbb{R} \to \mathbb{R}$ be a C^2 function such that $g(1) = -1$, $g'(1) = 2$ and $g''(1) = -2$. Define

$$f : \mathbb{R}^2 \to \mathbb{R}, \quad f(x,y) = g\left(\frac{x^2}{y} - 1\right).$$

Find the quadratic Taylor approximation of f at $(2,2)$.

Exercise 9. Let $f, g : \mathbb{R}^2 \to \mathbb{R}$ be two C^2 functions whose linear Taylor approximation at $(1, -1)$ is

$$L_f(x, y) = 3x + 4y + 3 \quad \text{and} \quad L_g(x, y) = 2x + y - 2.$$

Find the linear Taylor approximation at $(1, -1)$ of fg, $\frac{f}{g}$ and f^g.

Exercise 10. Use the linear or quadratic Taylor approximation to compute the following limits.

(i) $\displaystyle\lim_{(x,y) \to (0,0)} \frac{\sin(x^2 + xy) + y(y - x)}{x^2 + y^2}$;

(ii) $\displaystyle\lim_{(x,y) \to (0,0)} \frac{\ln(1 + xy - y) + y}{\sqrt{x^2 + y^2}}$;

(iii) $\displaystyle\lim_{(x,y) \to (0,0)} \frac{e^{x^2 + 2y^2} - 1 - y^2}{x^2 + y^2}$;

(iv) $\displaystyle\lim_{(x,y) \to (0,0)} \frac{\sin(2x + y^3) - 2x}{x^2 + y^2}$.

Exercise 11. (i) Give example of a continuously differentiable function $f : D \subset \mathbb{R}^m \to \mathbb{R}$ and $\mathbf{a} \in D$ such that

$$\lim_{\mathbf{x} \to \mathbf{a}} \frac{f(\mathbf{x}) - f(\mathbf{a})}{\|\mathbf{x} - \mathbf{a}\|} \quad \text{does not exist.}$$

(ii) Show that if $f : D \subset \mathbb{R}^m \to \mathbb{R}$ is continuously differentiable and $\mathbf{a} \in D$, then the above limit exists if and only if $\nabla f(\mathbf{a}) = 0$.
Hint: Use (8.4).

Exercise 12. The C^2 function $z = z(x, y) : \mathbb{R}^2 \to \mathbb{R}$ satisfies the implicit equation

$$z^2 + 2e^{xz} + 3y^2 = 6 \quad \text{and} \quad z(0, -1) = 1.$$

Find the quadratic Taylor approximation of z at $(0, -1)$.

Exercise 13. The continuously differentiable functions $u = u(x, y), v = v(x, y) : \mathbb{R}^2 \to \mathbb{R}$ satisfy

$$\begin{cases} xu + yv = \cos(yu) \\ yu + xv = \sin(xv) \\ u(0, 1) = 0, \ v(0, 1) = 1. \end{cases}$$

Find the linear Taylor approximation of u and v at $(0, 1)$.

9

Implicit Function Theorem

This chapter discusses the Implicit Function Theorem which has been partially used in Chapter 5 when the reader was introduced to implicit differentiation. In the current chapter a rigorous justification of differentiability is provided in the two- and three-variable setting and a general form of the Implicit Function Theorem is stated for any number of variables.

9.1   Preliminaries

It is now the right moment to justify the implicit differentiability we employed in Chapter 5 Section 5.2. More exactly, let $D \subset \mathbb{R}^2$ be an open set and $(a, b) \in D$. Assume $f : D \to \mathbb{R}$ is a continuously differentiable function such that $f(a, b) = m$. Then, the equality $f(x, y) = m$ defines y as a function of x in a neighbourhood $[a - \delta, a + \delta]$ of a provided $\frac{\partial f}{\partial y}(a, b) \neq 0$. The function ϕ has interesting properties which we applied (but not justified) in Section 5.2. In particular, ϕ is continuously differentiable and uniquely determined by the equality

$$f(x, \phi(x)) = m \quad \text{for all } x \in [a - \delta, a + \delta]. \tag{9.1}$$

Such a result is called the Implicit Function Theorem and we shall discuss it over the current chapter. We shall take small steps towards its full generality in order to get a clear understanding of it.

Let us note that the above equality shows, in particular, that the set

$$\{(x, y) : x \in [a - \delta, a + \delta], f(x, y) = m\} \subset L(f, m)$$

is the graph of the function $\phi : [a - \delta, a + \delta] \to \mathbb{R}$. Hence, the level curve $L(f, m)$ of f at level m is locally (or piecewise) the graph of a continuously differentiable function ϕ.

DOI: 10.1201/9781003449652-9

9.2 Two-Variable Case

We are now ready to state and prove the Implicit Function Theorem in the two-variables setting as announced above.

Theorem 9.1 *Let $D \subset \mathbb{R}^2$ be an open set, $(a, b) \in D$ and $f : D \to \mathbb{R}$ be a continuously differentiable function such that*

$$f(a, b) = m \quad and \quad \frac{\partial f}{\partial y}(a, b) \neq 0. \tag{9.2}$$

Then, there exist $\delta, \eta > 0$ such that $U := [a - \delta, a + \delta] \times [b - \eta, b + \eta] \subset D$ and the level curve $L(f, m) \cap U$ is the graph of a continuously differentiable function. That is, there exists $\phi : [a - \delta, a + \delta] \to [b - \eta, b + \eta]$ continuously differentiable such that:

(i) $\phi(a) = b$;

(ii) $f(x, \phi(x)) = m$ *for all $x \in [a - \delta, a + \delta]$*;

(iii) *whenever $(x, y) \in U$ and $f(x, y) = m$, then $y = \phi(x)$.*

Furthermore,

$$\phi'(x) = -\frac{\dfrac{\partial f}{\partial x}(x, \phi(x))}{\dfrac{\partial f}{\partial y}(x, \phi(x))} \quad for\ all\ \ x \in [a - \delta, a + \delta]. \tag{9.3}$$

Proof Replacing f by $f - m$ in the following, one may assume $f(a, b) = 0$. Also, replacing f by $-f$ one may also assume $\frac{\partial f}{\partial y}(a, b) > 0$.

Our argument is divided into three steps.
Step 1: Construction of ϕ.

Since $\frac{\partial f}{\partial y}$ is continuous, it follows that there exist $\eta, \sigma > 0$ such that

$$[a - \eta, a + \eta] \times [b - \eta, b + \eta] \subset D$$

and

$$\frac{\partial f}{\partial y} > \sigma \quad in \quad [a - \eta, a + \eta] \times [b - \eta, b + \eta]. \tag{9.4}$$

For any $(x, y) \in [a - \eta, a + \eta] \times [b - \eta, b + \eta]$ define $g_x(y) = h_y(x) = f(x, y)$.
Fix $x \in [a - \eta, a + \eta]$. Then,

$$g'_x(y) = \frac{\partial f}{\partial y}(x, y) > 0 \quad for\ all \quad y \in [b - \eta, b + \eta], \tag{9.5}$$

and so $g_x : [b - \eta, b + \eta] \to \mathbb{R}$ is increasing for all $x \in [a - \eta, a + \eta]$. Thus,

$$g_a(b - \eta) < g_a(b) = f(a, b) = 0 < g_a(b + \eta).$$

The above relation can be written as

$$h_{b-\eta}(a) < 0 < h_{b+\eta}(a).$$

Note that $h_{b-\eta}, h_{b+\eta} : [a - \eta, a + \eta] \to \mathbb{R}$ are continuous functions, which at a take a negative (resp. positive) value. Thus, there exists $\delta \in (0, \eta)$ such that

$$h_{b-\eta}(x) < 0 < h_{b+\eta}(x) \quad \text{for all } x \in [a - \delta, a + \delta].$$

Switching the above inequalities in terms of g_x one has

$$g_x(b - \eta) < 0 < g_x(b + \eta) \quad \text{for all } x \in [a - \delta, a + \delta].$$

Recall that by (9.5), g_x is increasing, so there exists a unique $y \in [b - \eta, b + \eta]$ so that $g_x(y) = 0$. We have thus constructed $\delta, \eta > 0$ such that $U := [a - \delta, a + \delta] \times [b - \eta, b + \eta] \subset D$ and for all $x \in [a - \delta, a + \delta]$ there exists a unique $y \in [b - \eta, b + \eta]$ so that $f(x, y) = 0$. We write

$$y = \phi(x) \quad \text{where} \quad \phi : [a - \delta, a + \delta] \to [b - \eta, b + \eta].$$

Step 2: $\phi : [a - \delta, a + \delta] \to [b - \eta, b + \eta]$ is continuous.

Let $x_0 \in [a - \delta, a + \delta]$. If ϕ was not continuous at x_0, then we could find two sequences $\{x_n^1\}, \{x_n^2\} \subset [a - \delta, a + \delta]$ so that $x_n^1, x_n^2 \to x_0$ and

$$\phi(x_n^1) \to \ell_1, \quad \phi(x_n^2) \to \ell_2 \quad \text{as} \quad n \to \infty \quad \text{and} \quad \ell_1 \neq \ell_2.$$

Note that

$$f(x_n^1, \phi(x_n^1)) = 0 \quad \text{and} \quad f(x_n^2, \phi(x_n^2)) = 0 \quad \text{for all} \quad n \geq 1.$$

Since f is continuous, we may pass to the limit in the above to deduce $f(x_0, \ell_1) = 0$ and $f(x_0, \ell_2) = 0$. By the definition of ϕ it follows that $\ell_1 = \phi(x_0) = \ell_2$, contradiction to the fact that $\ell_1 \neq \ell_2$. Hence, ϕ is continuous.

Step 3: $\phi : [a - \delta, a + \delta] \to [b - \eta, b + \eta]$ is differentiable.

Let as before $x_0 \in [a - \delta, a + \delta]$ and denote $y_0 = \phi(x_0)$.

Define $g : U \to \mathbb{R}$ by

$$g(x, y) = \begin{cases} \dfrac{f(x, y) - \nabla f(x_0, y_0) \bullet (x - x_0, y - y_0)}{\|(x - x_0, y - y_0)\|} & \text{if } (x, y) \neq (x_0, y_0) \\ 0 & \text{if } (x, y) = (x_0, y_0). \end{cases}$$

Since f is differentiable at (x_0, y_0), g is continuous at this point. Also, for $(x, y) \in U \setminus \{(x_0, y_0)\}$, we have

$$f(x, y) - f(x_0, y_0) = \frac{\partial f}{\partial x}(x_0, y_0)(x - x_0)$$

$$+ \frac{\partial f}{\partial y}(x_0, y_0)(y - y_0) + g(x, y)\|(x - x_0, y - y_0)\|.$$

We next let $y = \phi(x)$ in the above equality. Since

$$f(x, \phi(x)) = f(x_0, y_0) = 0$$

one has

$$\frac{\partial f}{\partial x}(x_0, y_0)(x - x_0) + \frac{\partial f}{\partial y}(x_0, y_0)(\phi(x) - \phi(x_0))$$

$$+ g(x, \phi(x))\|(x - x_0, \phi(x) - \phi(x_0))\| = 0,$$

for all $x \in [a - \delta, a + \delta]$. Since by (9.4), we have $\frac{\partial f}{\partial y}(x_0, y_0) > \sigma > 0$, the above equality entails

$$\left| \frac{\phi(x) - \phi(x_0)}{x - x_0} + \frac{\frac{\partial f}{\partial x}(x_0, y_0)}{\frac{\partial f}{\partial y}(x_0, y_0)} \right| = \frac{|g(x, \phi(x))|}{\frac{\partial f}{\partial y}(x_0, y_0)} \cdot \sqrt{1 + \left(\frac{\phi(x) - \phi(x_0)}{x - x_0} \right)^2}. \quad (9.6)$$

Claim: $\dfrac{\phi(x) - \phi(x_0)}{x - x_0}$ is bounded in a neighbourhood of x_0.

Indeed, since $g(x, \phi(x)) \to 0$ as $x \to x_0$, one can find an open interval that contains x_0 denoted $I \subset [a - \delta, a + \delta]$, such that

$$|g(x, \phi(x))| < \sigma/2 \quad \text{for all} \quad x \in I.$$

Then, from (9.4) it follows that

$$\frac{|g(x, \phi(x))|}{f_y(x_0, y_0)} < \frac{1}{2} \quad \text{for all } x \in I.$$

We use the above estimate in (9.6) together with the triangle inequality to deduce

$$\left| \left| \frac{\phi(x) - \phi(x_0)}{x - x_0} \right| - \left| \frac{\frac{\partial f}{\partial x}(x_0, y_0)}{\frac{\partial f}{\partial y}(x_0, y_0)} \right| \right| \le \frac{1}{2} \sqrt{1 + \left(\frac{\phi(x) - \phi(x_0)}{x - x_0} \right)^2}$$

$$\le \frac{1}{2}\left(1 + \left| \frac{\phi(x) - \phi(x_0)}{x - x_0} \right| \right)$$

so,

$$\left| \frac{\phi(x) - \phi(x_0)}{x - x_0} \right| \le 2 \left| \frac{\frac{\partial f}{\partial x}(x_0, y_0)}{\frac{\partial f}{\partial y}(x_0, y_0)} \right| + 1 \quad \text{for all} \quad x \in I$$

which proves our Claim. With this in mind, we can pass to the limit as $x \to x_0$ in (9.6). Using Corollary 3.12 (since $g(x, \phi(x)) \to 0$ as $x \to x_0$) one has

$$\lim_{x \to x_0} \frac{\phi(x) - \phi(x_0)}{x - x_0} = -\frac{\frac{\partial f}{\partial x}(x_0, y_0)}{\frac{\partial f}{\partial y}(x_0, y_0)}$$

which shows that ϕ is differentiable at x_0 and that (9.3) holds.   □

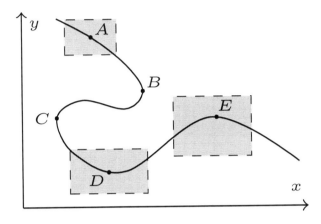

Figure 9.1
The level curve $L(f, m)$ which is locally the graph of a function.

Figure 9.1 illustrates the conclusion of Theorem 9.5 as follows. The curved line is the level curve $L(f, m)$ on which five points A, B, C, D and E are selected. We can find rectangles centred at A, D and E such that the intersection of these rectangles with the level curve $L(f, m)$ is the graph of a smooth function; one can employ the vertical line test to check this fact! Thus, the piece of the level curve inside a rectangle is the graph of a continuously differentiable function $y = \phi(x)$. By the same vertical line test, this property does not hold at B and C. In turn, we can express x as a function of y, that is, $x = \psi(y)$ around the points B and C. Indeed, if $\dfrac{\partial f}{\partial x}(a, b) \neq 0$, then the same argument from the proof of Theorem 9.5 implies that x can be cast as a function $x = \psi(y)$. In conclusion, the Implicit Function Theorem says that the level curve of a continuously differentiable function is the graph of a function around any regular point of it.

Example 9.2 *Consider the hyperbola* $x^2 - y^2 = 1$. *At what points* $(a, b) \in \mathbb{R}^2$ *on the hyperbola can we apply the Implicit Function Theorem?*

Solution. Let $f : \mathbb{R}^2 \to \mathbb{R}$, $f(x, y) = x^2 - y^2$. By the hypotheses of Theorem 9.5 we need to impose $\dfrac{\partial f}{\partial y}(a, b) \neq 0$, so $b \neq 0$. Thus, for any point (a, b), $b \neq 0$, on the hyperbola, one can apply the Implicit Function Theorem to deduce that $y = \phi(x)$ for some continuously differentiable function $\phi : [a - \delta, a + \delta] \to \mathbb{R}$. In fact, we can compute directly $\phi(x) = \pm\sqrt{x^2 - 1}$ which is differentiable as long as $|x| > 1$. Inspecting the graph of the hyperbola (by using the vertical line test) we clearly see that y is a local function of x whose graph passes through

(a, b) whenever $(a, b) \neq (\pm 1, 0)$. Take for instance $(a, b) = (-3, -2\sqrt{2})$. Then, by Theorem 9.5 there exists $\phi : I \to J$ a continuously differentiable function where I, J are intervals such that:

- $-3 \in I$ and $-2\sqrt{2} \in J$;

- $\phi(-3) = -2\sqrt{2}$;

- $x^2 - y^2 = 1$, $x \in I$, $y \in J \implies y = \phi(x)$.

Observe that $\phi(x) = -\sqrt{x^2 - 1}$ and the maximum interval on which ϕ is defined is $I = (-\infty, -1)$. By (9.3) one has $\phi'(-3) = \frac{3}{2\sqrt{2}}$; this can also be computed directly since we have the analytic expression of ϕ at hand.

Example 9.3 *Consider the implicit equation $x^3 + xy + y^3 = 5$. Show that there exists a unique continuously differentiable function $y = y(x)$ defined on an open interval that contains -1 and such that $y(-1) = 2$. Find $\frac{dy}{dx}(-1)$.*

Solution. We have seen in Section 5.2 how to compute the derivative $\frac{dy}{dx}(-1)$ and at that time we assumed (without proof) the existence of such continuously differentiable function $y(x)$. Let us see now how to prove its existence!
Consider the function $f : \mathbb{R}^2 \to \mathbb{R}$, $f(x, y) = x^3 + xy + y^3$. Hence, f is continuously differentiable and $\frac{\partial f}{\partial y}(-1, 2) = 11 \neq 0$. Thus, the hypotheses of Theorem 9.1 are fulfilled and in virtue of this result, there exist $\delta, \eta > 0$ and a continuously differentiable function $y : [-1 - \delta, -1 + \delta] \to [2 - \eta, 2 + \eta]$ such that $f(x, y(x)) = 5$ for all $x \in [-1 - \delta, -1 + \delta]$. By (9.3) one has

$$\frac{dy}{dx}(-1) = -\frac{\frac{\partial f}{\partial x}(-1, 2)}{\frac{\partial f}{\partial y}(-1, 2)} = -\frac{5}{11}.$$

Exercises

Exercise 1. Let $D \subset \mathbb{R}^2$ be an open set, $(a, b) \in D$, and let $f : D \subset \mathbb{R}^2 \to \mathbb{R}$ be a continuously differentiable function such that $\frac{\partial f}{\partial x}(a, b) \neq 0$.

Retake the proof of Theorem 9.1 to show that there exist $\delta, \eta > 0$ such that

$$[a - \delta, a + \delta] \times [b - \eta, b + \eta] \subset D$$

and a continuously differentiable function $\psi : [b - \eta, b + \eta] \to [a - \delta, a + \delta]$ such that:

(i) $\psi(b) = a$;

(ii) $f(\psi(y), y) = m$ for all $y \in [b - \eta, b + \eta]$;

(iii) whenever $(x, y) \in [a - \delta, a + \delta] \times [b - \eta, b + \eta]$ and $f(x, y) = m$, then
       $x = \psi(y)$.

Exercise 2. Can the equation $y(x^2 + y^2) = \cos(xy) - 1$ be solved near $(1, 0)$
by using the Implicit Function Theorem:

(i)   as a continuously differentiable function $y = y(x)$?

(ii) as a continuously differentiable function $x = x(y)$?

Exercise 3. For which values of $b > 0$ can the equation

$$x^y + y^x = b + 1$$

be solved as a continuously differentiable function $y = y(x)$ in a neighbourhood
of $x = b$ such that $y(b) = 1$?

Exercise 4. Let $f : D \subset \mathbb{R}^2 \to \mathbb{R}$ be a continuously differentiable function
such that $f(1, 2) = 1$. For which conditions on f can the functional equation

$$f\big(f(x, y), xy\big) = 1$$

be solved near $(1, 2)$ for y as a continuously differentiable function of x?

Exercise 5. Assume $f : D \subset \mathbb{R}^2 \to \mathbb{R}$ is a twice differentiable function that
satisfies (9.1). Show that the implicit function $\phi : [a - \delta, a + \delta] \to [b - \eta, b + \eta]$
from Theorem 9.1 is twice differentiable and for all $x \in [a - \delta, a + \delta]$ one has

$$\phi'(x) = \frac{2ABE - A^2 D - B^2 C}{B^3},$$

where

$$A = \frac{\partial f}{\partial x}(x\phi(x)), \quad B = \frac{\partial f}{\partial y}(x, \phi(x)),$$

$$C = \frac{\partial^2 f}{\partial x^2}(x, \phi(x)), \quad D = \frac{\partial^2 f}{\partial y^2}(x, \phi(x)), \quad E = \frac{\partial^2 f}{\partial x \partial y}(x, \phi(x)).$$

9.3   Three-Variable Case

In this section we make on step further and extend Theorem 9.1 to the case
of three variables.

Theorem 9.4 *Let $D \subset \mathbb{R}^3$ be an open set, $(a, b, c) \in D$ and $f : D \to \mathbb{R}$ be a continuously differentiable function such that*

$$f(a, b, c) = m \quad and \quad \frac{\partial f}{\partial z}(a, b, c) \neq 0. \tag{9.7}$$

Then, there exist $r, \eta > 0$ such that $U := \overline{B}_r(a, b) \times [c - \eta, c + \eta] \subset D$ and the level curve $L(f, m) \cap U$ is the graph of a continuously differentiable function. That is, there exists $\phi : \overline{B}_r(a, b) \to [c - \eta, c + \eta]$ continuously differentiable such that:

    (i) $\phi(a, b) = c$;

    (ii) $f(x, y, \phi(x, y)) = m$ *for all* $(x, y) \in \overline{B}_r(a, b)$;

    (iii) *whenever* $(x, y, z) \in U$ *and* $f(x, y, z) = m$, *then* $z = \phi(x, y)$.

Furthermore,

$$\begin{cases} \dfrac{\partial \phi}{\partial x}(x, y) = -\dfrac{\dfrac{\partial f}{\partial x}(x, y, \phi(x, y))}{\dfrac{\partial f}{\partial z}(x, y, \phi(x, y))} \\[4em] \dfrac{\partial \phi}{\partial y}(x, y) = -\dfrac{\dfrac{\partial f}{\partial y}(x, y, \phi(x, y))}{\dfrac{\partial f}{\partial z}(x, y, \phi(x, y))} \end{cases} \quad for\ all \quad (x, y) \in \overline{B}_r(a, b). \tag{9.8}$$

Proof The proof is very similar to that of Theorem 9.1 and we only point out the differences. Step 1 is as above; as in the proof of Theorem 9.1 we may assume $m = 0$ and $\frac{\partial f}{\partial z}(a, b, c) > 0$. We may then find a ball $B_r(a, b) \subset \mathbb{R}^2$ and $\eta > 0$ such that $U := \overline{B}_r(a, b) \times [c - \eta, c + \eta] \subset D$ and

$$\frac{\partial f}{\partial z}(x, y, z) > 0 \quad \text{for all} \quad (x, y, z) \in U.$$

For any $(x, y) \in \overline{B}_r(a, b)$ and any $z \in [c - \eta, c + \eta]$ we can now define

$$g_{x,y} : [c - \eta, c + \eta] \to \mathbb{R} \quad \text{and} \quad h_z : \overline{B}_r(a, b) \to \mathbb{R}$$

by

$$g_{x,y}(z) = h_z(x, y) = f(x, y, z).$$

As above, one deduces that the equation $g_{x,y}(z) = 0$ has a unique solution $z \in [c - \eta, c + \eta]$ which we denote $z = \phi(x, y)$. Step 2, in which one achieves the continuity of the mapping $\phi : \overline{B}_r(a, b) \to [c - \eta, c + \eta]$, is the same as in the proof of Theorem 9.1. Finally, one has to check that ϕ is continuously differentiable and that (9.8) holds. Instead of using the approach of the previous result, we can take advantage of Theorem 4.18 as follows. By now, we know that $z = \phi(x, y)$ satisfies

$$f(x, y, \phi(x, y)) = 0 \quad \text{for all } (x, y) \in \overline{B}_r(a, b). \tag{9.9}$$

For y fixed, one has by Theorem 9.1 that $\phi(\cdot, y)$ is continuously differentiable and then, for x fixed, the function $\phi(x, \cdot)$ is continuously differentiable. Thus, ϕ has continuous partial derivatives and by Theorem 4.18, we deduce that ϕ is continuously differentiable. Finally, the two equalities in (9.8) follow if we fix y (resp. x) and differentiate with respect to x (resp. y) in (9.9). $\qquad\square$

Example 9.5 *The implicit equation*

$$2x^2 y - 6yz + e^{x+z} = 1$$

has a unique continuously differentiable solution $z = z(x, y)$ *which satisfies* $z(-3, 1) = 3$. *Furthermore,* $\nabla z(-3, 1) = \left(-\frac{11}{5}, 0\right)$.

Solution. Let $f : \mathbb{R}^3 \to \mathbb{R}$, $f(x, y, z) = 2x^2 y - 6yz + e^{x+z}$. Then, f is continuously differentiable on $\mathbb{R}^3$ and $f(-3, 1, 3) = 1$. Observe that

$$\frac{\partial f}{\partial z}(-3, 1, 3) = -5 \neq 0,$$

so the hypotheses of Theorem 9.4 are fulfilled. Thus, there exist $r, \eta > 0$ and a continuously differentiable function $\phi : \overline{B}_r(-3, 1) \subset \mathbb{R}^2 \to \mathbb{R}$ defined on a closed disc in $\mathbb{R}^2$ centred at $(-3, 1)$ such that:

- $\phi(-3, 1) = 3$;

- $f(x, y, \phi(x, y)) = 1$ for all $(x, y) \in \overline{B}_r(-3, 1)$;

- whenever $(x, y, z) \in \overline{B}_r(-3, 1) \times [3 - \eta, 3 + \eta]$ and $f(x, y, z) = 1$, then $z = \phi(x, y)$;

- (9.8) holds for $(a, b) = (-3, 1)$.

In other words, $z = \phi(x, y)$ is a continuously differentiable function which satisfies the implicit equation $2x^2 y - 6yz + e^{x+z} = 1$ in a neighbourhood of $(-3, 1)$ and $z(-3, 1) = 3$. The partial derivatives of z at $(-3, 1)$ were computed in Example 5.8 by using implicit differentiation. Instead, we use here (9.8) as an alternative method. Thus,

$$\frac{\partial z}{\partial x}(-3, 1) = \frac{\partial \phi}{\partial x}(-3, 1) = -\frac{\dfrac{\partial f}{\partial x}(-3, 1, 3)}{\dfrac{\partial f}{\partial z}(-3, 1, 3)} = -\frac{11}{5},$$

and similarly,

$$\frac{\partial z}{\partial y}(-3, 1) = \frac{\partial \phi}{\partial y}(-3, 1) = -\frac{\dfrac{\partial f}{\partial y}(-3, 1, 3)}{\dfrac{\partial f}{\partial z}(-3, 1, 3)} = 0.$$

Exercises

Exercise 6. Show that the equation

$$e^{x+2y} + e^{y+2z} + e^{z+2x} = 3$$

can be solved in a neighbourhood of the point $(0,0,0)$ by using the Implicit Function Theorem in the form $z = z(x,y)$.

Exercise 7. For which $a, b \in \mathbb{R}$ can the Implicit Function Theorem 9.4 be employed to deduce the existence of a continuously differentiable function

$$z = z(x,y) : B_r(3,-2) \subset \mathbb{R}^2 \to \mathbb{R}, \ r > 0,$$

such that

$$xy + ay^2z + bz^3 = 2 \quad \text{and} \quad z(3,-2) = 1?$$

Exercise 8. Can the equation

$$xy^2 + yz + \sin(x - y + z) = 1$$

be solved near $(x,y,z) = (0, 0, \pi/2)$ by using the Implicit Function Theorem:

(i)   as a continuously differentiable function $z = z(x,y)$?

(ii) as a continuously differentiable function $y = y(x,z)$?

(iii) as a continuously differentiable function $x = x(y,z)$?

Exercise 9. Let $f : \mathbb{R} \to \mathbb{R}$ be a continuously differentiable function such that $f(1) \neq 0$. Using Theorem 9.4 show that the functional equation

$$f(xz) + zf(y - z) = yf(y - x)$$

can be solved near the point $(x,y,z) = (1,2,1)$ as a continuously differentiable function $z = z(x,y)$.

Exercise 10. Let $f : D \subset \mathbb{R}^3 \to \mathbb{R}$ be a continuously differentiable function such that $f(1,1,1) = 1$. For which conditions on f can the functional equation

$$f(xy, yz, zx) = 1$$

be solved near the point $(1,1,1)$ as a continuously differentiable function $z = z(x,y)$?

9.4   The General Case

In this section we formulate a general result that extends Theorem 9.1 and Theorem 9.4. Let $D \subset \mathbb{R}^m \times \mathbb{R}^p$, $m, p \geq 1$, be an open set. We prefer to denote the elements in D as two-component points $(\mathbf{x}, \mathbf{y})$ where $\mathbf{x} \in \mathbb{R}^m$ and $\mathbf{y} \in \mathbb{R}^p$. If $F : D \to \mathbb{R}^p$ is a continuously differentiable function, we denote by $J_{F(\mathbf{x}, \cdot)}$ and $J_{F(\cdot, \mathbf{y})}$ the Jacobian matrix of the functions $F(\mathbf{x}, \cdot)$ and $F(\cdot, \mathbf{y})$ respectively. We note that for $(\mathbf{x}, \mathbf{y}) \in D$, $J_{F(\mathbf{x}, \cdot)}(\mathbf{y})$ is a $p \times p$ matrix while $J_{F(\cdot, \mathbf{y})}(\mathbf{x})$ is a $p \times m$ matrix.

Theorem 9.6 *Let $D \subset \mathbb{R}^m \times \mathbb{R}^p$, $m, p \geq 1$, be an open set, $(\mathbf{a}, \mathbf{b}) \in D$ and $F : D \to \mathbb{R}^p$ be a continuously differentiable function such that*

$$F(\mathbf{a}, \mathbf{b}) = \mathbf{m} \quad and \quad J_{F(\mathbf{a}, \cdot)}(\mathbf{b}) \quad is\ invertible. \tag{9.10}$$

Then, there exist:

- *a ball $B_r(\mathbf{a}) \subset \mathbb{R}^m$ $(r > 0)$ and an open set $U \subset \mathbb{R}^p$ with $\mathbf{b} \in U$ and $\overline{B}_r(\mathbf{a}) \times \overline{U} \subset D$;*

- *a continuously differentiable function $\phi : \overline{B}_r(\mathbf{a}) \to \overline{U}$;*

such that:

   (i) *$\phi(\mathbf{a}) = \mathbf{b}$;*
   (ii) *$F(\mathbf{x}, \phi(\mathbf{x})) = \mathbf{m}$ for all $\mathbf{x} \in \overline{B}_r(\mathbf{a})$;*
   (iii) *whenever $(\mathbf{x}, \mathbf{y}) \in \overline{B}_r(\mathbf{a}) \times \overline{U}$ and $F(\mathbf{x}, \mathbf{y}) = \mathbf{m}$, then $\mathbf{y} = \phi(\mathbf{x})$;*
   (iv) *For all $\mathbf{x} \in B_r(\mathbf{a})$, we have*

$$J_\phi(\mathbf{x}) = -\left[J_{F(\mathbf{x}, \cdot)}(\phi(\mathbf{x})) \right]^{-1} J_{F(\cdot, \phi(\mathbf{x}))}(\mathbf{x}). \tag{9.11}$$

Theorem 9.6 can be used in systems of implicit equations when we want to study the interdependence of one set of data in terms of the other. Let us see next Theorem 9.6 at work on some examples.

Example 9.7 *Consider the simultaneous equations*

$$\begin{cases} 3x^2 - y^3 + 2z = 1 \\ xy + 2yz - z^2 = 5. \end{cases}$$

Show that there exist $\delta > 0$ and two continuously differentiable functions $y = y(x), z = z(x) : [1-\delta, 1+\delta] \to \mathbb{R}$ that satisfy the above equations and $y(1) = 2$, $z(1) = 3$. Find $y'(1)$ and $z'(1)$.

Solution. Let $F : \mathbb{R}^3 \to \mathbb{R}^2$ be given by

$$F(x, y, z) = (3x^2 - y^3 + 2z, xy + 2yz - z^2).$$

Then, F is continuously differentiable on $\mathbb{R}^3$ and

$$J_{F(x,\cdot,\cdot)}(y, z) = \begin{pmatrix} -3y^2 & 2 \\ x + 2z & 2y - 2z \end{pmatrix}.$$

For $(x, y, z) = (1, 2, 3)$ one has $\det J_{F(1,\cdot,\cdot)}(2, 3) = 10 \neq 0$, so $J_{F(1,\cdot,\cdot)}(2, 3)$ is invertible. We thus satisfy the hypotheses in Theorem 9.6 at $a = 1$ and $\mathbf{b} = (2, 3)$. Hence, there exist $\delta > 0$, an open set $U \subset \mathbb{R}^2$ that contains $(2, 3)$, and a continuously differentiable function $\phi : [1 - \delta, 1 + \delta] \to \overline{U}$ such that:

- $\phi(1) = (2, 3)$;

- $F(x, \phi(x)) = (1, 5)$ for all $x \in [1 - \delta, 1 + \delta]$;

- whenever $(x, y, z) \in [1 - \delta, 1 + \delta] \times \overline{U}$ and $F(x, y, z) = (1, 5)$ one has $(y, z) = \phi(x)$.

The last statement shows that we are entitled to write $(y, z) = \phi(x)$, so y and z are both functions of x which can be "inverted" from the above two simultaneous equations. Thus, $y = y(x)$, $z = z(x) : [1 - \delta, 1 + \delta] \to \mathbb{R}$ are continuously differentiable functions of x. From $F(x, y, z) = (1, 5)$ one has that $(x, y(x), z(x))$ is a solution of the coupled equations in the statement of our example. Finally, by (9.11) in Theorem 9.6(iv) one has

$$J_{(y,z)}(x) = -\left[J_{F(x,\cdot,\cdot)}(y, z)\right]^{-1} J_{F(\cdot,y,z)}(x).$$

At $x = 1$ we find

$$J_{(y,z)}(1) = -\left[J_{F(1,\cdot,\cdot)}(2, 3)\right]^{-1} J_{F(\cdot,2,3)}(1)$$

$$= -\begin{pmatrix} -12 & 2 \\ 7 & -2 \end{pmatrix}^{-1} \begin{pmatrix} 6 \\ 2 \end{pmatrix}$$

$$= \frac{1}{10} \begin{pmatrix} 2 & 2 \\ 7 & 12 \end{pmatrix} \begin{pmatrix} 6 \\ 2 \end{pmatrix} = \begin{pmatrix} \frac{8}{5} \\ \frac{33}{5} \end{pmatrix}$$

Hence

$$y'(1) = \frac{8}{5} \quad \text{and} \quad z'(1) = \frac{33}{5}. \tag{9.12}$$

Remark 9.8 To compute $y'(1)$ and $z'(1)$, we could use directly the implicit differentiation as in Section 5.2 of Chapter 5. Indeed, by the implicit differentiation with respect to z variable one has

$$\begin{cases} 6x - 3y^2y' + 2z' = 0 \\ y + xy' + 2y'z + 2yz' - 2zz' = 0. \end{cases}$$

Having in mind that for $x = 1$ one has $y(1) = 2$ and $z(1) = 3$ one obtains

$$\begin{cases} 6 - 12y'(1) + 2z'(1) = 0 \\ 2 + 7y'(1) - 2z'(1) = 0. \end{cases}$$

From here we deduce (9.12).

Example 9.9 *Consider the simultaneous equations*

$$\begin{cases} x \cos u - 1 = y \sin v \\ x \cos v + 1 = y \sin u. \end{cases}$$

(i) *If $u(1,1) = 0$ and $v(1,1) = \pi$, show that we can employ the Implicit Function Theorem 9.6 to express $u = u(x,y)$ and $v = v(x,y)$ as continuously differentiable functions in a ball centred at $(1,1)$. Find $\nabla u(1,1)$ and $\nabla v(1,1)$ in this case;*

(ii) *If $x(0,\pi) = 1$ and $y(0,\pi) = 1$, can we still apply Theorem 9.6 to deduce the existence of continuously differentiable functions $x = x(u,v)$ and $y = y(u,v)$ in a ball centred at $(0,\pi)$?*

Solution. (i) Let $F : \mathbb{R}^4 \to \mathbb{R}^2$ be defined as

$$F(x,y,u,v) = (x \cos u - 1 - y \sin v, x \cos v + 1 - y \sin u).$$

Then, F is continuously differentiable and for all $(x,y,u,v) \in \mathbb{R}^4$, we have

$$\det J_{F(x,y,\cdot,\cdot)}(u,v) = \det \begin{pmatrix} -x \sin u & -y \cos v \\ -y \cos u & -x \sin v \end{pmatrix}$$
$$= x^2 \sin u \sin v - y^2 \cos u \cos v.$$

At $(x,y,u,v) = (1,1,0,\pi)$ we have $\det J_{F(1,1,\cdot,\cdot)}(0,\pi) = 1$ so the matrix $J_{F(1,1,\cdot,\cdot)}(0,\pi)$ is invertible. This shows that we can apply Theorem 9.6 to deduce the existence of a ball $B_r(1,1) \subset \mathbb{R}^2$, $r > 0$, and two continuously differentiable functions $u = u(x,y), v = v(x,y) : B_r(1,1) \to \mathbb{R}^2$ that satisfy the above implicit equations and $u(1,1) = 0$, $v(1,1) = \pi$.

From Theorem 9.6(iv) we also have

$$J_\phi(1,1) = -\left[J_{F(1,1,\cdot,\cdot)}(0,\pi)\right]^{-1} J_{F(\cdot,\cdot,0,\pi)}(1,1)$$

$$= \begin{pmatrix} 0 & 1 \\ -1 & 0 \end{pmatrix}^{-1} \begin{pmatrix} 1 & 0 \\ -1 & 0 \end{pmatrix}$$

$$= \begin{pmatrix} 0 & -1 \\ 1 & 0 \end{pmatrix} \begin{pmatrix} 1 & 0 \\ -1 & 0 \end{pmatrix} = \begin{pmatrix} 1 & 0 \\ 1 & 0 \end{pmatrix}.$$

Hence, $\nabla u(1,1) = \nabla v(1,1) = (1,0)$. We could have used directly the implicit differentiation in the simultaneous equations as in Section 5.2 of Chapter 5 to compute $\nabla u(1,1)$ and $\nabla v(1,1)$.

(ii) For the same function F defined above, we have

$$\det J_{F(\cdot,\cdot,u,v)}(x,y) = \det \begin{pmatrix} \cos u & -\sin v \\ \cos v & -\sin u \end{pmatrix} = -\cos u \sin u + \cos v \sin v.$$

Thus, $\det J_{F(\cdot,\cdot,0,\pi)}(1,1) = 0$ so the conditions in Theorem 9.6 are not fulfilled to reach a similar conclusion as above.

Exercises

Exercise 11. Consider the simultaneous equations

$$\begin{cases} x + y + z = 3 \\ 2^x + 3^y + 4^z = 8. \end{cases}$$

Show that there exist $\delta > 0$ and two continuously differentiable functions $x = x(z), y = y(z) : [-\delta, \delta] \to \mathbb{R}$ that satisfy the above equations and $x(0) = 2$, $y(0) = 1$. Find $x'(0)$ and $y'(0)$.

Exercise 12. Consider the simultaneous equations

$$\begin{cases} x^3 + y^3 = u^2 + v^2 \\ x^4 + y^4 = u^3 - v^3. \end{cases}$$

Can we solve the above system near $(x, y, u, v) = (1, 1, 1, -1)$:

(i) as a continuously differentiable function (u, v) depending of (x, y)?
(ii) as a continuously differentiable function (x, y) depending of (u, v)?

Exercise 13. For which $a \in \mathbb{R}$ can Theorem 9.6 be applied to deduce that the system
$$\begin{cases} ax^2v + uy^3 - e^{xu-v} = a - 1 \\ xu^2 + yv^3 - e^{yv} = 0 \end{cases}$$
has a continuously differentiable solution $u = u(x, y)$, $v = v(x, y)$ defined in a disc centred at $(1, 0)$ and $u(1, 0) = v(1, 0) = 1$?

Exercise 14. What condition should be imposed on the real numbers a, b and c such that the simultaneous equations
$$\begin{cases} xy^2 + zv - ve^{uz} = 0 \\ yv + u^2vx - ze^{uv} = 0 \end{cases}$$
can be solved in a neighbourhood of $(x, y, z) = (a, b, 0)$ by the Implicit Function Theorem and thus obtain continuously differentiable functions $u = u(x, y, z)$ and $v = v(x, y, z)$ that satisfy the above equations and $u(a, b, 0) = 1$, $v(a, b, 0) = c$?

Exercise 15. Use the Implicit Function Theorem to show that the system of the simultaneous equations
$$\begin{cases} xu + yv = w\sin(u + v) \\ yv + xw = u\sin(v + w) \\ xu + yw = v\sin(w + u). \end{cases}$$
can be solved in a neighbourhood of $(x, y) = (0, 0)$ and derive the existence of continuously differentiable functions $u = u(x, y)$, $v = v(x, y)$ and $w = w(x, y)$ that satisfy the above equations and $u(0, 0) = \pi$, $v(0, 0) = \pi$, $w(0, 0) = 2\pi$.

Exercise 16. Let $f : \mathbb{R} \to \mathbb{R}$ be a continuously differentiable function and consider the system
$$\begin{cases} xf(y) + yf(z) = c \\ f(xy) + xf(yz) = c. \end{cases}$$
Show that if $c = 2f(1) \neq 0$, then there exist two continuously differentiable functions $x = x(z)$, $y = y(z)$ that satisfy the above equations in neighbourhood of 1 and such that $x(1) = y(1) = 1$. Find $x'(1)$ and $y'(1)$.

Exercise 17. Let $f : \mathbb{R} \to \mathbb{R}$ be a continuously differentiable function which is monotone. For $(x_0, y_0) \in \mathbb{R}^2$ consider the system
$$\begin{cases} u = f(x) + y \\ v = x - f(y) \\ x(0, 0) = x_0, y(0, 0) = y_0. \end{cases}$$
Show that there exist $r > 0$ and $x = x(u, v), y = y(u, v) : \overline{B}_r(0, 0) \to \mathbb{R}$ continuously differentiable that satisfy the above above system.

10

Local and Global Extrema

This chapter discusses the local and global extrema for functions of several variables. The theory is built around the Second Order Derivative Test which allows to classify the critical points into local minima, local maxima and saddle points. In such a context, the second order derivatives and the properties of the Hessian matrix play an important role.

10.1 Extreme Values and Critical Points

In this chapter we are interested in finding the largest and the smallest value of a function $f : D \subset \mathbb{R}^m \to \mathbb{R}$ whenever these exist. To this aim, we formalise the concept of local and global extrema in the definition below.

Definition 10.1 *Let* $f : D \subset \mathbb{R}^m \to \mathbb{R}$ *and* $\mathbf{a} \in D$ *which is not an isolated point of* D. *We say that*

(i) $\mathbf{a}$ *is a local minimum point of* f *if* $f(\mathbf{x}) \geq f(\mathbf{a})$ *for all* $\mathbf{x} \in B_r(\mathbf{a}) \cap D$, *for some* $r > 0$;

(ii) $\mathbf{a}$ *is a local maximum point of* f *if* $f(\mathbf{x}) \leq f(\mathbf{a})$ *for all* $\mathbf{x} \in B_r(\mathbf{a}) \cap D$, *for some* $r > 0$;

(iii) $\mathbf{a}$ *is a local extreme point of* f *if it is either a local minimum or a local maximum point of* f. *In this case* $f(\mathbf{a})$ *is called an extreme value of* f.

(iv) $\mathbf{a}$ *is a global (or absolute) minimum point of* f *if* $f(\mathbf{x}) \geq f(\mathbf{a})$ *for all* $\mathbf{x} \in D$;

(v) $\mathbf{a}$ *is a global (or absolute) maximum point of* f *if* $f(\mathbf{x}) \leq f(\mathbf{a})$ *for all* $\mathbf{x} \in D$;

(vi) $\mathbf{a}$ *is a global (or absolute) extreme point of* f *if it is either a global minimum or a global maximum point of* f. *In this case* $f(\mathbf{a})$ *is called a global extreme value of* f.

As the reader has already noted, the local extreme points satisfy the required inequality $f(\mathbf{x}) \geq f(\mathbf{a})$, or $f(\mathbf{x}) \leq f(\mathbf{a})$, only in a ball centred at $\mathbf{a}$ as opposed

DOI: 10.1201/9781003449652-10

to global (or absolute) extreme points which satisfy the inequality at all points in the domain of the definition of f. Also, global extreme points are local extreme points. A point $\mathbf{a} \in D$ is called a *strict* local (respectively, global) maximum, or minimum, or extreme point if the above inequalities are strict and are required to hold for all $\mathbf{x} \in B_r(\mathbf{a}) \setminus \{\mathbf{a}\}$ (respectively, for all $\mathbf{x} \in D \setminus \{\mathbf{a}\}$).

The result below states that continuous functions defined on compact sets of $\mathbb{R}^m$ achieve their global extreme values.

Theorem 10.2 (Extreme Value Theorem)
Let $D \subset \mathbb{R}^m$ be a compact set and $f : D \to \mathbb{R}$ be a continuous function. Then, f achieves a global minimum value and a global maximum value on D. That is, there exist $\underline{\mathbf{a}}$, $\overline{\mathbf{a}} \in D$ such that

$$f(\underline{\mathbf{a}}) \leq f(\mathbf{x}) \leq f(\overline{\mathbf{a}}) \quad \text{for all } \mathbf{x} \in D.$$

Proof We divide the proof into two steps.
Step 1: f is bounded on D, that is, for some $M > 0$ we have

$$|f(\mathbf{x})| \leq M \quad \text{for all } \mathbf{x} \in D.$$

If f was not bounded, then $|f(\mathbf{x})|$ could be arbitrarily large for some $\mathbf{x} \in D$. This means that for any $n \geq 1$ there exists $\mathbf{a}_n \in D$ such that $|f(\mathbf{a}_n)| \geq n$. Since D is compact, by Theorem 1.22, $\{\mathbf{a}_n\}$ has a convergent subsequence to some $\mathbf{a} \in D$. Replacing in the following $\{\mathbf{a}_n\}$ with this subsequence, one may assume that $\{\mathbf{a}_n\}$ converges to $\mathbf{a} \in D$ (this also saves us from using complicated notations). Since f is continuous at $\mathbf{a}$, one has that $\{f(\mathbf{a}_n)\}$ converges to $f(\mathbf{a})$ as $n \to \infty$. On the other hand $|f(\mathbf{a}_n)| \geq n$ and putting these facts together we find

$$|f(\mathbf{a})| = \lim_{n\to\infty} |f(\mathbf{a}_n)| \geq \lim_{n\to\infty} n = \infty, \quad \text{contradiction.}$$

Hence, f is bounded on D.

Step 2: f has a global minimum point and a global maximum point on D.
By the above Step 1, we know that there exists $M > 0$ such that

$$f(\mathbf{x}) \leq M \quad \text{for all } \mathbf{x} \in D. \tag{10.1}$$

We choose the infimum of all $M > 0$ with the above property. This means that if M is replaced with $M - \dfrac{1}{n}$, then the above inequality does not hold for all $\mathbf{x} \in D$. Thus, by our choice, $M > 0$ satisfies (10.1) and

$$\text{for all } n \geq 1 \text{ there exists } \mathbf{a}_n \in D \text{ so that } f(\mathbf{a}_n) \geq M - \frac{1}{n}. \tag{10.2}$$

As above, $\{\mathbf{a}_n\}$ has a convergent subsequence to some $\bar{\mathbf{a}} \in D$. Replacing further $\{\mathbf{a}_n\}$ with this subsequence, one may assume that $\{\mathbf{a}_n\}$ converges to $\bar{\mathbf{a}}$. Combining (10.1) and (10.2), we have

$$M - \frac{1}{n} \leq f(\mathbf{a}_n) \leq M \quad \text{for all } n \geq 1.$$

This shows that $f(\mathbf{a}_n) \to M$ as $n \to \infty$. Using the fact that $\mathbf{a}_n \to \bar{\mathbf{a}}$ and f is continuous, it follows that $f(\bar{\mathbf{a}}) = \lim_{n \to \infty} f(\mathbf{a}_n) = M$. Hence, by (10.1) it follows that $f(\mathbf{x}) \leq M = f(\bar{\mathbf{a}})$ for all $\mathbf{x} \in D$. This shows that $\bar{\mathbf{a}}$ is a global maximum point of f. To show the existence of a global minimum point $\underline{\mathbf{a}}$ we proceed similarly, or apply verbatim the above argument to $-f$ whose global maximum point is a global minimum point of f. $\qquad\square$

If the domain of definition $D \subset \mathbb{R}^m$ of a function is open, then the local extrema are found among the critical points. This is stated and proved next.

Theorem 10.3 (Critical Point Theorem)
  Let $D \subset \mathbb{R}^m$ be an open set, $f : D \to \mathbb{R}$ and $\mathbf{a} \in D$ be such that:

  (i) $\mathbf{a}$ is a local extremum point for f;

  (ii) f has partial derivatives at $\mathbf{a}$.

Then, $\mathbf{a}$ is a critical point of f; that is, $\nabla f(\mathbf{a}) = \mathbf{0}$.

Proof Replacing f by $-f$ we may assume that $\mathbf{a}$ is a local maximum point of f. Let $1 \leq i \leq m$; we want to show that $\dfrac{\partial f}{\partial x_i}(\mathbf{a}) = 0$. Recall that

$$\frac{\partial f}{\partial x_i}(\mathbf{a}) = \lim_{t \to 0} \frac{f(\mathbf{a} + t\mathbf{e}_i) - f(\mathbf{a})}{t}.$$

Since $\mathbf{a}$ is a local maximum point, the top line of the above quotient is negative for t close to zero. Hence, considering sided limits one has

$$\lim_{\substack{t \to 0 \\ t > 0}} \frac{f(\mathbf{a} + t\mathbf{e}_i) - f(\mathbf{a})}{t} \leq 0 \quad \text{and} \quad \lim_{\substack{t \to 0 \\ t < 0}} \frac{f(\mathbf{a} + t\mathbf{e}_i) - f(\mathbf{a})}{t} \geq 0.$$

Since both the above limits exist and are equal, they must be zero and this yields $\dfrac{\partial f}{\partial x_i}(\mathbf{a}) = 0$. $\qquad\square$

10.2 Second-Order Derivative Test

Before we discuss the main result of this section, we introduce a new concept which is peculiar to functions of several variables.

Definition 10.4 *Let $f : D \subset \mathbb{R}^m \to \mathbb{R}$ and $\mathbf{a} \in D$ be a critical point of f. We say that $\mathbf{a}$ is a saddle point of f if for any $r > 0$ the set $B_r(\mathbf{a}) \cap D$ contains points $\mathbf{x}$ such that $f(\mathbf{x}) < f(\mathbf{a})$ and points $\mathbf{y}$ such that $f(\mathbf{y}) > f(\mathbf{a})$.*

In other words, a critical point $\mathbf{a}$ is a saddle point if there is a curve passing through $\mathbf{a}$ along which $\mathbf{a}$ is a maximum point and there exists another curve passing through $\mathbf{a}$ along which $\mathbf{a}$ is a minimum point.

Example 10.5 *Let $f : \mathbb{R}^2 \to \mathbb{R}$, $f(x,y) = x^2 - y^2$. Then, $(0,0)$ is a saddle point of f.*

Solution. The surface $z = f(x,y)$ is a quadric and was discussed in Section 2.4. At that time, we were not interested in studying particular points on the surface. At this stage, the origin plays a special role since it is a critical point of f. Next, the graph of f which is pictured in Figure 10.2 is called a hyperbolic paraboloid. Observe that along the line $(x,y) = (t,0)$, $t \in \mathbb{R}$ one

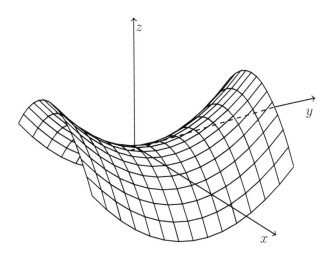

Figure 10.1
The hyperbolic paraboloid $z = x^2 - y^2$ and the saddle point $(0,0)$.

has
$$f(x,y) = f(t,0) = t^2 > 0 = f(0,0) \quad \text{for all } t \neq 0.$$
On the other hand, along the line $(x,y) = (0,t)$, $t \in \mathbb{R}$ one has
$$f(x,y) = f(t,0) = -t^2 < 0 = f(0,0) \quad \text{for all } t \neq 0.$$
Thus, in any disc centred at $(0,0)$ the function f takes both larger and smaller values than $f(0,0)$. This shows that $(0,0)$ is a saddle point of f.

Theorem 10.6 (Second Order Derivative Test)

   Let $f : D \subset \mathbb{R}^m \to \mathbb{R}$ be a function of class C^2 and let $\mathbf{a} \in D^\circ$ be a critical point of f.

   (i) *If the Hessian matrix $H_f(\mathbf{a})$ is positive definite, then $\mathbf{a}$ is a strict local minimum point of f;*

   (ii) *If the Hessian matrix $H_f(\mathbf{a})$ is negative definite, then $\mathbf{a}$ is a strict local maximum point of f;*

   (iii) *If there exist two vectors $\mathbf{u}, \mathbf{v} \in \mathbb{R}^m$ such that*

$$\mathbf{u}H_f(\mathbf{a})\mathbf{u}^T > 0 \quad and \quad \mathbf{v}H_f(\mathbf{a})\mathbf{v}^T > 0, \tag{10.3}$$

   then $\mathbf{a}$ is a saddle point of f.

The definition and properties of positive and negative matrices are provided in Section A.5 of the Appendix.

Proof Let $r > 0$ be such that $B_r(\mathbf{a}) \subset D$. By Quadratic Taylor's Approximation Theorem 8.10, we have

$$f(\mathbf{x}) = f(\mathbf{a}) + (\mathbf{x} - \mathbf{a})H_f(\mathbf{a})(\mathbf{x} - \mathbf{a})^T + g(\mathbf{x}) \quad \text{for all } \mathbf{x} \in B_r(\mathbf{a}),$$

where $g : B_r(\mathbf{a}) \to \mathbb{R}$ satisfies

$$\lim_{\mathbf{x} \to \mathbf{a}} \frac{g(\mathbf{x})}{\|\mathbf{x} - \mathbf{a}\|^2} = 0.$$

Letting

$$h(\mathbf{x}) = \begin{cases} \dfrac{g(\mathbf{x})}{\|\mathbf{x} - \mathbf{a}\|^2} & \text{if } \mathbf{x} \in B_r(\mathbf{a}) \setminus \{\mathbf{a}\}, \\ 0 & \text{if } \mathbf{x} = \mathbf{a}, \end{cases}$$

we have that h is continuous at $\mathbf{a}$ and

$$f(\mathbf{x}) = f(\mathbf{a}) + (\mathbf{x}-\mathbf{a})H_f(\mathbf{a})(\mathbf{x}-\mathbf{a})^T + \|\mathbf{x}-\mathbf{a}\|^2 h(\mathbf{x}) \quad \text{for all } \mathbf{x} \in B_r(\mathbf{a}). \tag{10.4}$$

   (i) Assume $H_f(\mathbf{a})$ is positive definite. Then, by Proposition A.14 there exists a constant $c > 0$ such that

$$\mathbf{u}H_f(\mathbf{a})\mathbf{u}^T > c\|\mathbf{u}\|^2 \quad \text{for all } \mathbf{u} \in \mathbb{R}^m \setminus \{\mathbf{0}\}. \tag{10.5}$$

Since h is continuous at $\mathbf{a}$ and $h(\mathbf{a}) = 0$, one may find $\delta \in (0, r)$ such that

$$|h(\mathbf{x})| < \frac{c}{2} \quad \text{for all } \mathbf{x} \in B_\delta(\mathbf{a}). \tag{10.6}$$

We now combine (10.4), (10.5) and (10.6) to derive

$$\begin{aligned} f(\mathbf{x}) &\geq f(\mathbf{a}) + \|\mathbf{x} - \mathbf{a}\|^2(c + h(\mathbf{x})) \\ &\geq f(\mathbf{a}) + \frac{c}{2}\|\mathbf{x} - \mathbf{a}\|^2 \\ &> f(\mathbf{a}) \quad \text{for all } \mathbf{x} \in B_\delta(\mathbf{a}) \setminus \{\mathbf{a}\}. \end{aligned}$$

This shows that **a** is a strict local minimum point of f.

(ii) Replacing f by $-f$ we have that $H_{-f}(\mathbf{a})$ is positive definite, so **a** is a strict local minimum of $-f$, that is, **a** is a strict local maximum of f.

(iii) Normalizing the vectors **u** and **v** that satisfy (10.3), we may assume $\|u\| = \|v\| = 1$. Let us start again from (10.4) in which we take $\mathbf{x} = \mathbf{a} + t\mathbf{u}$ and $\mathbf{x} = \mathbf{a} + t\mathbf{v}$, $|t| < r$. We have

$$f(\mathbf{a} + t\mathbf{u}) = f(\mathbf{a}) + t^2 \left\{ \mathbf{u}H_f(\mathbf{a})\mathbf{u}^T + h(\mathbf{a} + t\mathbf{u}) \right\}$$
$$f(\mathbf{a} + t\mathbf{v}) = f(\mathbf{a}) + t^2 \left\{ \mathbf{v}H_f(\mathbf{a})\mathbf{v}^T + h(\mathbf{a} + t\mathbf{v}) \right\}.$$

Now, from (10.3) and the continuity of h at **a** we may find $\delta \in (0, r)$ such that

$$\begin{aligned} \mathbf{u}H_f(\mathbf{a})\mathbf{u}^T + h(\mathbf{a} + t\mathbf{u}) &> 0 \\ \mathbf{v}H_f(\mathbf{a})\mathbf{v}^T + h(\mathbf{a} + t\mathbf{v}) &< 0 \end{aligned} \qquad \text{for all } |t| < \delta.$$

This shows that $f(\mathbf{x}) > f(\mathbf{a})$ along the segment line $\mathbf{a} + t\mathbf{u}$, $0 < t < \delta$ and $f(\mathbf{x}) < f(\mathbf{a})$ along the segment line $\mathbf{a} + t\mathbf{v}$, $0 < t < \delta$. Hence, **a** is a saddle point of f.   □

Using the characterization of positive and negative definite matrices we derive:

Corollary 10.7 *Let* $f : D \subset \mathbb{R}^m \to \mathbb{R}$ *be a function of class* C^2 *and let* $\mathbf{a} \in D°$ *be a critical point of* f. *Denote by* $A_1, A_2, \ldots, A_m$ *the leading principal submatrices of* $H_f(\mathbf{a})$; *that is,* A_k *is the* $k \times k$ *matrix whose entries are located at the intersection of the first* k *rows and the first* k *columns of* $H_f(\mathbf{a})$.

   (i) *If* $\det(A_k) > 0$ *for all* $1 \le k \le m$, *then* **a** *is a strict local minimum point of* f;

   (ii) *If* $(-1)^k \det(A_k) > 0$ *for all* $1 \le k \le m$, *then* **a** *is a strict local maximum point of* f;

   (iii) *If* $\det(A_{2k}) < 0$ *for some integer* $1 \le k \le m/2$, *then* **a** *is a saddle point of* f.

There are a number of cases in Corollary 10.7 which are left inconclusive. For instance, we have no classification of the critical points if $\det(A_k) = 0$ for some $1 \le k \le m$. The most relevant situation in this framework is when the Hessian matrix $H_f(\mathbf{a})$ is singular, that is, $\det(H_f(\mathbf{a})) = 0$. Critical points with this property is called *degenerate* and one has to determine their nature (local minimum, local maximum or saddle point) on a case by case basis. We shall illustrate this fact on some specific examples in the next section.

Proof Part (i) and (ii) follow directly from Theorem 10.6, Theorem A.15 and Corollary A.16 in the Appendix.

   (iii) We know that all eigenvalues of the symmetric matrix A_{2k} are real numbers and $\det(A_{2k}) < 0$. Then, there is at least one positive eigenvalue, say $\lambda > 0$, and at least one negative eigenvalue $\mu < 0$ of A_{2k}. Let

$\mathbf{u} = (u_1, u_2, \ldots, u_{2k})$ and $\mathbf{v} = (v_1, v_2, \ldots, v_{2k})$ be an eigenvector of A_{2k} corresponding to λ and μ respectively. Denote

$$\tilde{\mathbf{u}} = (u_1, u_2, \ldots, u_{2k}, 0, 0, \ldots, 0) \in \mathbb{R}^m$$
$$\tilde{\mathbf{v}} = (v_1, v_2, \ldots, v_{2k}, 0, 0, \ldots, 0) \in \mathbb{R}^m.$$

Then

$$\tilde{\mathbf{u}} H_f(\mathbf{a}) \tilde{\mathbf{u}}^T = \mathbf{u} A_{2k} \mathbf{u}^T = \lambda \|\mathbf{u}\|^2 > 0$$

and similarly,

$$\tilde{\mathbf{v}} H_f(\mathbf{a}) \tilde{\mathbf{v}}^T = \mathbf{v} A_{2k} \mathbf{v}^T = \mu \|\mathbf{v}\|^2 < 0.$$

Using now Theorem 10.6(iii) one has that $\mathbf{a}$ is a saddle point of f. □
   In the case of two-variables functions, Corollary 10.7 reads:

Corollary 10.8 *Let* $f : D \subset \mathbb{R}^2 \to \mathbb{R}$ *be a function of class* C^2 *and let* $(a, b) \in D^\circ$ *be a critical point of* f.

   (i) *If*

$$\det(H_f(a, b)) > 0 \quad and \quad \frac{\partial^2 f}{\partial x^2}(a, b) > 0,$$

   then (a, b) *is a strict local minimum point of* f;
   (ii) *If*

$$\det(H_f(a, b)) > 0 \quad and \quad \frac{\partial^2 f}{\partial x^2}(a, b) < 0,$$

   then (a, b) *is a strict local maximum point of* f;
   (iii) *If* $\det(H_f(a, b)) < 0$, *then* (a, b) *is a saddle point of* f.

Since $H_f(a, b)$ is symmetric, it can be shown that the situation where

$$\det(H_f(a, b)) > 0 \quad and \quad \frac{\partial^2 f}{\partial x^2}(a, b) = 0$$

cannot occur. Thus, the only inconclusive case in Corollary 10.8 is $\det(H_f(a, b)) = 0$.

Example 10.9 *Find and classify the critical points of the function*

$$f : \mathbb{R}^2 \to \mathbb{R}, \quad f(x, y) = x^4 - 2xy^3 - 3xy^2.$$

Solution. The gradient of f is $\nabla f(x, y) = (4x^3 - 2y^3 - 3y^2, -6xy^2 - 6xy)$. The critical points are found by solving

$$\nabla f(x, y) = (0, 0) \implies \begin{cases} 4x^3 - 2y^3 - 3y^2 = 0 \\ xy(y + 1) = 0. \end{cases} \tag{10.7}$$

From the second equation of (10.7), we have $x = 0$ or $y = 0$ or $y = -1$.

(i) If $x = 0$, then from the first equation of (10.7) we find

$$-2y^3 - 3y^2 = 0 \Longrightarrow y = 0 \text{ or } y = -\frac{3}{2}.$$

(ii) If $y = 0$, then from the first equation of (10.7) we find $x = 0$.

(iii) If $y = -1$, then from the first equation of (10.7) we find $4x^3 - 1 = 0$, that is, $x = \frac{1}{\sqrt[3]{4}}$.

Thus, the critical points of $f(x, y)$ are

$$(x, y) = \left(0, -\frac{3}{2}\right), \ \left(\frac{1}{\sqrt[3]{4}}, -1\right), \ (0, 0).$$

The Hessian matrix is

$$H_f(x, y) = \begin{pmatrix} 12x^2 & -6y(y+1) \\ -6y(y+1) & -6x(2y+1) \end{pmatrix}.$$

• For $(x, y) = \left(0, -\frac{3}{2}\right)$, we have

$$\det H_f\left(0, -\frac{3}{2}\right) = -\frac{81}{4} < 0 \quad \text{so} \quad \left(0, -\frac{3}{2}\right) \text{ is a saddle point.}$$

• For $(x, y) = \left(\frac{1}{\sqrt[3]{4}}, -1\right)$, we have

$$\det H_f\left(\frac{1}{\sqrt[3]{4}}, -1\right) = 18 > 0 \quad \text{and} \quad \frac{\partial^2 f}{\partial x^2}\left(\frac{1}{\sqrt[3]{4}}, -1\right) = \frac{12}{\sqrt[3]{16}} > 0.$$

Thus, $\left(\frac{1}{\sqrt[3]{4}}, -1\right)$ is a local minimum point.

• For $(x, y) = (0, 0)$, we have $\frac{\partial^2 f}{\partial x^2}(0, 0) = 0$ and $\det H_f(0, 0) = 0$ so the second order derivative test is inconclusive. We see that along the line $y = x$, we have $f(x, y) = -x^3(x + 3)$ which takes both positive values (if $x < 0$) and negative values (if $x > 0$) in any disc centred at the origin. This means that $(0, 0)$ is a saddle point.

Example 10.10 *Find and classify the critical points of the function*

$$f : \mathbb{R}^3 \to \mathbb{R}, \quad f(x, y, z) = x^2 + y^2 z + x^2 y - 4z^2.$$

Solution. To find the critical points of f we solve

$$\nabla f(x, y, z) = (0, 0, 0) \Longrightarrow \begin{cases} 2x + 2xy = 0 \\ 2yz + x^2 = 0 \\ y^2 - 8z = 0. \end{cases} \tag{10.8}$$

This first equation of (10.8) implies $x(1 + y) = 0$, so $x = 0$ or $y = -1$.

(i) If $x = 0$, then the last two equations of (10.8) yield

$$\begin{cases} yz = 0 \\ y^2 - 8z = 0 \end{cases} \implies y = z = 0.$$

(ii) If $y = -1$, then from (10.8) we derive

$$\begin{cases} -2z + x^2 = 0 \\ 1 - 8z = 0 \end{cases} \implies z = \frac{1}{8}, x = \pm\frac{1}{2}.$$

Thus, the critical points of f are

$$(x, y) = \left(\frac{1}{2}, -1, \frac{1}{8}\right), \left(-\frac{1}{2}, -1, \frac{1}{8}\right), (0, 0, 0).$$

The Hessian matrix is

$$H_f(x, y, z) = \begin{pmatrix} 2 + 2y & 2x & 0 \\ 2x & 2z & 2y \\ 0 & 2y & -8 \end{pmatrix}.$$

- For $(x, y, z) = \left(\frac{1}{2}, -1, \frac{1}{8}\right)$, we have the leading principal submatrices

$$A_1 = (0), A_2 = \begin{pmatrix} 0 & 1 \\ 1 & \frac{1}{4} \end{pmatrix}, A_3 = \begin{pmatrix} 0 & 1 & 0 \\ 1 & \frac{1}{4} & -2 \\ 0 & -2 & -8 \end{pmatrix}.$$

Since $\det(A_2) = -1 < 0$ it follows from Corollary 10.7 that $\left(\frac{1}{2}, -1, \frac{1}{8}\right)$ is a saddle point of f.

- For $(x, y, z) = \left(-\frac{1}{2}, -1, \frac{1}{8}\right)$ we proceed exactly in the same manner as above to obtain that $\left(-\frac{1}{2}, -1, \frac{1}{8}\right)$ is a saddle point of f.

- For $(x, y, z) = (0, 0, 0)$ we find

$$H_f(0, 0, 0) = \begin{pmatrix} 2 & 0 & 0 \\ 0 & 0 & 0 \\ 0 & 0 & -8 \end{pmatrix}.$$

In this case $(0, 0, 0)$ is a degenerate critical point. We observe that:

(i) along the line $(x, y, z) = (t, t, -t)$, $t \in \mathbb{R}$, we have
$$f(x, y, z) = f(t, t, -t) = -3t^2 < 0 \quad \text{for all } t \neq 0;$$

(ii) along the line $(x, y, z) = (t, t, 0)$, we have $f(x, y, z) = t^2(1 + t) > 0$ for $t \neq 0$ close to zero.

According to Definition 10.4 this means that $(0, 0, 0)$ is a saddle point.

Exercises

Exercise 1. Find and classify the critical points of the functions below.

(i) $f : \mathbb{R}^2 \to \mathbb{R}$, $f(x,y) = x^4 + y^4 - 36xy$;

(ii) $f : \mathbb{R}^2 \to \mathbb{R}$, $f(x,y) = x^3 y + 3xy^3 + xy$;

(iii) $f : \mathbb{R}^2 \to \mathbb{R}$, $f(x,y) = (x - y)e^{x-y^2}$;

(iv) $f : \mathbb{R} \times (0, \infty) \to \mathbb{R}$, $f(x,y) = 3\ln(x^2 + y) - xy$;

Exercise 2. Find and classify the critical points of the functions below.

(i) $f : \mathbb{R}^3 \to \mathbb{R}$, $f(x,y,z) = xy^3 + 4yz^2 + 2xz - 7y$;

(ii) $f : \mathbb{R}^2 \to \mathbb{R}$, $f(x,y) = xye^{-x^2-y^2-z^2}$;

(iii) $f : \mathbb{R}^3 \to \mathbb{R}$, $f(x,y,z) = e^{xy+z} - xyz - z$.

Exercise 3. Let $f : D \subset \mathbb{R}^2 \to \mathbb{R}$ be a function of class C^2 and let $(a,b) \in D^\circ$ be a critical point of f such that

$$\det(H_f(a,b)) \neq 0 \quad \text{and} \quad \frac{\partial^2 f}{\partial y^2}(a,b) = 0.$$

Check that (a,b) is a saddle point of f.

Exercise 4. Find $a, b, c \in \mathbb{R}$ so that the function

$$f : \mathbb{R}^2 \to \mathbb{R}, \quad f(x,y) = ax^4 + by^4 + cxy^2 + 3$$

has an extreme value 2 at $(1, -1)$. Is $(1, -1)$ a local or a global extreme point? A maximum or a minimum point?

Exercise 5. The Taylor quadratic approximation at $(1,0)$ of a C^2 function $f : \mathbb{R}^2 \to \mathbb{R}$ is

$$Q(x,y) = -1 + 3(x - 1)^2 + 4(x - 1)y + 2y^2.$$

(i) Show that $(1,0)$ is a critical point of f.

(ii) Is $(1,0)$ a local maximum, a local minimum or a saddle point of f?

Exercise 6. The C^2 function $z : B_r(0,0) \subset \mathbb{R}^2 \to \mathbb{R}$, $r > 0$, satisfies the implicit equation

$$z^3 + 2xy + \sin z = 0 \quad \text{and} \quad z(0,0) = 0.$$

Show that $(0,0)$ is a critical point of f and find its nature.

Exercise 7.

(i) Show that there exists a unique function $z = z(x,y) : \mathbb{R}^2 \to \mathbb{R}$ such that

$$e^z + x^2 + z = \sin y \quad \text{for all} \quad (x,y) \in \mathbb{R}^2.$$

(ii) Assuming that z is twice differentiable (which is true, by using the Implicit Function Theorem), show that $\left(0, \dfrac{\pi}{2}\right)$ is a critical point of z and find its nature.

Exercise 8. The C^2 functions $u = u(x,y), v = v(x,y) : \mathbb{R}^2 \to \mathbb{R}$ satisfy

$$\begin{cases} 2e^{u+y} - e^{v-x} = x + 2y + 1 \\ 3e^{v+y} - 2e^{u-x} = 2x + 3y + 1. \end{cases}$$

(i) Find $u(0,0)$ and $v(0,0)$;

(ii) Show that $(0,0)$ is a critical point for both u and v and find its nature.

10.3   The Inconclusive Case

In this section we analyze the degenerate critical points on specific examples. As already mentioned in this chapter, there is no general method to classify these points. Rather, one has to work according to the particular features related to each situation.

Example 10.11 *Let $f : \mathbb{R}^3 \to \mathbb{R}$, $f(x,y,z) = x\cos(y+z) - \sin x$.*

(i) *Check that all critical points of f are degenerate;*

(ii) *Show that $(0,0,0)$ is a saddle point of f.*

Solution. (i) To find the critical points of f we solve

$$\nabla f(x,y,z) = 0 \implies \begin{cases} \cos(y+z) - \cos x = 0 \\ -x\sin(y+z) = 0. \end{cases} \tag{10.9}$$

- If $x = 0$, then the first equation of (10.9) yields $\cos(y+z) = 1$ so $y+z = 2k\pi$, for some integer k.

- If $x \neq 0$, then the second equation of (10.9) implies $\sin(y+z) = 0$ so $y + z = \ell\pi$. Now, the first equation of (10.9) yields $x = p\pi$, where ℓ, p are integers and $p - \ell$ is even.

Observe that

$$\frac{\partial^2 f}{\partial x^2}(x,y,z) = \sin x, \quad \frac{\partial^2 f}{\partial x \partial y}(x,y,z) = \frac{\partial^2 f}{\partial x \partial z}(x,y,z) = -\sin(y+z).$$

If $x = p\pi$ and $y + z = \ell\pi$ where p, ℓ are integers, we see that

$$\frac{\partial^2 f}{\partial x^2}(x,y,z) = \frac{\partial^2 f}{\partial x \partial y}(x,y,z) = \frac{\partial^2 f}{\partial x \partial z}(x,y,z) = 0,$$

so the entries in the first row of the Hessian matrix $H_f(x,y,z)$ are all zero. Thus, $\det H_f(x,y,z) = 0$ which shows that all critical points of f are degenerate.

(ii) Note that along the line $(x,y,z) = (t,t,-t)$, we have

$$f(x,y,z) = t - \sin t > 0 = f(0,0,0) \quad \text{for all } t > 0;$$

$$f(x,y,z) = t - \sin t < 0 = f(0,0,0) \quad \text{for all } t < 0.$$

Thus, in any ball centred at $(0,0,0)$ the function f takes both larger and smaller values than $f(0,0,0)$. This shows that $(0,0,0)$ is a saddle point.

Example 10.12 *Classify the critical points of $f : \mathbb{R}^3 \to \mathbb{R}$, $f(x,y,z) = x^4 + e^{y^2 - 2y - z^2}$.*

Solution. Since

$$\nabla f(x,y,z) = (4x^3, 2(y-1)e^{y^2-2y-z^2}, -2ze^{y^2-2y-z^2}),$$

the only critical point of f is $(0,1,0)$. We observe that the Hessian matrix of f at this point is

$$H_f(0,1,0) = \begin{pmatrix} 0 & 0 & 0 \\ 0 & 2e^{-1} & 0 \\ 0 & 0 & -2e^{-1} \end{pmatrix}$$

and $\det H_f(0,1,0) = 0$ so that $(0,1,0)$ is a degenerate critical point of f; this renders the Second Order Derivative Test inapplicable. At this stage we need to observe that the 2×2 minor formed with the last two rows and the last two columns of $H_f(0,1,0)$ is

$$A = \begin{pmatrix} 2e^{-1} & 0 \\ 0 & -2e^{-1} \end{pmatrix} \quad \text{and} \quad \det(A) = -4e^{-2} < 0.$$

Letting $g(x,y,z) = f(y,z,x)$ we may apply the above computations to deduce that the 2×2 minor formed with the first two rows and two columns of $H_g(1,0,0)$ has negative determinant. We are thus entitled to apply Corollary 10.7(iii) to deduce that $(1,0,0)$ is a saddle point of g, hence $(0,1,0)$ is a saddle point of f.

An alternative way to classify the degenerate critical point $(0,1,0)$ is to observe that $f(x,y,z) = x^2 + e^{(y-1)^2 - z^2 - 1}$ and that:

- along the line $(x, y, z) = (t, t+1, 0)$, $t \in \mathbb{R}$, we have

$$f(x, y, z) = t^2 + e^{t^2-1} > e^{-1} = f(0, 1, 0) \quad \text{for all } t \neq 0;$$

- along the line $(x, y, z) = (0, t+1, 2t)$, $t \in \mathbb{R}$, we have

$$f(x, y, z) = e^{-3t^2-1} < e^{-1} = f(0, 1, 0) \quad \text{for all } t \neq 0.$$

Thus, in light of Definition 10.4, we have that $(0, 1, 0)$ is a saddle point of f.

We should point our that not all degenerate critical points yield saddle points. This is illustrated in the next example.

Example 10.13 *Classify the critical points of* $f : \mathbb{R}^3 \to \mathbb{R}$, $f(x, y, z) = e^{-x^4-y^2-2z^2}$.

Solution. Searching first for critical points of f, we are led to solve $\nabla f(x, y, z) = (0, 0, 0)$, where

$$\nabla f(x, y, z) = (-4x^3 e^{-x^4-y^2-2z^2}, -2ye^{-x^4-y^2-2z^2}, -4ze^{-x^4-y^2-2z^2}).$$

It is easily seen that $(0, 0, 0)$ is the only critical point of f and its Hessian matrix $H_f(0, 0, 0)$ is given by

$$H_f(0, 0, 0) = \begin{pmatrix} 0 & 0 & 0 \\ 0 & -2 & 0 \\ 0 & 0 & -4 \end{pmatrix}.$$

Hence, $(0, 0, 0)$ is a degenerate critical point of f. Let us notice that

$$-x^4 - y^2 - 2z^2 \leq 0 \Longrightarrow f(x, y, z) = e^{-x^4-y^2-2z^2} \leq e^0 = f(0, 0, 0),$$

for all $(x, y, z) \in \mathbb{R}^3$. This tells us that $(0, 0, 0)$ is a global maximum point of f. We tried hopelessly to employ Theorem 10.6 in order to decide whether $(0, 0, 0)$ is a *local* extreme point of f but with a simple observation we derived that $(0, 0, 0)$ is in fact a *global* extreme point!

In the next example we employ the implicit differentiation in order to find and discuss the nature of critical points.

Example 10.14 *The differentiable function* $z = z(x, y) : \mathbb{R}^2 \to \mathbb{R}$ *satisfies the implicit equation*

$$z^3 + (y^2 + 1)z = 10 + x^2 y.$$

(i) *Show that* $z(x, 0) = 2$ *for all* $x \in \mathbb{R}$;

(ii) *Show that* $(x, y) = (0, 0)$ *is the only critical point of* f;

(iii) *Show that* $(0, 0)$ *is a saddle point of* z.

Solution. (i) For $y = 0$ we find that $z(x, 0)$ satisfies

$$z^3 + z = 10 \implies (z - 2)(z^2 + 2z + 5) = 0 \implies z = 2 \text{ the only real solution.}$$

(ii) We differentiate implicitly to deduce

$$\begin{cases} 3z^2 \dfrac{\partial z}{\partial x} + (y^2 + 1)\dfrac{\partial z}{\partial x} = 2xy \\ 3z^2 \dfrac{\partial z}{\partial y} + 2yz + (y^2 + 1)\dfrac{\partial z}{\partial y} = x^2 \end{cases} \implies \begin{cases} \dfrac{\partial z}{\partial x} = \dfrac{2xy}{3z^2 + y^2 + 1} \\ \dfrac{\partial z}{\partial y} = \dfrac{x^2 - 2yz}{3z^2 + y^2 + 1}. \end{cases} \quad (10.10)$$

If (x, y) is a critical point of $z = z(x, y)$, then the equalities (10.10) and the implicit equation yield $(x, y) = (0, 0)$.

(iii) From (10.10) we deduce that $\dfrac{\partial z}{\partial x}$ and $\dfrac{\partial z}{\partial y}$ are differentiable and

$$\begin{cases} \dfrac{\partial^2 z}{\partial x^2} = 2y \cdot \dfrac{3z^2 + y^2 + 1 - 6xz\dfrac{\partial z}{\partial x}}{(3z^2 + y^2 + 1)^2} \\ \dfrac{\partial^2 z}{\partial x \partial y} = 2x \cdot \dfrac{3z^2 - y^2 + 1 - 6yz\dfrac{\partial z}{\partial y}}{(3z^2 + y^2 + 1)^2} \\ \dfrac{\partial^2 z}{\partial y^2} = -2 \cdot \dfrac{3z^3 - y^2 z + z + x^2 y + \dfrac{\partial z}{\partial y}(y^3 + y + 3x^2 z - 3yz^2)}{(3z^2 + y^2 + 1)^2} \end{cases}.$$

Hence,

$$H_z(0, 0) = \begin{pmatrix} 0 & 0 \\ 0 & -\frac{4}{13} \end{pmatrix} \text{ and } (0, 0) \text{ is a degenerate critical point.}$$

Observe that for $(x, y) \in \mathbb{R}^2$ the function

$$F(x, y, \cdot) : \mathbb{R} \to \mathbb{R}, \quad F(x, y, z) = z^3 + (y^2 + 1)z - x^2 y - 10 \text{ is increasing in } z.$$

Also,

$$F(t, t^2, 2) = t^4 > 0 \quad \text{for all } t \in \mathbb{R} \setminus \{0\}.$$

This yields $z(t, t^2) < 2 = z(0, 0)$ for all $t \neq 0$. Further, we have

$$F\left(t, \frac{t^2}{3}, 2\right) = -\frac{t^4}{9} < 0 \quad \text{for all } t \in \mathbb{R} \setminus \{0\}.$$

Hence, $z\left(t, \frac{t^2}{3}\right) > 2 = z(0, 0)$ for all $t \neq 0$. Thus, $(0, 0)$ is a saddle point of z.

Exercises

Exercise 9. Find the nature of $(0,0)$ (as a local maximum, local minimum or saddle point) for each of the following functions:

(i) $f(x,y) = \sin(x^2 y)$;

(ii) $f(x,y) = (x+y)^2 - x^4$;

(iii) $f(x,y) = y^2 \ln(x^2 + 1)$.

Exercise 10. Show that for all $a, b \in \mathbb{R} \setminus \{0\}$ the function

$$f : \mathbb{R}^3 \to \mathbb{R}, \quad f(x,y,z) = x^2 y + ay^2 z + bxz$$

has a unique critical point which is a saddle point.

Exercise 11. Let $f : \mathbb{R}^2 \to \mathbb{R}$, $f(x,y) = \cos(x+y) + \sin(x-y)$.

(i)   Check that all critical points of f are not degenerate;

(ii)   Show that $(\frac{5\pi}{4}, \frac{3\pi}{4})$ is a strict local maximum point of f.

Exercise 12. Let $f : \mathbb{R}^3 \to \mathbb{R}$, $f(x,y,z) = (x^2 + 2y^2 + z) \sin z$.

(i)   Check that f has infinitely many degenerate critical points;

(ii)   Check that f has infinitely many local maximum points;

(iii)   Show that $(0,0,0)$ is a saddle point of f.

Exercise 13. Let $f : \mathbb{R}^3 \to \mathbb{R}$ be a C^2 function and define

$$g : \mathbb{R}^3 \to \mathbb{R}, \quad g(x,y,z) = xyf(x,y,z).$$

(i)   Show that if $f(0,0,0) \neq 0$, then $(0,0,0)$ is a saddle point of g;

(ii)   Give an example of a function f with $f(0,0,0) = 0$, so that $(0,0,0)$ is not a saddle point of g.

Exercise 14. Let $f : \mathbb{R}^3 \to \mathbb{R}$, $f(x,y,z) = \sin(y+z) - \cos(x-y)$.

(i)   Check that all critical points of f are degenerate;

(ii)   Check that f has infinitely many global minimum and global maximum points;

(iii)   Check that f has infinitely saddle points.

Exercise 15. The C^2 function $z = z(x, y) : B_r(1, 1) \to \mathbb{R}$, $0 < r < 1$, satisfies the implicit equation

$$xz + (y - 1)^2 + 2e^{yz} = 2 \quad \text{and} \quad z(1, 1) = 0.$$

(i)   Check that $(1, 1)$ is a degenerate critical point of f;

(ii) Show that $(1, 1)$ is a global maximum point of z.

11

Constrained Optimization

This chapter introduces the Lagrange Multipliers Method for finding the extrema of functions subject to certain constraints. The method is further applied to various settings such as: finding extrema of functions defined on compact sets, derive various inequalities involving real numbers, modelling problems in business and economics.

11.1 Motivation

To motivate the presentation in this chapter, let us start with a real life example. Suppose we want to find the maximum area of a computer screen whose diagonal is 60 cm (this roughly corresponds to 24 inches). Thus, denoting by $x \geq 0$ and $y \geq 0$ the screen sides, one has to maximize the area function $f(x, y) = xy$ knowing that

$$x^2 + y^2 = 3600 \quad \text{and} \quad x, y \geq 0.$$

Hence, rephrasing the problem in similar terms to those of the previous chapter, our task is to find extrema of $f : D \to \mathbb{R}$, $f(x, y) = xy$ where

$$D = \{(x, y) \in \mathbb{R}^2 : x^2 + y^2 = 3600, x, y \geq 0\}.$$

Let us note that geometrically, D is a quarter of a circle centred at the origin and having radius 60; D is compact (being bounded and closed) so the extrema of f always exist on D. However, since D has empty interior, neither the Critical Point Theorem nor the Second Order Derivative Test can be used here. This is a significant difference from what we have seen so far!

Consider the level sets $L(f, c)$ of f where $c > 0$. Geometrically, we want to find the largest value of $c > 0$ such that the level set $L(f, c)$ is nonempty. We observe that the larger is $c > 0$, the larger is also the distance from the origin $(0, 0)$ to the level curve $L(f, c)$. Thus, the maximum value of the function f is achieved when the level set $L(f, c)$ is tangent to D. Let us note that D is an arc of the level set $L(g, 3600)$ where $g(x, y) = x^2 + y^2$. At the maximum point of f on D, say (a, b), the two level sets $L(f, c)$ and $L(g, 3600)$ have a common tangent line T; see Figure 11.1. This is one valuable fact we want to

DOI: 10.1201/9781003449652-11

exploit next! Another important piece of information is provided by Theorem 6.11 which says that the two gradients $\nabla f(a,b)$ and $\nabla g(a,b)$, as long as they are not zero, must be orthogonal to T, so, they are parallel, see Figure 11.1.

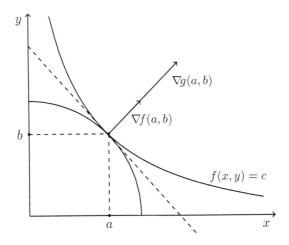

Figure 11.1
The gradient vectors $\nabla f(a,b)$ and $\nabla g(a,b)$ are orthogonal to the tangent T.

This yields
$$\nabla f(a,b) = \lambda g(a,b), \tag{11.1}$$
for some real number λ. Recall that (a,b) also satisfies
$$g(a,b) = 3600. \tag{11.2}$$

Thus, (11.1)–(11.2) consist of three equations involving three unknowns: a, b and λ. At this stage, we achieved our goal to introduce the new method in the above simple setting: this approach is called the *Lagrange Multipliers Method* and the parameter λ is called the *Lagrange multiplier*. We shall investigate this technique in various degrees of generality throughout the present chapter. Just to finish our example, we see that (11.1)–(11.2) yield

$$\begin{cases} b = 2\lambda a \\ a = 2\lambda b \\ a^2 + b^2 = 3600. \end{cases}$$

Squaring the first two equations and using the third one, we find

$$3600 = a^2 + b^2 = 4\lambda^2(a^2 + b^2) \implies 4\lambda^2 = 1.$$

Hence, $\lambda = \pm\frac{1}{2}$ and since $a, b \geq 0$, we deduce $a = b = 30\sqrt{2}$. Thus, the maximum area of the computer screen is $f(30\sqrt{2}, 30\sqrt{2}) = 1800$ cm^2.

There are a few things left to clear up here:

- How do we know that $(a, b) = (30\sqrt{2}, 30\sqrt{2})$ is indeed a global maximum and not a global minimum point of f?

- We were looking for a global maximum but how about the global minimum of f on D?

Well, by the Extreme Value Theorem 10.2 we know that f achieves a global maximum on the compact set D. By the above arguments we found that the only candidate is $(30\sqrt{2}, 30\sqrt{2})$; one simply has to compare the value achieved by f at this point with any other value taken by f on D. Let us take the value of f at the endpoints of the domain, namely $(60, 0)$ and $(0, 60)$. We find that $f(60, 0) = f(0, 60) = 0$ and thus, since $f(30\sqrt{2}, 30\sqrt{2}) > f(60, 0)$ it follows that $(30\sqrt{2}, 30\sqrt{2})$ is indeed a global maximum point of f on D. As a matter of fact, $(60, 0)$ and $(0, 60)$ are both global minimum points of f since

$$f(x, y) \geq 0 = f(60, 0) = f(0, 60) \quad \text{for all } (x, y) \in D.$$

As a rule of thumb, always check the values taken by f at the endpoints (if any) of the domain!

11.2 Lagrange Multipliers Method, Part I

We are now ready to present the Lagrange Multipliers Method in a more formal setting. Suppose we want to find the extrema of a three-variable function $f : D \subset \mathbb{R}^3 \to \mathbb{R}$ subject to the condition $g(x, y, z) = m$, where $m \in \mathbb{R}$. The latter condition is called the *constraint* while f is called the *objective function*. We assume that f and g are continuously differentiable functions; this enables us to use the properties of the gradient relative to the level sets as discussed in the previous section. Let us further note that the set $\{(x, y, z) \in D : g(x, y, z) = m\}$ is compact whose interior, just as before, may be empty. Our main result is the following.

Theorem 11.1 (Lagrange Multipliers Method with One Constraint)
    Let $D \subset \mathbb{R}^3$ *be an open set and* $f, g : D \to \mathbb{R}$ *be two continuously differentiable functions. Let* $E = \{(x, y, z) \in D : g(x, y, z) = m\}$, *where* $m \in \mathbb{R}$, *and suppose that* $(a, b, c) \in E$ *satisfies:*

(i) (a, b, c) *is a local extreme point of* $f\big|_E : E \to \mathbb{R}$;
(ii) $\nabla g(a, b, c) \neq (0, 0, 0)$.

Then, there exists $\lambda \in \mathbb{R}$ *such that* $\nabla f(a, b, c) = \lambda \nabla g(a, b, c)$.

We should note that if D is bounded and E is nonempty, then E is compact (one can check that E is indeed bounded and closed). Thus, $f\big|_E$ has extreme points on E. Also, the point (a, b, c) in Theorem 11.1 is a local extreme point of f on E (which is a smaller subset of D) and not necessarily on the whole domain of definition D of f.

The parameter λ is called the *Lagrange multiplier* and the quantity

$$\mathcal{L}(x, y, z) = f(x, y, z) - \lambda g(x, y, z) \quad \text{is called the } \textit{Lagrangian.}$$

The above theorem states that if (a, b, c) is an extreme point of $f\big|_E$ and $\nabla g(a, b, c) \neq (0, 0, 0)$, then there exists $\lambda \in \mathbb{R}$ so that (a, b, c) is a critical point of the Lagrangian $\mathcal{L}$.

Proof Since $\nabla g(a, b, c) \neq (0, 0, 0)$, by swapping eventually the variables x, y and z, one may assume that

$$\frac{\partial g}{\partial z}(a, b, c) \neq 0.$$

We are now in the frame of the Implicit Function Theorem 9.1 which yields the existence of $r, \eta > 0$ such that $U := \overline{B}_r(a, b) \times [c - \eta, c + \eta] \subset D$ and the existence of a continuously differentiable function $\phi : \overline{B}_r(a, b) \to [c - \eta, c + \eta]$ such that:

(i) $\phi(a, b) = c$;

(ii) for all $(x, y) \in \overline{B}_r(a, b)$ one has

$$g(x, y, \phi(x, y)) = m; \tag{11.3}$$

(iii) whenever $(x, y, z) \in U$ and $g(x, y, z) = m$, then $z = \phi(x, y)$.

Let now define $H : B_r(a, b) \to \mathbb{R}$ by $H(x, y) = f(x, y, \phi(x, y))$. Then, (a, b) is a local extremum of H and by the Critical Point Theorem 10.3 one has $\nabla H(a, b) = (0, 0)$. Hence,

$$\begin{cases} \dfrac{\partial f}{\partial x}(a, b, c) + \dfrac{\partial f}{\partial z}(a, b, c)\dfrac{\partial \phi}{\partial x}(a, b) = 0 \\[3mm] \dfrac{\partial f}{\partial y}(a, b, c) + \dfrac{\partial f}{\partial z}(a, b, c)\dfrac{\partial \phi}{\partial y}(a, b) = 0. \end{cases} \tag{11.4}$$

We next differentiate implicitly with respect to x and then with respect to y variable in (11.3) to obtain

$$\begin{cases} \dfrac{\partial g}{\partial x}(a, b, c) + \dfrac{\partial g}{\partial z}(a, b, c)\dfrac{\partial \phi}{\partial x}(a, b) = 0 \\[3mm] \dfrac{\partial g}{\partial y}(a, b, c) + \dfrac{\partial g}{\partial z}(a, b, c)\dfrac{\partial \phi}{\partial y}(a, b) = 0. \end{cases} \tag{11.5}$$

Let now

$$\lambda = \frac{\dfrac{\partial f}{\partial z}(a,b,c)}{\dfrac{\partial g}{\partial z}(a,b,c)}. \tag{11.6}$$

Then, (11.4), (11.5) and (11.6) yield $\nabla f(a,b,c) = \lambda \nabla g(a,b,c)$. □

Let us next illustrate the method through several examples.

Example 11.2 *Find the extrema of $f : \mathbb{R}^2 \to \mathbb{R}$, $f(x,y) = x^3 - x^2 + xy^2$ subject to the constraint $x^2 + y^2 = 1$.*

Solution. The constraint is $g(x,y) = 1$ where $g(x,y) = x^2 + y^2$. Geometrically, the constraint is the unit circle D in the plane and we want to find the extrema of $f : D \to \mathbb{R}$. Note that D is compact and as in the first section of this chapter, D has empty interior.

In order to employ the Lagrange multipliers method one has first to check that $\nabla g(x,y) \neq (0,0)$ on the constraint $\{g(x,y) = 1\}$. This is the case indeed, since $\nabla g(x,y) = (2x, 2y)$ and then $\nabla g(x,y) = (0,0)$ implies $(x,y) = (0,0)$ which does not fulfil $g(x,y) = 1$.

We are now entitled to apply the Lagrange Multipliers Method and hence, we are prompted to solve

$$\begin{cases} \nabla f(x,y) = \lambda g(x,y) \\ g(x,y) = 1 \end{cases} \implies \begin{cases} 3x^2 - 2x + y^2 = 2\lambda x \\ 2xy = 2\lambda y \\ x^2 + y^2 = 1. \end{cases} \tag{11.7}$$

From the second equation of (11.7) one has $y(\lambda - x) = 0$, so either $y = 0$ or $\lambda = x$.

- If $y = 0$, then from the third equation of (11.7) we deduce $x^2 = 1$, so $(x,y) = (\pm 1, 0)$.

- If $\lambda = x$ it follows from the first equation of (11.7) that $3x^2 - 2x + y^2 = 2x^2$, so $y^2 = -x^2 + 2x$. We use this last equality in the third equation of (11.7) to obtain $2x = 1$ and then $(x,y) = \left(\frac{1}{2}, \pm \frac{\sqrt{3}}{2}\right)$.

We have thus obtained four candidates for the extrema of f subject to the constraint $g(x,y) = 1$, namely

$$(x,y) = (\pm 1, 0), \qquad (x,y) = \left(\frac{1}{2}, \pm \frac{\sqrt{3}}{2}\right).$$

We gather the values of f at the above points in order to select the largest and the smallest value. We compute

$$f(1,0) = 0, \quad f(-1,0) = -2 \quad \text{and} \quad f\left(\frac{1}{2}, \pm \frac{\sqrt{3}}{2}\right) = \frac{1}{4}.$$

Thus, the global minimum of f on the constraint $\{g(x,y) = 1\}$ is $f(-1,0) = -2$ and the global maximum on this constraint is $f\left(\frac{1}{2}, \pm \frac{\sqrt{3}}{2}\right) = \frac{1}{4}$.

Example 11.3 *Find the shortest distance from the origin to the curve* $x^3 + 3xy^2 = 32$.

Solution. The implicit curve $x^3 + 3xy^2 = 32$ is depicted in Figure 11.2.

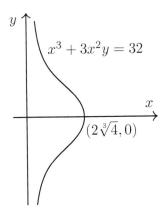

Figure 11.2
The implicit curve $x^3 + 3xy^2 = 32$.

The constraint is $g(x, y) = 32$, where $g(x, y) = x^3 + 3xy^2$. We want to analyze the distance from the origin $(0,0)$ to a general point (x, y) on the curve $g(x, y) = 32$. From the above picture we see that the distance becomes arbitrarily large as $x > 0$ is getting closer to zero while $|y|$ is very large; thus, there is no maximum value of this distance as a real number. We also see that there are two points which minimize the distance; these points are symmetric one from the other with respect to the x axis. This fact will also follow from our analysis below.

The function we want to maximize is $f(x, y) = \sqrt{x^2 + y^2}$. It is more convenient to work instead with the function $F(x, y) = f(x, y)^2 = x^2 + y^2$.

Observe that $\nabla g(x, y) = (3x^2 + 3y^2, 6xy)$ and $\nabla g(x, y) = (0,0)$ implies $x = y = 0$. However, the point $(0,0)$ does not lie on the curve, so $\nabla g(x, y) \neq (0,0)$ whenever $g(x, y) = 32$. Thus, we fulfil the conditions in the Lagrange Multipliers Theorem.

We solve

$$\begin{cases} \nabla F(x, y) = \lambda \nabla g(x, y) \\ g(x, y) = 32 \end{cases} \implies \begin{cases} 2x = 3\lambda(x^2 + y^2) \\ 2y = 6\lambda xy \\ x^3 + 3xy^2 = 32 \end{cases} \tag{11.8}$$

From the second condition in (11.8) one has $y(1 - 3\lambda x) = 0$, so either $y = 0$ or $3\lambda x = 1$.

- If $y = 0$, we obtain from the third equation of (11.8) that $x^3 = 32$ so $(x, y) = (2\sqrt[3]{4}, 0)$.

- If $3\lambda x = 1$, then $x \neq 0$, $\lambda \neq 0$ and the first two equations in (11.8) yield

$$\lambda = \frac{1}{3x} = \frac{2x}{3(x^2 + y^2)} \implies x^2 = y^2 \implies x = \pm y.$$

Using $x^2 = y^2$ we obtain from the third equation of (11.8) that $4x^3 = 32$ so $x = 2$ and then $(x, y) = (2, \pm 2)$.

We found $(x, y) = (2\sqrt[3]{4}, 0)$, $(2, 2)$, $(2, -2)$ and then

$$f(2\sqrt[3]{4}, 0) = 2\sqrt[3]{4} \quad \text{and} \quad f(2, \pm 2) = 2\sqrt{2}.$$

Thus, the shortest distance from the origin to the curve is $2\sqrt{2}$. As we observed from Figure 11.2, the shortest distance is achieved at two points on the curve, namely $(x, y) = (2, 2)$ and $(x, y) = (2, -2)$.

We should also note that without taking into account the shape of the implicit curve $g(x, y) = 32$ we may be tempted to say that $2\sqrt[3]{4}$ is the largest distance. However, this is not correct as explained at the beginning when we agreed by analysing the graph in Figure 11.2 that the distance between the origin and the points on the curve becomes arbitrarily large as $x > 0$ is small while $|y|$ is large.

Example 11.4 *Find the extrema of $f : \mathbb{R}^2 \to \mathbb{R}$, $f(x, y) = \cos(x + y)$ subject to the constraint $\sin x + \sin y = 1$.*

Solution. Let $g(x, y) = \sin x + \sin y$ and observe that $\nabla g(x, y) = (0, 0)$ yields $\cos x = \cos y = 0$. In such a case one has $\sin x$, $\sin y \in \{-1, 1\}$. This is not possible to hold on the constraint $g(x, y) = 1$. Hence,

$$\nabla g(x, y) \neq (0, 0) \quad \text{for all } (x, y) \text{ that satisfy } g(x, y) = 1.$$

We are now entitled to use the Lagrange multipliers and solve

$$\begin{cases} \nabla f(x, y) = \lambda g(x, y) \\ g(x, y) = 1 \end{cases} \implies \begin{cases} -\sin(x + y) = \lambda \cos x \\ -\sin(x + y) = \lambda \cos y \\ \sin x + \sin y = 1. \end{cases} \quad (11.9)$$

From the first two equations in (11.9) we find $\lambda \cos x = \lambda \cos y$.

- If $\lambda = 0$, then (11.9) yields $\sin(x + y) = 0$ so $x + y = k\pi$ for some integer k. This further implies $y = k\pi - x$ and the constraint $g(x, y) = 1$ yields

$$1 = \sin x + \sin y = \sin x + \sin(k\pi - x) = \begin{cases} 0 & \text{if } k \text{ is even} \\ 2\sin x & \text{if } k \text{ is odd.} \end{cases}$$

Thus, k must be odd and then $f(x, y) = \cos(x + y) = \cos(k\pi) = -1$.

- If $\cos x = \cos y$, then $y = 2k\pi \pm x$, for some integer k. If $y = 2k\pi x - x$, then the constraint $g(x, y) = 1$ reads

$$1 = \sin x + \sin y = \sin x + \sin(2k\pi - x) = 0, \quad \text{impossible.}$$

Hence, it remains to discuss the case $y = 2k\pi + x$ and from the constraint condition one has $2 \sin x = 1$, so

$$x = \frac{\pi}{6} + 2\ell\pi \quad \text{or} \quad x = \frac{5\pi}{6} + 2\ell\pi \quad \text{for some integer } \ell.$$

From the above values of x and y one finds

$$f(x, y) = \cos(x + y) \in \left\{ \cos\frac{\pi}{3}, \cos\frac{5\pi}{3} \right\} = \left\{ \frac{1}{2} \right\}.$$

The global minimum of f is -1 and the global maximum value is $\frac{1}{2}$.

Example 11.5 *Let $f : \mathbb{R}^3 \to \mathbb{R}$, $f(x, y, z) = x^2 - 3xy + 3yz$ and let E be the infinite cylinder in $\mathbb{R}^3$ of equation $y^2 + z^2 = 20$.*

    (i) *Show that the function f has a minimum value but not maximum value on E;*

    (ii) *Using the Lagrange Multipliers Method find the minimum value of f on E.*

Solution. (i) The set E is closed but not bounded, so not compact. Observe that $y^2 + z^2 = 20$ implies $|y|, |z| \leq \sqrt{20} = 2\sqrt{5}$. As f is a quadratic function in x and the set E is independent on the x values, we have that f takes a minimum value in terms of x which is achieved at

$$x = -\frac{b}{2a} = \frac{3y}{2} \quad \text{and} \quad f\left(\frac{3y}{2}, y, z\right) = -\frac{9y^2}{4} + 3yz,$$

and $-2\sqrt{5} \leq y, z \leq 2\sqrt{5}$, so f has a minimum value on E. Again, because f is quadratic in x, it becomes arbitrarily large for x large; thus f has no maximum value (as a real number).

    (ii) Let $g(x, y, z) = y^2 + z^2$. Then $\nabla g(x, y) = (0, 2y, 2z)$ and $\nabla g(x, y, z) \neq (0, 0, 0)$ on E. Thus, by Lagrange Multipliers Method we solve

$$\begin{cases} \nabla f(x, y, z) = \lambda g(x, y, z) \\ g(x, y, z) = 20 \end{cases} \implies \begin{cases} 2x - 3y = 0 \\ -3x + 3z = 2\lambda y \\ 3y = 2\lambda z \\ y^2 + z^2 = 20. \end{cases} \quad (11.10)$$

From the first and the third equation of (11.10) we find $y = \frac{2\lambda}{3}z$ and $x = \lambda z$. We use these two equalities in the second equation of (11.10) to find $z = 0$ or $4\lambda^2 + 9\lambda - 9 = 0$ with the solutions $\lambda = -3$, $\lambda = \frac{3}{4}$.

(ii1) If $z = 0$, then the first three equation in (11.10) yield $x = y = z = 0$ which contradict the last equation in (11.10). There are no solutions in this case.

(ii2) If $\lambda = -3$, then $x = -3z$, $y = -2z$. Also, the fourth equation in (11.10) yields $y^2 + z^2 = 20$, so $5z^2 = 20$ and then $(x, y, z) = (-6, -4, 2)$, $(6, 4, -2)$.

(ii3) If $\lambda = \frac{3}{4}$, then $x = \frac{3}{4}z$, $y = \frac{z}{2}$. As above, from $y^2 + z^2 = 20$ we deduce $5z^2 = 80$ and then $(x, y, z) = (3, 2, 4)$, $(-3, -2, -4)$.

The minimum value of f is $f(-6, -4, 2) = f(6, 4, -2) = -60$.

The Lagrange Multipliers Method can be used to prove some important inequalities in algebra and trigonometry. We illustrate two such examples in the following.

Example 11.6 (AM-GM Inequality)

Let $m \geq 2$ be an integer. Then, for any real numbers $x_1, x_2, \ldots, x_m \geq 0$, we have

$$\frac{x_1 + x_2 + \cdots + x_m}{m} \geq \sqrt[m]{x_1 x_2 \ldots x_m}. \tag{11.11}$$

The fraction on the left-hand side of (11.11) represents the *arithmetic mean* while the right-hand side of (11.11) represents the *geometric mean* of the numbers $x_1, x_2, \ldots, x_m > 0$. The inequality (11.11) bears the name of Arithmetic-Geometric Mean (in short AM-GM) inequality. We have already seen a proof of (11.11) in the particular case $m = 2$ (see inequality (3.4) in Chapter 3) when we discussed the existence of the limit to a function of two variables.

Solution. We trust the reader has already noticed that Theorem 11.1 holds in a more general setting of functions of m variables.

If any of the nonnegative numbers $x_1, x_2, \ldots, x_m$ is zero, then the inequality (11.11) clearly holds. We shall assume $x_1, x_2, \ldots, x_m > 0$. Let us note that the inequality (11.11) is homogeneous, in the sense that if $(x_1, x_2, \ldots, x_m)$ satisfies (11.11), then for any $\lambda > 0$ one has that $(\lambda x_1, \lambda x_2, \ldots, \lambda x_m)$ satisfies (11.11) too. Thus, it is enough to prove (11.11) for all systems $(x_1, x_2, \ldots, x_m)$ with $x_1 + x_2 + \cdots + x_m = 1$. Indeed, if (11.11) holds for all such systems $(x_1, x_2, \ldots, x_m)$ with sum 1 and if $y_1, y_2, \ldots, y_m$ are positive numbers with sum $S > 0$, we apply (11.11) for $(y_1/S, y_2/S, \ldots, y_m/S)$ and deduce

$$\frac{y_1 + y_2 + \cdots + y_m}{mS} \geq \sqrt[m]{\frac{y_1 y_2 \cdots y_m}{S^m}}.$$

Multiplying the above inequality by S we find

$$\frac{y_1 + y_2 + \cdots + y_m}{m} \geq \sqrt[m]{y_1 y_2 \cdots y_m}$$

and this is equivalent to (11.11) for the system $(y_1, y_2, \ldots, y_m)$.

Hence, it is enough to assume $x_1 + x_2 + \cdots + x_m = 1$ in (11.11), case in which inequality (11.11) reads

$$\frac{1}{m} \geq \sqrt[m]{x_1 x_2 \cdots x_m} \quad \text{or equivalently,} \quad x_1 x_2 \cdots x_m \leq m^{-m}. \qquad (11.12)$$

We have until now obtained an equivalent form of (11.11) that is suitable for Lagrange Multipliers Method. Precisely, let

$$D = \{(x_1, x_2, \ldots, x_m) \in \mathbb{R}^m : x_i \geq 0 \text{ for all } 1 \leq i \leq m\}.$$

We want to find the maximum of

$$f : D \to \mathbb{R} \quad \text{defined by} \quad f(x_1, x_2, \ldots, x_m) = x_1 x_2 \cdots x_m,$$

subject to the constraint

$$g(x_1, x_2, \ldots, x_m) = 1, \quad \text{where} \quad g(x_1, x_2, \ldots, x_m) = x_1 + x_2 + \cdots + x_m = 1.$$

Note that $\nabla g(x_1, x_2, \ldots, x_m) \neq \mathbf{0}$ and we can proceed with the Lagrange Multipliers Method. Thus, for an extreme point $(x_1, x_2, \ldots, x_m) \in D$ there exists $\lambda \in \mathbb{R}$ such that

$$\begin{cases} x_2 x_3 \cdots x_m = 2\lambda \\ x_1 x_3 \cdots x_m = 2\lambda \\ \cdots \quad \cdots \quad \cdots \\ x_1 x_2 \cdots x_{m-1} = 2\lambda \\ x_1 + x_2 + \cdots + x_m = 1. \end{cases}$$

If $\lambda = 0$, then at least one of $x_1, x_2, \ldots, x_m$ must be zero, case in which $f(x_1, x_2, \ldots, x_m) = 0$ and this leads to a minimum value of f. Assume next $\lambda \neq 0$ which means that all x_i are positive and let $P = x_1 x_2 \cdots x_m$. From the above equations we find

$$x_1 = x_2 = \cdots = x_m = \frac{P}{2\lambda}$$

and from the constraint $x_1 + x_2 + \cdots + x_m = 1$ we deduce $x_i = 1/m$ for all $1 \leq i \leq m$. Thus, for any $(x_1, x_2, \ldots, x_m) \in D$ with sum 1 one has

$$f(x_1, x_2, \ldots, x_m) \leq f\left(\frac{1}{m}, \frac{1}{m}, \ldots, \frac{1}{m}\right) = m^{-m}.$$

This shows that (11.12) holds which also implies (11.11).

Example 11.7 *In any triangle ABC one has*

$$1 < \cos A + \cos B + \cos C \leq \frac{3}{2}. \qquad (11.13)$$

Also, the upper bound $\frac{3}{2}$ is achieved if and only if $A = B = C = \frac{\pi}{3}$, that is, when the triangle is equilateral.

Apparently there is no connection between trigonometric inequalities like (11.13) and the Lagrange Multipliers Method. However, a deeper thought reveals the function we want to maximize and the corresponding constraint.
Solution. Let $D = \{(x, y, z) \in \mathbb{R}^3 : x, y, z \geq 0\}$ and

$$f : D \to \mathbb{R}, \quad f(x, y, z) = \cos x + \cos y + \cos z. \qquad (11.14)$$

We want to maximize f subject to the constraint

$$g(x, y, z) = \pi \quad \text{where} \quad g(x, y, z) = x + y + z. \qquad (11.15)$$

The conditions of Lagrange Multipliers Method are fulfilled and at an extreme point $(x, y, z) \in D$ there exists $\lambda \in \mathbb{R}$ such that

$$\sin x = \sin y = \sin z = -\lambda \quad \text{and} \quad x + y + z = \pi.$$

Since $0 \leq y, z \leq \pi$ it follows that $y, z \in \{x, \pi - x\}$. Hence, three main situations and all permutations of these may occur:

$$\begin{cases} x = y \\ z = \pi - x \end{cases} \quad \text{or} \quad z = y = \pi - x \quad \text{or} \quad x = y = z.$$

We find

$$(x, y, z) = (\pi, 0, 0), (0, \pi, 0), (0, 0, \pi), \left(\frac{\pi}{3}, \frac{\pi}{3}, \frac{\pi}{3}\right).$$

The values of f at the above points are

$$f(\pi, 0, 0) = f(0, \pi, 0) = f(0, 0, \pi) = 1 \quad \text{and} \quad f\left(\frac{\pi}{3}, \frac{\pi}{3}, \frac{\pi}{3}\right) = \frac{3}{2}.$$

Hence

$$1 \leq f(x, y, z) \leq \frac{3}{2} \quad \text{for all } (x, y, z) \in D, \, x + y + z = \pi.$$

By the above arguments, the lower bound 1 is achieved if and only if $(x, y, z) = (\pi, 0, 0)$, $(0, \pi, 0)$, $(0, 0, \pi)$ which is not possible in case of a general triangle ABC. Also, the upper bound $\frac{3}{2}$ is achieved if and only if $(x, y, z) = \left(\frac{\pi}{3}, \frac{\pi}{3}, \frac{\pi}{3}\right)$ which in our case means the triangle ABC is equilateral.

We finish this example with a recap of our approach. We wanted to prove the inequality (11.13) in the triangle ABC, which means the constraints are

$$0 < A, B, C < \pi \quad \text{and} \quad A + B + C = \pi. \qquad (11.16)$$

Instead, we worked with the function f defined in (11.14) on the closed set D and we found the extrema of f subject to the weaker constraint (11.15). Next, we used the Lagrange Multipliers Method to find the extrema of f subject to (11.15) and then discarded the extreme points which did not fulfil the stronger constraint (11.16). This led us to the strict inequality in the lower bound of (11.13).

Exercises

Exercise 1. Find the extrema of $f : \mathbb{R}^2 \to \mathbb{R}$, $f(x, y) = x^3 + 3xy^2 - 3xy$ on the circle $x^2 + y^2 = 5$.

Exercise 2. Find the shortest and the largest distance from the point $(2, 0)$ to the ellipse $x^2 + 2y^2 = 88$.

Exercise 3. Find $a \in \mathbb{R}$ such the function $f(x, y) = axy + 3y^2$ subject to the constraint $x^2 + y^2 = 5$ has an extreme point at $(2, -1)$. Is this a global maximum or a global minimum of f subject to the constraint?

Exercise 4. The continuously differentiable function $f : \mathbb{R}^2 \to \mathbb{R}$ increases the most rapidly at $(3, 4)$ in the direction $(1, -2)$. Is it possible that $f(x, y)$ subject to the constraint $x^2 + y^2 = 25$ has a local extreme point at $(3, 4)$?

Exercise 5. Find the extrema of $f : \mathbb{R}^2 \to \mathbb{R}$, $f(x, y) = \cos(x - y)$ subject to the constraint $\sin x + \sin y = 1$.

Exercise 6. Let $f : \mathbb{R}^3 \to \mathbb{R}$, $f(x, y, z) = x^2 - 6y^2 + 8yz$. Find the extrema of f on the sphere $x^2 + y^2 + z^2 = 5$.

Exercise 7. Let $f : \mathbb{R}^3 \to \mathbb{R}$, $f(x, y, z) = xy + 3yz + 3zx$. Find the extrema of f on the ellipsoid $x^2 + y^2 + 3z^2 = 20$.

Exercise 8. In the following we want to prove the Cauchy-Schwarz inequality using the Lagrange Multipliers Method. Let $\mathbf{a} = (a_1, a_2, \ldots, a_m) \in \mathbb{R}^m$. We want to show that

$$(a_1x_1 + a_2x_2 + \cdots + a_mx_m)^2 \leq (a_1^2 + a_2^2 + \cdots + a_m^2)(x_1^2 + x_2^2 + \cdots + x_m^2)$$

for all $(x_1, x_2, \ldots, x_m) \in \mathbb{R}^m$.

(i) Notice first that if $\mathbf{a} = \mathbf{0}$, then the above inequality holds. Assume in the following that $\mathbf{a} \neq \mathbf{0}$;

(ii) Let $b_i = \frac{a_i}{\|\mathbf{a}\|}$ and $\mathbf{b} = (b_1, b_2, \ldots, b_m)$. Check that the original inequality is equivalent to

$$(\mathbf{b} \bullet \mathbf{x})^2 \leq \|\mathbf{x}\|^2 \quad \text{for all } \mathbf{x} \in \mathbb{R}^m;$$

(iii) Check that it is enough to prove the above inequality only for $\|\mathbf{x}\| = 1$. Thus, the proof of the initial inequality is reduced to finding the maxima of $f : \mathbb{R}^n \to \mathbb{R}$, $f(\mathbf{x}) = (\mathbf{b} \bullet \mathbf{x})^2$ subject to $\|\mathbf{x}\| = 1$.

Exercise 9. Show that in any triangle ABC one has

$$\sin(2A) + \sin(2B) + \sin(2C) \leq \frac{3\sqrt{3}}{2},$$

with equality if and only if ABC is equilateral.

Exercise 10. Show that in any quadrilateral $ABCD$ one has

$$\cos A + \cos B + \cos C + \cos D \geq 0.$$

When does the equality hold?

11.3   Lagrange Multipliers Method, Part II

We are now ready to take one step further and expand the Lagrange Multipliers Method to the case of two constraints.

Theorem 11.8 (Lagrange Multipliers Method with Two Constraints)
   Let $D \subset \mathbb{R}^3$ be an open set and $f, g_1, g_2 : D \to \mathbb{R}$ be continuously differentiable functions. Let $E = \{(x, y, z) \in D : g_1(x, y, z) = m_1, g_2(x, y, z) = m_2\}$, where $m_1, m_2 \in \mathbb{R}$, and suppose that $(a, b, c) \in E$ satisfies:

(i) *(a, b, c) is a local extreme point of $f\big|_E : E \to \mathbb{R}$;*

(ii) *The Jacobian matrix of $G = (g_1, g_2) : D \to \mathbb{R}^2$ at (a, b, c), namely*

$$J_G(a, b, c) = \begin{pmatrix} \dfrac{\partial g_1}{\partial x}(a, b, c) & \dfrac{\partial g_1}{\partial y}(a, b, c) & \dfrac{\partial g_1}{\partial z}(a, b, c) \\[2mm] \dfrac{\partial g_2}{\partial x}(a, b, c) & \dfrac{\partial g_2}{\partial y}(a, b, c) & \dfrac{\partial g_2}{\partial z}(a, b, c) \end{pmatrix} \quad \text{has rank 2.}$$

Then, there exist $\lambda_1, \lambda_2 \in \mathbb{R}$ such that

$$\nabla f(a, b, c) = \lambda_1 \nabla g_1(a, b, c) + \lambda_2 \nabla g_2(a, b, c). \tag{11.17}$$

As in Theorem 11.1, the above result says that local extreme points of $f\big|_E$ are critical points of the Lagrangian

$$\mathcal{L}(x, y, z) = f(x, y, z) - \lambda_1 g_1(x, y, z) - \lambda_2 g_2(x, y, z).$$

Proof Since the Jacobian matrix $J_G(a, b, c)$ has rank 2, one may relabel the variables x, y, z if necessary and assume that

$$A = \begin{pmatrix} \dfrac{\partial g_1}{\partial y}(a, b, c) & \dfrac{\partial g_1}{\partial z}(a, b, c) \\[2mm] \dfrac{\partial g_2}{\partial y}(a, b, c) & \dfrac{\partial g_2}{\partial z}(a, b, c) \end{pmatrix} \quad \text{is invertible.} \tag{11.18}$$

Thus, by Theorem 9.6 there exists $\delta > 0$ and a continuously differentiable function

$$\phi = (\phi_1, \phi_2) : (a - \delta, a + \delta) \to \mathbb{R}^2$$

such that $\phi(a) = (b, c)$ and

$$G(x, \phi_1(x), \phi_2(x)) = (m_1, m_2) \quad \text{for all } x \in (a - \delta, a + \delta). \tag{11.19}$$

On the other hand, (11.19) entails

$$G(x, \phi_1(x), \phi_2(x)) = \big(g_1(x, \phi_1(x), \phi_2(x)), g_2(x, \phi_1(x), \phi_2(x))\big) \in E$$

for all $x \in (a - \delta, a + \delta)$. This further implies that $x = a$ is a local extreme point of

$$\psi : (a - \delta, a + \delta) \to \mathbb{R}, \quad \psi(x) = f(x, \phi_1(x), \phi_2(x)). \tag{11.20}$$

At this stage, there are two pieces of information to take advantage of. Firstly, from the Critical Point Theorem 10.3 one has $\psi'(a) = 0$. Secondly, we get further knowledge from (11.19) by differentiating it with respect to x variable.

From $\psi'(a) = 0$ and (11.20) we find

$$\frac{\partial f}{\partial x}(a, b, c) + \frac{\partial f}{\partial y}(a, b, c)\frac{d\phi_1}{dx}(a) + \frac{\partial f}{\partial z}(a, b, c)\frac{d\phi_2}{dx}(a) = 0,$$

which yields

$$\frac{\partial f}{\partial x}(a, b, c) = -\Big(\frac{\partial f}{\partial y}(a, b, c), \frac{\partial f}{\partial z}(a, b, c)\Big)\phi'(a)^T, \tag{11.21}$$

where $\phi'(a) = \Big(\frac{d\phi_1}{dx}(a), \frac{d\phi_2}{dx}(a)\Big)$. We next differentiate in (11.19) with respect to x and then let $x = a$. We obtain

$$\begin{cases} \frac{\partial g_1}{\partial x}(a, b, c) + \frac{\partial g_1}{\partial y}(a, b, c)\frac{d\phi_1}{dx}(a) + \frac{\partial g_1}{\partial z}(a, b, c)\frac{d\phi_2}{dx}(a) = 0 \\ \frac{\partial g_2}{\partial x}(a, b, c) + \frac{\partial g_2}{\partial y}(a, b, c)\frac{d\phi_1}{dx}(a) + \frac{\partial g_2}{\partial z}(a, b, c)\frac{d\phi_2}{dx}(a) = 0, \end{cases}$$

which we rewrite

$$\begin{pmatrix} \frac{\partial g_1}{\partial x}(a, b, c) \\ \frac{\partial g_2}{\partial x}(a, b, c) \end{pmatrix} = -A\phi'(a)^T,$$

where A is the invertible matrix defined in (11.18). Thus, the above equality yields

$$\phi'(a)^T = -A^{-1}\begin{pmatrix} \frac{\partial g_1}{\partial x}(a, b, c) \\ \frac{\partial g_2}{\partial x}(a, b, c) \end{pmatrix}.$$

We use the above formula in (11.21) to derive

$$\frac{\partial f}{\partial x}(a, b, c) = \Big(\frac{\partial f}{\partial y}(a, b, c), \frac{\partial f}{\partial z}(a, b, c)\Big)A^{-1}\begin{pmatrix} \frac{\partial g_1}{\partial x}(a, b, c) \\ \frac{\partial g_2}{\partial x}(a, b, c) \end{pmatrix}. \tag{11.22}$$

Next, we denote

$$(\lambda_1, \lambda_2) = \left(\frac{\partial f}{\partial y}(a, b, c), \frac{\partial f}{\partial z}(a, b, c)\right) A^{-1},$$

which yields

$$\left(\frac{\partial f}{\partial y}(a, b, c), \frac{\partial f}{\partial z}(a, b, c)\right) = (\lambda_1, \lambda_2) \begin{pmatrix} \frac{\partial g_1}{\partial y}(a, b, c) & \frac{\partial g_1}{\partial z}(a, b, c) \\ \frac{\partial g_2}{\partial y}(a, b, c) & \frac{\partial g_2}{\partial z}(a, b, c) \end{pmatrix}, \quad (11.23)$$

and from (11.22) we also have

$$\frac{\partial f}{\partial x}(a, b, c) = (\lambda_1, \lambda_2) \begin{pmatrix} \frac{\partial g_1}{\partial x}(a, b, c) \\ \frac{\partial g_2}{\partial x}(a, b, c) \end{pmatrix}. \quad (11.24)$$

Now, (11.23) and (11.24) are equivalent to (11.17) and conclude our proof. $\square$

Example 11.9 *The plane $y + z = 4$ intersects the infinite cylinder $x^2 + y^2 = 9$ in an ellipse, see Figure 11.3. Find the points on this ellipse which are closest to, and farthest from the origin.*

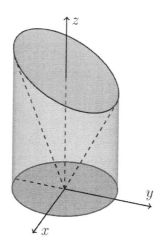

Figure 11.3
The cylinder $x^2 + y^2 = 9$ and its intersection with the plane $y + z = 4$.

Solution. We want to maximize and minimize the distance function $\sqrt{x^2 + y^2 + z^2}$ subject to the constraints $g_1(x, y, z) = 9$ and $g_2(x, y, z) = 4$, where

$$g_1, g_2 : \mathbb{R}^3 \to \mathbb{R}, \quad g_1(x, y, z) = x^2 + y^2 \quad \text{and} \quad g_2(x, y, z) = y + z.$$

As already explained in Example 11.3 it is more convenient to work with the squared distance function $f(x, y, z) = x^2 + y^2 + z^2$. Let us first note that the Jacobian matrix of $G = (g_1, g_2)$ is

$$J_G(x, y, z) = \begin{pmatrix} 2x & 2y & 0 \\ 0 & 1 & 1 \end{pmatrix}$$

and on the constraint set $E = \{g_1(x, y, z) = 9\} \cap \{g_2(x, y, z) = 4\}$ one has $(x, y) \neq (0, 0)$ which implies the rank of J_G is always 2 on E. Hence, by Lagrange Multipliers Method there exists $\lambda_1, \lambda_2 \in \mathbb{R}$ such that

$$\begin{cases} \nabla f(x, y, z) = \lambda_1 \nabla g_1(x, y, z) + \lambda_2 \nabla g_2(x, y, z) \\ g_1(x, y, z) = 9 \\ g_2(x, y, z) = 4, \end{cases}$$

which yields

$$\begin{cases} x = \lambda_1 x \\ 2y = 2\lambda_1 y + \lambda_2 \\ 2z = \lambda_2 \\ x^2 + y^2 = 9 \\ y + z = 4. \end{cases} \tag{11.25}$$

From the first equation of (11.25) one has $x = 0$ or $\lambda_1 = 1$.

(i) If $x = 0$, then the last two equations of (11.25) imply $(x, y, z) = (0, 3, 1)$, $(0, -3, 7)$.

(ii) If $\lambda_1 = 1$, then (11.25) yields $\lambda_2 = 0$, $z = 0$, $y = 4$ and $x^2 = -7 < 0$ which is not possible.

Comparing the values of the distance function at the above points we find that $(0, 3, 1)$ is the closest point to the origin while $(0, -3, 7)$ is the farthest point on the ellipse from the origin.

Example 11.10 *Find the extrema of $f : \mathbb{R}^3 \to \mathbb{R}$, $f(x, y, z) = x + y + z$ subject to*

$$x^2 + y^2 + z^2 = 3 \quad and \quad x + yz = 2.$$

Solution. Define $g_1(x, y, z) = x^2 + y^2 + z^2$ and $g_2(x, y, z) = x + yz$. The constraint set $E = \{g_1(x, y, z) = 3\} \cap \{g_2(x, y, z) = 2\}$ is a closed subset of the sphere of radius $\sqrt{3}$ centred at the origin, so E is compact. Thus, f achieves a global maximum and a global minimum on E. Also, it is easy to check that the Jacobian matrix of $G = (g_1, g_2)$ is

$$J_G(x, y, z) = \begin{pmatrix} 2x & 2y & 2z \\ 1 & z & y \end{pmatrix}$$

and has rank 2 for all $(x, y, z) \in E \setminus \{(1, 1, 1), (1, -1, -1)\}$. If $(x, y, z) \in E \setminus \{(1, 1, 1), (1, -1, -1)\}$, we can use the Lagrange Multipliers Method and solve

$$\begin{cases} \nabla f(x, y, z) = \lambda_1 \nabla g_1(x, y, z) + \lambda_2 \nabla g_2(x, y, z) \\ g_1(x, y, z) = 4 \\ g_2(x, y, z) = 1, \end{cases}$$

which yields

$$\begin{cases} 1 = 2\lambda_1 x + \lambda_2 \\ 1 = 2\lambda_1 y + \lambda_2 z \\ 1 = 2\lambda_1 z + \lambda_2 y \\ x^2 + y^2 + z^2 = 3 \\ x + yz = 2. \end{cases} \tag{11.26}$$

From the second and third equation of (11.26) we find

$$2\lambda_1 y + \lambda_2 z = 2\lambda_1 z + \lambda_2 y \implies (2\lambda_1 - \lambda_2)(y - z) = 0,$$

hence $2\lambda_1 = \lambda_2$ or $y = z$.

(i) If $2\lambda_1 = \lambda_2$, then the first two equations of (11.26) yield

$$1 = 2\lambda_1(x + 1) = 2\lambda_1(y + z)$$

and thus $\lambda_1 \neq 0$ and $x + 1 = y + z$. We now exploit the last two equations of (11.26) as follows: we double the last equation of (11.26) and add it to the fourth equation of (11.26) to obtain

$$x^2 + 2x + (y^2 + 2yz + z^2) = 7 \implies x^2 + 2x + (x + 1)^2 = 7$$
$$\implies x^2 + 2x - 3 = 0.$$

From here we find $x = 1$ and $x = -3$.

If $x = 1$, then $y + z = 2$ and $yz = 1$ so $(x, y, z) = (1, 1, 1)$.

If $x = -3$, then $y + z = -2$ and $yz = 5$ with no real solutions.

(ii) If $y = z$, then the last two equations of (11.26) imply $x^2 + 2y^2 = 3$ and $x + y^2 = 2$ which yield $(x, y, z) = (1, 1, 1), (1, -1, -1)$.

We found that f has no extreme points on $E \setminus \{(1, 1, 1), (1, -1, -1)\}$. Thus the extreme values of f on E are achieved at $(1, 1, 1)$ and $(1, -1, -1)$. By direct calculation we have $f(1, 1, 1) = 3$ and $f(1, -1, -1) = -1$.

The global maximum of f on E is $f(1, 1, 1) = 3$ and the global minimum of f on E is $f(1, -1, -1) = -1$.

Exercises

Exercise 11. Find the extrema of $f : \mathbb{R}^2 \to \mathbb{R}$, $f(x, y) = x + y + z$ subject to the constraints $x^2 + y^2 + z^2 = 3$ and $y(x + z) = 2$.

Exercise 12. Let $g_1, g_2 : \mathbb{R}^3 \to \mathbb{R}$ be given by

$$g_1(x, y, z) = x^2 + z^2 \quad \text{and} \quad g_2(x, y, z) = x + y - z.$$

(i)   Explain why the set $E = \{g_1(x, y, z) = 10\} \cap \{g_2(x, y, z) = 3\}$ is compact. You may use a geometric reason;

(ii) Find the extreme values of $f(x, y, z) = x^2 - y^2 + 2yz$ on E.

Exercise 13. The plane $x + z = -2$ intersects the infinite cylinder $x^2 + y^2 = 10$ in an ellipse. Find the points on this ellipse which are closest to, and farthest from $(0, 1, -2)$.

Exercise 14. The plane of equation $x + y + 2z = 6$ intersects the paraboloid $z = x^2 + y^2$ in an ellipse. Find the points of the ellipse which are closest to and farthest from the origin.

Exercise 15. The plane of equation $-x + y + z = 3$ intersects the hyperboloid $x^2 - y^2 + z^2 = 9$ in a curve $\mathcal{C}$.

(i)   Show that $\mathcal{C}$ is unbounded;

(ii)   Find the points on $\mathcal{C}$ which are closest to the origin;

(iii) Are there points on $\mathcal{C}$ which are farthest from the origin?

11.4   Extrema of Functions on General Compact Sets

Let $f : D \subset \mathbb{R}^m \to \mathbb{R}$ be a continuously differentiable function on the compact set D. By the Extreme Value Theorem 10.2 we know that f has a global maximum point and a global minimum point on D. Let $\mathbf{a} \in D$ be one such a global extreme point of f.

   If $\mathbf{a} \in D^\circ$, then by the Critical Point Theorem 10.3, we have that $\mathbf{a}$ is a critical point of f, that is, $\nabla f(\mathbf{a}) = \mathbf{0}$. If $\mathbf{a} \in \partial D$, then $\mathbf{a}$ is an extreme point of $f|_{\partial D}$. Assuming now that ∂D can be described as the level set of a continuously differentiable function g, one may use the Lagrange Multipliers Method to identify all extrema of $f|_{\partial D}$ and hence, find $\mathbf{a}$. We therefore deduce that all extrema of f are obtained along three steps as follows:

(i)   Find all critical points of f in D°;

(ii) Find all extreme points of f on ∂D;

(iii) Compare the values taken by f at the points found in Steps (i)–(ii) above to select the global extreme points.

We should point out that we need not employ the Second Order Derivative Test at Step (i) above. Indeed, since we do know that global extreme points of f exist, these extrema are found by selecting those points obtained in Step (i) and Step (ii) at which f assumes the largest/smallest value.

Example 11.11 *Let $D = \{(x, y) \in \mathbb{R}^2 : x^2 + y^2 \leq 4, y \geq 0\}$. Find the global extrema of $f : D \to \mathbb{R}$, $f(x, y) = x^2 - 4xy + y^2$.*

Solution. Observe that D is the closed upper half disc centred at the origin and having radius 2. We need first to determine the critical points of f in D°. Solving $\nabla f(x, y) = (0, 0)$ we find $(x, y) = (0, 0) \notin D^\circ$, so we discard this point.

Next we want to find the extrema of f on ∂D. Since the boundary of D consists of two regions, a half circle and a segment line, we shall proceed separately.

We first find the extrema of f subject to $x^2 + y^2 = 4$ and $y \geq 0$. To this aim, we use the Lagrange Multipliers Method (first one has to check that the hypotheses in Theorem 11.1 are fulfilled) and solve

$$
\begin{cases}
2x - 4y = 2\lambda x \\
-4x + 2y = 2\lambda y \\
x^2 + y^2 = 4 \\
y \geq 0
\end{cases}
\implies
\begin{cases}
(1 - \lambda)x = 2y \\
(1 - \lambda)y = 2x \\
x^2 + y^2 = 4 \\
y \geq 0.
\end{cases}
\tag{11.27}
$$

If $x = 0$, then the first equation of (11.27) yields $y = 0$ which contradicts $x^2 + y^2 = 4$. Hence $x \neq 0$ and similarly $y \neq 0$. We may now divide the first two equations of (11.27) to deduce $x^2 = y^2$ and thus $x = \pm y$. Using this fact and the last two equations in (11.27) one gets $(x, y) = (\pm\sqrt{2}, \sqrt{2})$.

It remains to find the extrema of f on the segment line that joins $(-2, 0)$ with $(2, 0)$. On this segment line, we have $x \in [-2, 2]$ and $y = 0$. Thus, we find the extrema of $h : [-2, 2] \to \mathbb{R}$, $h(x) = f(x, 0) = x^2$. The extrema are $(0, 0)$, $(-2, 0)$ and $(2, 0)$.

We now compare the values taken by f at all the above points. We find that the global maximum of f is $f(-\sqrt{2}, \sqrt{2}) = 12$ while the global minimum of f in D is $f(\sqrt{2}, \sqrt{2}) = -4$.

Example 11.12 *Find the extreme values of $f : D \to \mathbb{R}$, $f(x, y) = (x - y)e^{xy}$ where D is the closed disc defined by $x^2 + y^2 \leq 8$.*

Solution. We first determine the critical points of f on D°. From $\nabla f(x, y) = (0, 0)$ one gets

$$
\begin{cases}
y(x - y) = -1 \\
x(x - y) = 1
\end{cases}
\implies x^2 = y^2 \implies x = \pm y.
\tag{11.28}
$$

There are no solutions of the above simultaneous equations if $x = y$, while if $x = -y$ one finds $(x, y) = \left(\frac{1}{\sqrt{2}}, -\frac{1}{\sqrt{2}}\right)$, $\left(-\frac{1}{\sqrt{2}}, \frac{1}{\sqrt{2}}\right)$. In this case

$$f\left(\frac{1}{\sqrt{2}}, -\frac{1}{\sqrt{2}}\right) = \sqrt{\frac{2}{e}}, \quad f\left(-\frac{1}{\sqrt{2}}, \frac{1}{\sqrt{2}}\right) = -\sqrt{\frac{2}{e}}.$$

Note that $\partial D = \{(x, y) \in \mathbb{R}^2 : g(x, y) = 8\}$ where $g(x, y) = x^2 + y^2$. Also, $\nabla g(x, y) \neq (0, 0)$ on ∂D and thus, by Lagrange Multipliers Method we solve

$$\begin{cases} \left(1 + y(x - y)\right)e^{xy} = 2\lambda x \\ \left(-1 + x(x - y)\right)e^{xy} = 2\lambda y \\ x^2 + y^2 = 8. \end{cases} \tag{11.29}$$

If $\lambda = 0$, then the above equations lead to (11.28) which has no solutions on ∂D. If $x = 0$, then the first equation in (11.29) yields $y = \pm 1$ but $(x, y) = (0, \pm 1)$ is not an element of ∂D. Thus, $x \neq 0$ and similarly $y \neq 0$. We can now divide the two equations of (11.29) to obtain

$$\frac{1 + y(x - y)}{-1 + x(x - y)} = \frac{x}{y} \implies (x + y)\left(1 - (x - y)^2\right) = 0.$$

From the above we find $x + y = 0$ or $x - y = \pm 1$.

(i)   If $x + y = 0$, then $x^2 + y^2 = 8$ yields $(x, y) = (2, -2)$, $(-2, 2)$ and $f(2, -2) = 4e^{-4}$, $f(-2, 2) = -4e^{-4}$.

(ii) If $x - y = 1$, then $x^2 + y^2 = 8$ yields $2y^2 + 2y = 7$. Note that one needs not solve this quadratic equation in y since we can compute directly $f(x, y) = e^{y^2 + y} = e^{7/2}$.

(iii) If $x - y = -1$, then from $x^2 + y^2 = 8$ we find $2y^2 - 2y = 7$ and similar to above one finds $f(x, y) = e^{y^2 - y} = -e^{7/2}$.

Thus, the global maximum of f is $e^{7/2}$ and the global minimum is $-e^{7/2}$.

The last example in this section concerns the extrema of a function of three variables.

Example 11.13 *Let $D = \{(x, y) \in \mathbb{R}^3 : x^2 + y^2 + z^2 \leq 2\}$. Find the extreme values of $f : D \to \mathbb{R}$, $f(x, y, z) = 4yz - x^3$.*

Solution. To determine the critical points of f on D° we solve $\nabla f(x, y, z) = (0, 0, 0)$ and obtain $(x, y, z) = (0, 0, 0)$. Next, if $g(x, y, z) = x^2 + y^2 + z^2$, then $\nabla g(x, y, z) \neq (0, 0, 0)$ on ∂D and by Lagrange Multipliers Method we solve

$$\begin{cases} \nabla f(x, y, z) = \lambda \nabla g(x, y, z) \\ g(x, y, z) = 2 \end{cases} \implies \begin{cases} -3x^2 = 2\lambda x \\ 4z = 2\lambda y \\ 4y = 2\lambda z \\ x^2 + y^2 + z^2 = 2. \end{cases} \tag{11.30}$$

The second and third equation of (11.30) yield

$$4z = \lambda(2y) = \lambda^2 z \implies z = 0 \text{ or } \lambda = \pm 2.$$

(i) If $z = 0$, then the last two equations in (11.30) imply $y = 0$ and $x^2 = 2$, so $(x, y, z) = (\pm\sqrt{2}, 0, 0)$.

(ii) If $\lambda = 2$, then the first two equations in (11.30) imply $y = z$ and $-3x^2 = 4x$ and then

$$(x, y, z) = (0, 1, 1), \ (0, -1, -1), \ \left(-\frac{4}{3}, \frac{1}{3}, \frac{1}{3}\right), \ \left(-\frac{4}{3}, -\frac{1}{3}, -\frac{1}{3}\right).$$

(iii) If $\lambda = -2$, similar to part (ii) above one has $y = -z$ and $3x^2 = 4x$ and then

$$(x, y, z) = (0, 1, -1), \ (0, -1, 1), \ \left(\frac{4}{3}, \frac{1}{3}, -\frac{1}{3}\right), \ \left(\frac{4}{3}, -\frac{1}{3}, \frac{1}{3}\right).$$

The global maximum value of f is $f(0, 1, 1) = f(0, -1, -1) = 4$ and the global minimum value of f is $f(0, 1, -1) = f(0, -1, 1) = -4$.

Exercises

Exercise 16. Find the extrema of $f : D \to \mathbb{R}$, $f(x, y) = x^3 - 2xy^2$ where $D = \{(x, y) \in \mathbb{R}^2 : x^2 + 2y^2 \le 24\}$.

Exercise 17. Let $D = \{(x, y) \in \mathbb{R}^2 : x^2 + 3y^2 \le 4\}$. Find the extrema of $f : D \to \mathbb{R}$, $f(x, y) = x + y^3$.

Exercise 18. Find the extrema of $f(x, y) = xye^{x+y}$ on the closed disc $x^2 + y^2 \le 8$.

Exercise 19. Find the extrema of $f(x, y, z) = xy + 2z$ on the closed ball $x^2 + y^2 + z^2 \le 12$.

Exercise 20. Find the extrema of $f(x, y, z) = x^2 - y^2 + yz^2$ on the closed ball $x^2 + y^2 + z^2 \le 5$.

11.5   Some Applications to Business and Economics

In Economics, utility functions are used to describe consumers' relative preference for two or more goods or services. One simple example is the preference for certain types of foods, clothes or internet providers.

Consider a consumer purchasing x units of one good and y units of another good, with prices per unit r Euro and s Euro respectively. The contentment of the consumer with each choice of (x, y) is given by a *utility function* $U(x, y)$. Note that $x \geq 0$ and $y \geq 0$.

Our goal is to maximize the utility function $U(x, y)$ given that the consumer wants to spend exactly t Euro, that is, we want to find for what values of x and y the consumer will get maximum contentment of his purchase given that he spends t Euro. This last condition means that

$$g(x, y) = t, \quad \text{where} \quad g(x, y) = rx + sy, \quad (11.31)$$

and is called the *budget constraint*. We have thus a mathematical model suitable for Lagrange Multipliers Method. Since $g(x, y)$ is a linear function and $x, y \geq 0$, the geometric graph of the constraint is a segment line. Suppose we found a maximum value (a, b) of U subject to the above constraint and that now, we want our consumer to spend a new amount, say $\bar{t}$ Euro. Thus, we want to find the maximum value of U subject to the new budget constraint $g(x, y) = \bar{t}$. This will result in a new (but similar) optimization problem whose output the Economists prefer to approximate as follows. Assume we have already found a maximum $(\bar{a}, \bar{b})$ for the constraint $g(x, y) = \bar{t}$.

By linear Taylor approximation, we have

$$U(x, y) \simeq U(a, b) + \nabla U(a, b) \bullet (x - a, y - b). \quad (11.32)$$

Recall that by the Lagrange Multipliers Method, one has

$$\nabla U(a, b) = \lambda \nabla g(a, b) = \lambda(r, s).$$

We use this last equality in (11.32) in which we take $(x, y) = (\bar{a}, \bar{b})$. We find

$$U(\bar{a}, \bar{b}) \simeq U(a, b) + \lambda(r, s) \bullet (\bar{a}, \bar{b}) - \lambda(r, s) \bullet (a, b),$$

that is,

$$U(\bar{a}, \bar{b}) \simeq U(a, b) + \lambda g(\bar{a}, \bar{b}) - \lambda g(a, b).$$

We write the above approximation as

$$\text{New} \, U \simeq \text{Old} \, U + \lambda \big(\text{New} \, g - \text{Old} \, g \big). \quad (11.33)$$

In utility function problems the Lagrange multiplier λ is called the *marginal utility of money*. By (11.33), we have that the marginal utility of money λ is approximately the change in the utility function U caused by one unit increase in the budget constraint g. In other words, for each additional Euro spent, an additional λ units are added to the utility.

Example 11.14 *A consumer is choosing between Product A priced at €5 per unit and Product B priced at €4 per unit. The utility function is*

$$U(x, y) = xy(2x + y)$$

where x and y denote the units of Product A and Product B respectively.

(i) *If the consumer spends exactly €90, how many units of Product A and Product B should he purchase in order to maximize the utility?*

(ii) *What is the increase in utility if the consumer wishes to spend €100 instead?*

Solution. (i) We use the Method of Lagrange Multipliers. Let $U(x, y) = xy(2x + y)$ and $g(x, y) = 5x + 4y$. The budget constraint is $g(x, y) = 90$. Note that $\nabla g(x, y) = (5, 4) \neq (0, 0)$. Thus, we solve

$$\begin{cases} \nabla U(x, y) = \lambda \nabla g(x, y) \\ g(x, y) = 90 \end{cases} \implies \begin{cases} 4xy + y^2 = 5\lambda \\ 2x^2 + 2xy = 4\lambda \\ 5x + 4y = 90. \end{cases} \tag{11.34}$$

From the first two equations of (11.34), we have

$$\lambda = \frac{4xy + y^2}{5} = \frac{x^2 + xy}{2} \implies 5x^2 - 3xy - 2y^2 = 0. \tag{11.35}$$

If $y = 0$ we obtain from the third equation of (11.34) that $5x = 90$ so $x = 18$. Assume in the following $y \neq 0$. We divide in (11.35) by y^2 and let $t = \frac{x}{y}$. We thus find the quadratic equation

$$5t^2 - 3t - 2 = 0 \implies t = 1 \text{ and } t = -\frac{2}{5}.$$

The solution $t = \frac{x}{y} = -\frac{2}{5}$ is not admissible since $x, y \geq 0$. We thus impose $t = 1$ so $x = y$. From the third equation of (11.34) we deduce $9x = 90$ hence $x = y = 10$.

For $(x, y) = (18, 0)$ the utility function is $U(x, y) = 0$.

For $(x, y) = (10, 10)$ the utility function is $U(x, y) = 3000$.

Thus, the consumer should buy 10 units of Product A and 10 units of Product B.

(ii) From (11.34) we deduce $\lambda = 100$. Thus,

$$\text{New } U \simeq \text{Old } U + \lambda(\text{New } g - \text{Old } g)$$
$$\simeq 3000 + 100(100 - 90)$$
$$\simeq 4000.$$

If we were to solve the optimization problem under the new constraint $g(x, y) = 100$ we would find $x = y = 100/9$ and then the maximum utility $U = 4115.23$.

The outcome of manufacturing systems is modelled by Economists using production functions that have many in common to utility functions. The

Cobb-Douglas production function which models the relationship between production input and output is one example in this sense. Its mathematical expression is

$$q(x,y) = Cx^\alpha y^\beta,$$

where the variable x represents the number of units of capital while the variable y represents the number of units of labour. Further, $C > 0$ is a constant and the exponent α, β satisfy $0 < \alpha, \beta < 1$.

We should note that if $\alpha + \beta = 1$, then the production function q is homogeneous, namely

$$q(\lambda x, \lambda y) = \lambda q(x,y) \quad \text{for all } \lambda > 0.$$

In particular, this means that doubling the usage of capital x and the labour y will also double the output q.

If the cost of capital is r Euro per unit, the cost of labour is s Euro per unit and the total available budget is t Euro, then the budget constraint takes the form (11.31).

It is worth noting that for any Cobb-Douglas production function q, its value at the endpoints of the line segment determined by the budget constraint $g(x,y) = t$, $x, y \geq 0$ is always zero! As a consequence, we are only interested in finding the maximum value of q. Since the endpoints occur when $x = 0$, respectively $y = 0$, we may as well assume that $x > 0$ and $y > 0$.

In Cobb-Douglas production function problems, the Lagrange multiplier λ is called the *marginal productivity of money*. It is used in the same way as the marginal utility of money in utility function problems, as described above.

Example 11.15 *Consider the Cobb-Douglas production function*

$$q = 18x^{2/3}y^{1/2}$$

of a firm whose unit capital and labour costs are €6 and €3, respectively.

   (i) *Find the values of x and y that maximise the output if the total input costs are fixed at €2,268;*

   (ii) *Approximate the new maximum output if the input costs are increased to €2,300.*

Solution. (i) We have the constraint $g(x,y) = 2268$ and $x,y \geq 0$, where $g(x,y) = 6x + 3y$. Since $\nabla g(x,y) = (6,3) \neq (0,0)$, we may use the Lagrange Multipliers Method and thus solve

$$\begin{cases} \nabla q(x,y) = \lambda \nabla g(x,y) \\ g(x,y) = 2268 \end{cases} \implies \begin{cases} 12x^{-1/3}y^{1/2} = 6\lambda \\ 9x^{2/3}y^{-1/2} = 3\lambda \\ 6x + 3y = 2268. \end{cases} \quad (11.36)$$

We recall the assumption $x, y > 0$ and now we divide the first two equations of (11.36) to obtain $y = \frac{3}{2}x$. Using this fact in the last equation of (11.36) one gets $x = 216$ and then $y = 324$. We also find $\lambda = 6$ and the maximum value of the production function is $q = 11,664$.

(ii) Using the formula

$$\text{New } q \simeq \text{Old } q + \lambda(\text{New } g - \text{Old } g),$$

which is the counterpart of (11.33) for production function, we have

$$\text{New } q \simeq 11,664 + 6(2300 - 2268)$$
$$\simeq 11,856.$$

Exercises

Exercise 21. A consumer is choosing between Product A priced at €6 per unit and Product B priced at €8 per unit. The utility function is

$$U(x,y) = y^3 \sqrt{x}$$

where x and y denote the units of Product A and Product B respectively. If the consumer spends exactly €168, use the Lagrange Multipliers Method to find how many units of Product A and Product B should he purchase in order to maximize the utility.

Exercise 22. A consumer is choosing between Product A priced at €4 per unit and Product B priced at €8 per unit. The utility function is

$$U(x,y) = xye^{x+2y}$$

where x and y denote the units of Product A and Product B respectively.

(i) If the consumer spends exactly €80, how many units of Product A and Product B should he purchase in order to maximize the utility?

(ii) What is the increase in utility if the consumer wishes to spend €100 instead?

Exercise 23. The Cobb-Douglas production function of a company is given by

$$q = Cx^\alpha y^\beta$$

where x represents the number of units of capital and y represents the number of units of labour. The constant $C > 0$ and exponents $0 < \alpha, \beta < 1$ are given. Suppose that a company's unit capital and labour costs are €r and €s respectively and that the company's total budget is €t.

Show that the maximum value of the production function is

$$C\left(\frac{t}{\alpha + \beta}\right)^{\alpha+\beta} \left(\frac{\alpha}{r}\right)^\alpha \left(\frac{\beta}{s}\right)^\beta.$$

12

Solutions

12.1 Solutions Chapter 1

1. Let
$$\mathbf{u} = (u_1, u_2, \ldots, u_m)$$
$$\mathbf{v} = (v_1, v_2, \ldots, v_m)$$
$$\mathbf{w} = (w_1, w_2, \ldots, w_m)$$

be three vectors in $\mathbb{R}^m$. Part (i) follows from the associativity property of the real numbers:

$$\mathbf{u} \bullet (\mathbf{v} + \mathbf{w}) = (u_1, u_2, \ldots, u_m) \bullet (v_1 + w_1, v_2 + w_2, \ldots, v_m + w_m)$$
$$= u_1(v_1 + w_1) + u_2(v_2 + w_2) + \cdots + u_m(v_m + w_m)$$
$$= (u_1 v_1 + u_2 v_2 + \cdots + u_m v_m)$$
$$\quad + (u_1 w_1 + u_2 w_2 + \cdots + u_m w_m)$$
$$= \mathbf{u} \bullet \mathbf{v} + \mathbf{u} \bullet \mathbf{w}.$$

Part (ii) can be proved similarly.

(iii) We have

$$\mathbf{u} \bullet \mathbf{v} = u_1 v_1 + u_2 v_2 + \cdots + u_m v_m$$
$$= v_1 u_1 + v_2 u_2 + \cdots + v_m u_m$$
$$= \mathbf{v} \bullet \mathbf{u}.$$

(iv) $\mathbf{u} \bullet \mathbf{u} = u_1^2 + u_2^2 + \cdots + u_m^2 = \|\mathbf{u}\|^2$. If $\mathbf{u} \bullet \mathbf{u} = 0$, then $u_1^2 + u_2^2 + \cdots + u_m^2 = 0$. The sum of squares is zero if and only if each square is zero, so $\mathbf{u} \bullet \mathbf{u} = 0$ if and only if $u_1 = u_2 = \cdots = u_m = 0$, that is, $\mathbf{u} = \mathbf{0}$.

2. (iii) Squaring both sides of the inequality, we have to check that $\|\mathbf{u} + \mathbf{v}\|^2 \leq (\|\mathbf{u}\| + \|\mathbf{v}\|)^2$. This is equivalent to

$$(\mathbf{u} + \mathbf{v}) \bullet (\mathbf{u} + \mathbf{v}) \leq \|\mathbf{u}\|^2 + 2\|\mathbf{u}\| \cdot \|\mathbf{v}\| + \|\mathbf{v}\|^2. \qquad (12.1)$$

Expanding the scalar product on the left-hand side of (12.1) we see that

$$(\mathbf{u} + \mathbf{v}) \bullet (\mathbf{u} + \mathbf{v}) = \mathbf{u} \bullet \mathbf{u} + 2(\mathbf{u} \bullet \mathbf{v}) + \mathbf{v} \bullet \mathbf{v}$$
$$= \|\mathbf{u}\|^2 + 2(\mathbf{u} \bullet \mathbf{v}) + \|\mathbf{v}\|^2.$$

DOI: 10.1201/9781003449652-12

Thus, (12.1) is further equivalent to $\mathbf{u} \bullet \mathbf{v} \leq \|\mathbf{u}\| \cdot \|\mathbf{u}\|$ which follows from the Cauchy-Schwarz inequality (1.3).

3. (i) By triangle inequality, we have

$$\|\mathbf{x} + \mathbf{y}\| \leq \|\mathbf{x}\| + \|\mathbf{y}\| \quad \text{for all vectors } \mathbf{x}, \mathbf{y} \in \mathbb{R}^m. \qquad (12.2)$$

Letting $\mathbf{x} = \mathbf{v}$ and $\mathbf{y} = \mathbf{u} - \mathbf{v}$ in (12.2), we find $\|\mathbf{u}\| - \|\mathbf{v}\| \leq \|\mathbf{u} - \mathbf{v}\|$. Letting now $\mathbf{x} = \mathbf{u}$ and $\mathbf{y} = \mathbf{v} - \mathbf{u}$ in (12.2), we find

$$\|\mathbf{v}\| - \|\mathbf{u}\| \leq \|\mathbf{u} - \mathbf{v}\|.$$

Thus, $\pm(\|\mathbf{u}\| - \|\mathbf{v}\|) \leq \|\mathbf{u} - \mathbf{v}\|$ which yields $\big|\|\mathbf{u}\| - \|\mathbf{v}\|\big| \leq \|\mathbf{u} - \mathbf{v}\|$.

(ii) By the properties of the scalar product, we have

$$\|\mathbf{u} + \mathbf{v}\|^2 = (\mathbf{u} + \mathbf{v}) \bullet (\mathbf{u} + \mathbf{v}) = \|\mathbf{u}\|^2 + 2(\mathbf{u} \bullet \mathbf{v}) + \|\mathbf{v}\|^2,$$
$$\|\mathbf{u} - \mathbf{v}\|^2 = (\mathbf{u} - \mathbf{v}) \bullet (\mathbf{u} - \mathbf{v}) = \|\mathbf{u}\|^2 - 2(\mathbf{u} \bullet \mathbf{v}) + \|\mathbf{v}\|^2.$$

Adding the above equalities we obtain (ii) while subtracting them we derive (iii).

4. (i) From $\big|\mathbf{u} \bullet \mathbf{v}\big| = \|\mathbf{u}\| \cdot \|\mathbf{v}\|$, we have either $\mathbf{u} \bullet \mathbf{v} = \|\mathbf{u}\| \cdot \|\mathbf{v}\|$ or $\mathbf{u} \bullet \mathbf{v} = -\|\mathbf{u}\| \cdot \|\mathbf{v}\|$. The latter case can be written as $\mathbf{u} \bullet (-\mathbf{v}) = \|\mathbf{u}\| \cdot \|-\mathbf{v}\|$. Thus, replacing $\mathbf{v}$ with $-\mathbf{v}$, we may always assume $\mathbf{u} \bullet \mathbf{v} = \|\mathbf{u}\| \cdot \|\mathbf{v}\|$.

(ii) From (1.5) we deduce $E(t) = 0$ through a direct calculation.

(iii) Using now (1.4) we find $\|t\mathbf{u} + \mathbf{v}\| = 0$. By Property (iv) of the norm, it follows that $t\mathbf{u} + \mathbf{v} = 0$ from which we easily derive $\frac{\mathbf{u}}{\|\mathbf{u}\|} = \frac{\mathbf{v}}{\|\mathbf{v}\|}$. Thus, $\mathbf{u} = \lambda \mathbf{v}$ where $\lambda = \frac{\|\mathbf{u}\|}{\|\mathbf{v}\|} \in \mathbb{R}$.

5. Direct calculations, by using the components of vectors $\mathbf{u}$, $\mathbf{v}$ and $\mathbf{w}$.

6. One can use direct calculations. A more elegant approach is to see that if one of the vectors $\mathbf{u}$ and $\mathbf{v}$ is zero, then the equality is clearly true. If none of $\mathbf{u}$ and $\mathbf{v}$ are zero, then denoting by θ the angle between them, we have (see (1.2) and Theorem 1.4(iv)):

$$\cos \theta = \frac{\mathbf{u} \bullet \mathbf{v}}{\|\mathbf{u}\| \cdot \|\mathbf{v}\|} \quad \text{and} \quad \sin \theta = \frac{\|\mathbf{u} \times \mathbf{v}\|}{\|\mathbf{u}\| \cdot \|\mathbf{v}\|}.$$

The required identity follows now from $\cos^2 \theta + \sin^2 \theta = 1$.

7. Let D be another point in the plane determined by A, B, C such that $ABCD$ is a parallelogram. Then, by Theorem 1.4(iv), we have

$$\text{Area}(ABCD) = \|\mathbf{u} \times \mathbf{v}\| = BC \cdot \text{dist}(A, BC) = AB \cdot \text{dist}(C, AB),$$

from which we deduce both (i) and (ii).

8. (i) As shown in Figure 12.1, we have the base area $A = \|\mathbf{u} \times \mathbf{v}\|$ and the height of the parallelepiped is $h = \|\mathbf{w}\| \cos \theta$. Then

$$V = A h = \|\mathbf{u} \times \mathbf{v}\| \cdot \|\mathbf{w}\| \cos \theta = |(\mathbf{u} \times \mathbf{v}) \bullet \mathbf{w}|.$$

(ii) The vectors are coplanar if and only if the parallelepiped is degenerate, that is, its volume is zero.

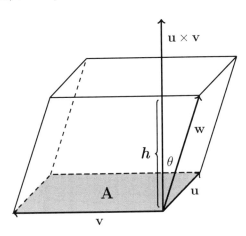

Figure 12.1
The parallelepiped constructed with the vectors $\mathbf{u}$, $\mathbf{v}$ and $\mathbf{w}$.

9. (i) From Theorem 1.4(iv), we have

$$\|(\mathbf{u} \times \mathbf{v}) \times \mathbf{w}\| \le \|\mathbf{u} \times \mathbf{v}\| \cdot \|\mathbf{w}\| \le \|\mathbf{u}\| \cdot \|\mathbf{v}\| \cdot \|\mathbf{w}\| = 1$$

and equality holds if and only if $\mathbf{w}$ and $\mathbf{u} \times \mathbf{v}$ are orthogonal and $\mathbf{u}$ and $\mathbf{v}$ are orthogonal. This also means $\mathbf{u}, \mathbf{v}$ and $\mathbf{w}$ lie in the plan orthogonal to $\mathbf{u} \times \mathbf{v}$, so they are coplanar.

(ii) From the above argument, we have that $\mathbf{u}$, $\mathbf{v}$, $\mathbf{w}$ are coplanar and $\mathbf{u} \perp \mathbf{v}$, $\mathbf{v} \perp \mathbf{w}$, so $\mathbf{w} = \pm\mathbf{u}$.

(iii) This means that $\mathbf{u} \perp \mathbf{v}$, $\mathbf{v} \perp \mathbf{w}$ and $\mathbf{w} \perp \mathbf{u}$ which is not possible, since the three vectors are coplanar.

10. The direction of the line ℓ is given by the normal vector $\mathbf{n} = (2, -3, 1)$ to the plane. Thus, the parametric equation of ℓ is

$$(x, y, z) = (1, 0, -1) + t(2, -3, 1), \quad t \in \mathbb{R}.$$

11. To find the equation of the intersection line we solve $x = 1 - 2y$, $z = x + y - 3 = -y - 2$. Letting for instance $y = t \in \mathbb{R}$, one gets $(x, y, z) = (1 - 2t, t, -2 - t)$, $t \in \mathbb{R}$. Let now $\sigma : ax + by + cz = d$ be

the equation of the plane that contains the line ℓ and is orthogonal to the plane $2x - 3y + 2z = 5$. Thus,

$$a(1 - 2t) + bt + c(-2 - t) = d \Longrightarrow \begin{cases} -2a + b - c = 0 \\ a - 2c = d. \end{cases}$$

On the other hand, σ is orthogonal to the plane $2x - 3y + 2z = 5$, so the corresponding normal vectors (a, b, c) and $(2, -3, 2)$ must be orthogonal. This yields $2a - 3b + 2c = 0$. From the above equalities we find $b = -2a$, $c = -4a$ and $d = 9a$. We may take $a = 1$ (any other nonzero value of a would do!) and deduce $\sigma : x - 2y - 4z = 9$.

12. Let $\sigma : ax + by + cz = d$ be the equation of the required plane. Since σ passes through A and B, one has $a - b - c = d$ and $-a - 6b + 2c = d$. Since σ is orthogonal to the plane $2x + 3y - z = 1$, their normal vectors (a, b, c) and $(2, 3, -1)$ must be orthogonal too. Thus, $2a + 3b - c = 0$. Combining the above equalities we get $b = c = -a$, $d = 3a$. Taking $a = 1$ we deduce the equation of the plane $\sigma : x - y - z = 3$.

13. The vectors $\mathbf{n}_1 = (0, 1, 1)$ and $\mathbf{n}_2 = (-1, m, 1)$ are the normal vectors to the two planes. Apply the formula for the cosine of the angle given in Corollary 1.8(iii) to deduce $m = 2$.

14. The line has the direction $\mathbf{u} = (-1, 1, p)$, the normal vector to the plane is $\mathbf{n} = (-1, 0, 1)$ and they form an angle of measure $\pi/6$. With formula (1.2) we find

$$\frac{\sqrt{3}}{2} = \cos\frac{\pi}{6} = \frac{|p - 1|}{\sqrt{2(p^2 + 2)}}.$$

Note that one must use the absolute value in the above equality since the angle is less than $\pi/2$. We find $p = -2$.

15. (i)–(iii) If $\{\mathbf{a}_n\}$ and $\{\mathbf{b}_n\}$ are convergent sequences to $\mathbf{a}$ and $\mathbf{b}$ respectively, then for each $1 \le i \le m$ the sequences of real numbers $\{a_{i,n}\}$, $\{b_{i,n}\}$ are convergent to a_i and b_i. This implies that $\{a_{i,n} \pm b_{i,n}\}$, $\{\lambda a_{i,n}\}$ and $\{a_{i,n} b_{i,n}\}$ converge to $a_i \pm b_i$, λa_i and $a_i b_i$ respectively.

(iv) From Exercise 3(i), we have $\big| \|\mathbf{a}_n\| - \|\mathbf{a}\| \big| \le \|\mathbf{a}_n - \mathbf{a}\| \to 0$ as $n \to \infty$. This shows that $\|\mathbf{a}_n\| \to \|\mathbf{a}\|$ as $n \to \infty$.

16. The statement is false. To construct a counterexample, let

$$\mathbf{a}_n = ((-1)^n, 0, 0, \ldots, 0) \in \mathbb{R}^m.$$

Then $\|\mathbf{a}_n\| = 1$, so the sequence $\{\|\mathbf{a}_n\|\}$ is convergent. However, $\{\mathbf{a}_n\}$ is not convergent in $\mathbb{R}^m$ because this means that each component of $\mathbf{a}_n$ forms a convergent sequence in $\mathbb{R}$. In particular, this would mean that $a_{1,n} = (-1)^n$ is convergent, contradiction.

17. (i) Observe that A is a strip along the x axis in the plane. Let $(a, b) \in A$. Then $-1 < b < 2$. Take $0 < r < \min\{b + 1, 2 - b\}$ and then the open ball $B_r(a, b)$ is contained in A. This shows that $(a, b) \in A^\circ$, so A is open.

(ii) Let $(a, b) \in B$. Then $b > a$ and we take $0 < r < (b - a)/2$. We claim that $B_r(a, b) \subset B$. Indeed, if $(x, y) \in B_r(a, b)$, then $x < a + r$ and $y > b - r$ (draw a picture!) and thus $y > b - r > a + r > x$ which shows that $(x, y) \in B$, so $B_r(a, b) \subset B$. This means that B is an open set.

(iii) The set C is the upper half space in $\mathbb{R}^m$. Take $\mathbf{a} = (a_1, a_2, \ldots, a_m) \in C$. Then $a_m > 0$ and by taking $0 < r < a_m$, we have that $B_r(\mathbf{a}) \subset C$, so $\mathbf{a} \in C^\circ$, which yields C is open.

18. (i) Let $\mathbf{x} \in \overline{B}_\rho(\mathbf{b})$, so that $\|\mathbf{x} - \mathbf{b}\| \le \rho$. We want to show $\|\mathbf{x} - \mathbf{a}\| < r$. Indeed, by the triangle inequality, we have

$$\|\mathbf{x} - \mathbf{a}\| \le \|\mathbf{x} - \mathbf{b}\| + \|\mathbf{b} - \mathbf{a}\| \le \rho + \|\mathbf{b} - \mathbf{a}\| < r,$$

which shows that $\mathbf{x} \in B_r(\mathbf{a})$, so $\overline{B}_\rho(\mathbf{b}) \subset B_r(\mathbf{a})$.

(ii) Let $\mathbf{b} \in B_r(\mathbf{a})$. Then $\|\mathbf{b} - \mathbf{a}\| < r$. Take $0 < \rho < r - \|\mathbf{a} - \mathbf{b}\|$ and by the above arguments one has $B_\rho(\mathbf{b}) \subset B_r(\mathbf{a})$. Hence, $B_r(\mathbf{a})$ is open in $\mathbb{R}^m$.

19. (i) Let $\mathbf{a} \in G_1 \cup G_2$. We want to show that $B_r(\mathbf{a}) \subset G_1 \cup G_2$ for some $r > 0$. Without loosing the generality we may assume $\mathbf{a} \in G_1$. Since G_1 is open, there exists $r > 0$ such that $B_r(\mathbf{a}) \subset G_1$ which further yields $B_r(\mathbf{a}) \subset G_1 \cup G_2$.

Let now $\mathbf{a} \in G_1 \cap G_2$. We want to show that $B_r(\mathbf{a}) \subset G_1 \cap G_2$ for some $r > 0$. Indeed, since $\mathbf{a} \in G_1$, $\mathbf{a} \in G_2$ and G_1, G_2 are open sets, there exist $r_1, r_2 > 0$ so that $B_{r_1}(\mathbf{a}) \subset G_1$ and $B_{r_2}(\mathbf{a}) \subset G_2$. Taking $r = \min\{r_1, r_2\} > 0$ we find that $B_r(\mathbf{a}) \subset G_1 \cap G_2$.

(ii) We proceed similarly, by showing that $(D_1 \cup D_2)^c$ and $(D_1 \cap D_2)^c$ are open sets. Useful equalities are $(D_1 \cup D_2)^c = D_1^c \cap D_2^c$ and $(D_1 \cap D_2)^c = D_1^c \cup D_2^c$.

20. Recall that $\overline{D} = D^\circ \cup \partial D$ and $D^\circ \subset D \subset D \cup \partial D$. Using the fact that D° and ∂D are disjoint, we have:

D is open $\Leftrightarrow D = D^\circ \Leftrightarrow D \cap \partial D = \emptyset \Leftrightarrow D$ has no boundary points.

D is closed $\Leftrightarrow D = \overline{D} \Leftrightarrow D = D \cup \partial D \Leftrightarrow \partial D \subset D$.

21. (i) $\mathbf{a} \in (D^c)^\circ$ if and only if there exists $r > 0$ such that $B_r(\mathbf{a}) \subset D^c$ $\Leftrightarrow \mathbf{a}$ is an interior point of $D^c \Leftrightarrow \mathbf{a} \notin D^\circ$ and $\mathbf{a} \notin \partial D \Leftrightarrow \mathbf{a} \notin \overline{D} \Leftrightarrow \mathbf{a} \in (\overline{D})^c$.

(ii) We can proceed as before, or we simply replace D by D^c in (i) above. We have $(\overline{D^c})^c = D^\circ$ and by taking the complement we derive $\overline{D^c} = (D^\circ)^c$.

(iii) $a \in \partial D$ if and only if $a \notin D^\circ$ and $a \notin (D^c)^\circ$. Similarly, $a \in \partial D^c$ if and only if $a \notin (D^c)^\circ$ and $a \notin \left[(D^c)^c\right]^\circ$. Since $(D^c)^c = D$, the conclusion follows.

22. (i) We have $A^\circ = (-2,6) \times (1,8)$, $\overline{A} = [-2,6] \times [1,8]$ and

$$\partial A = \left(\{-2,6\} \times [1,8]\right) \cup \left([-2,6] \times \{1,8\}\right).$$

(ii) We have $B^\circ = (-\infty,1) \times (-1,\infty)$, $\overline{B} = (-\infty,1] \times [-1,\infty)$ and

$$\partial B = \left(\{1\} \times [-1,\infty)\right) \cup \left((-\infty,1] \times \{-1\}\right).$$

(iii) We have

$$C^\circ = \{(x,y) \in \mathbb{R}^2 : -1 < x < 2\,, x^2 < y < 5\}$$
$$\overline{C} = \{(x,y) \in \mathbb{R}^2 : -1 \leq x \leq 2\,, x^2 \leq y \leq 5\}$$
$$\partial C = C_1 \cup C_2 \text{ where}$$
$$C_1 = \left(\{-1\} \times [1,5]\right) \cup \left([-1,2] \times \{5\}\right) \cup \left(\{2\} \times [4,5]\right)$$
$$C_2 = \{(x,y) \in \mathbb{R}^2 : -1 \leq x \leq 2\,, y = x^2\}.$$

(iv) $D^\circ = \{(x,y) \in \mathbb{R}^2 : xy^2 \neq 1\}$, $\partial D = \{(x,y) \in \mathbb{R}^2 : xy^2 = 1\}$ and $\overline{D} = \mathbb{R}^2$.

(v) We have

$$E^\circ = \{(x_1,x_2,x_3) \in \mathbb{R}^3 : x_1^2 + x_2^2 < x_3^2 < 1\}$$
$$\overline{E} = \{(x_1,x_2,x_3) \in \mathbb{R}^3 : x_1^2 + x_2^2 \leq x_3^2 \leq 1\}$$
$$\partial E = E_1 \cup E_2$$
$$E_1 = \{(x_1,x_2,x_3) \in \mathbb{R}^3 : x_1^2 + x_2^2 = x_3^2\,, |x_3| < 1\}$$
$$E_2 = \{(x_1,x_2,x_3) \in \mathbb{R}^3 : x_1^2 + x_2^2 \leq 1\,, |x_3| = 1\}.$$

23. (i) Let $M > 0$ be such that $\|\mathbf{a}\| \leq M$ for all $\mathbf{a} \in D$. We want to show that the above inequality also holds for all $\mathbf{a} \in \partial D$. Let $\mathbf{a} \in \partial D$. By Theorem 1.19 there exists a sequence $\{\mathbf{a}_n\} \subset D$ such that $\mathbf{a}_n \to \mathbf{a}$ as $n \to \infty$. By Theorem 1.14(iv), we have $\|\mathbf{a}_n\| \to \|\mathbf{a}\|$ as $n \to \infty$. This yields $\|\mathbf{a}\| = \lim_{n \to \infty}\|\mathbf{a}_n\| \leq M$. Hence, $\overline{D}$ is bounded.

(ii) Since D is bounded, we have $\overline{D}$ and ∂D are also bounded. They are also closed sets and thus compact.

24. Let $D = \mathbb{R}^m \setminus B_1(\mathbf{0})$ (the exterior of the open unit ball). Then D is unbounded and $\partial D = \partial B_1(\mathbf{0})$ is compact.

25. Since D_1, D_2 are compact sets, they are closed and bounded. By Exercise 19(ii) it follows that $D_1 \cup D_2$ and $D_1 \cap D_2$ are also closed. The boundedness of $D_1 \cup D_2$ and $D_1 \cap D_2$ is obvious, so they are compact.

12.2   Solutions Chapter 2

1. (i) Impose $3x^2 + y^2 \neq 0$; we find

$$D = \{(x,y) \in \mathbb{R}^2 : (x,y) \neq (0,0)\} = \mathbb{R}^2 \setminus \{(0,0)\}.$$

The domain is the whole plane except the origin.

(ii) Impose $16 - x^2 - y^2 > 0$; we find

$$D = \{(x,y) \in \mathbb{R}^2 : x^2 + y^2 < 16\} = B_4(0,0).$$

The domain is the open disc of radius 4 centred at the origin.

(iii) Impose $x^2 + y^2 - 4 > 0$. Hence,

$$D = \{(x,y) \in \mathbb{R}^2 : x^2 + y^2 > 4\} = \mathbb{R}^2 \setminus \overline{B}_2(0,0).$$

The domain is the exterior of the closed disc of radius 2 centred at the origin.

(iv) Impose $y - x \geq 0$, $D = \{(x,y) \in \mathbb{R}^2 : y \geq x\}$. The domain is a half-plane bounded by the line $y = x$.

(v) Impose $x^2 - y^2 > 0$, $D = \{(x,y) \in \mathbb{R}^2 : |x| > |y|\}$. The domain consists of the interior of two regions bounded by the lines $y = x$ and $y = -x$.

(vi) Impose $1 - \frac{e^x}{y} > 0 \iff \frac{e^x}{y} < 1 \iff y < 0$ or $y > e^x$.

In set notation, we have $D = \{(x,y) \in \mathbb{R}^2 | \ y < 0 \text{ OR } y > e^x\}$.

(vii) Impose $1 + xy^2 \geq 0 \iff xy^2 \geq -1 \iff y = 0$ or $x \geq -\dfrac{1}{y^2}$.

In set notation, we have $D = \left\{(x,y) \in \mathbb{R}^2 | \ y = 0 \text{ OR } x \geq -\frac{1}{y^2}\right\}$.

2. (i) We impose $x, y \geq 0$ and then $x - \sqrt{y} > 0$, $y - \sqrt{x} > 0$. We find $x, y > 0$ and $x^2 > y$, $y^2 > x$, so $x^4 > y^2 > x > 0$. In set notation $D = \{(x,y) \in \mathbb{R}^2 : x^4 > y^2 > x > 0, \ y > 0\}$. Geometrically, D is the unbounded open region between the curves $y = x^2$ and $y = \sqrt{x}$ with $x, y > 1$.

(ii) We have to impose that (x,y,z) does not lie in the planes of equations $y = 0$, $y + z = 0$ and $x + y + z = 0$. Thus, $D = \{(x,y,z) \in \mathbb{R}^3 : y \neq 0, \ x + y \neq 0, \ x + y + z \neq 0\}$. Geometrically D is the space $\mathbb{R}^3$ except the three planes described above.

3. (i) $\Psi(x,y,z) = \left(\dfrac{x}{1+z}, \dfrac{y}{1+z}\right)$.

(ii) We find $\Psi^{-1} : \mathbb{R}^2 \to \mathbb{S}^2 \setminus \{S\}$,

$$\Psi^{-1}(X,Y) = \left(\frac{2X}{X^2 + Y^2 + 1}, \frac{2Y}{X^2 + Y^2 + 1}, -\frac{X^2 + Y^2 - 1}{X^2 + Y^2 + 1}\right).$$

By direct computation one obtains for all $(X, Y) \in \mathbb{R}^2 \setminus \{(0,0)\}$ that

$$(\Phi \circ \Psi^{-1})(X, Y) = \left(\frac{X}{X^2 + Y^2}, \frac{Y}{X^2 + Y^2} \right).$$

4. Let $\mathbf{u} = (u_1, u_2, \ldots, u_m) \in \mathbb{R}^{m-1}$. We show that the equation $\Phi(\mathbf{x}) = \mathbf{u}$ has a unique solution $\mathbf{x} = (x_1, x_2, \ldots, x_m) \in \mathbb{S}^{m-1} \setminus \{N\}$. Indeed, $\Phi(\mathbf{x}) = \mathbf{u}$ yields $u_i = \frac{x_i}{1 - x_m}$ and then

$$\|\mathbf{u}\|^2 = \sum_{i=1}^{m-1} u_i^2 = \frac{1}{(1 - x_m)^2} \sum_{i=1}^{m-1} x_i^2 = \frac{1 - x_m^2}{(1 - x_m)^2} = \frac{1 + x_m}{1 - x_m}.$$

From here one has

$$x_m = \frac{\|\mathbf{u}\|^2 - 1}{\|\mathbf{u}\|^2 + 1} \quad \text{and then} \quad x_i = \frac{2u_i}{\|\mathbf{u}\|^2 + 1}, \, 1 \leq i \leq m - 1.$$

Thus, we found a unique solution $\mathbf{x} \in \mathbb{S}^{m-1} \setminus \{N\}$ which shows that Φ is bijective.

5. If $k = -1$, then $\|F(\mathbf{x})\| = 1$ for all $\mathbf{x} \in \mathbb{R}^m$, so F cannot be surjective. Take now $k \neq -1$ and we show that F is bijective. To this aim, one has to prove that for all $\mathbf{u} \in \mathbb{R}^m$, the equation $F(\mathbf{x}) = \mathbf{u}$ has a unique solution $\mathbf{x} \in \mathbb{R}^m$. If $\mathbf{u} = \mathbf{0}$, then $\mathbf{x} = \mathbf{0}$. Assume $\mathbf{u} \neq \mathbf{0}$ and thus $F(\mathbf{x}) = \mathbf{u}$ implies $\mathbf{x}\|\mathbf{x}\|^k = \mathbf{u}$ and $\|\mathbf{x}\|^{k+1} = \|\mathbf{u}\|$. Then $\|\mathbf{x}\| = \|\mathbf{u}\|^{1/(k+1)}$ from which we find $\mathbf{x} = \mathbf{u}\|\mathbf{u}\|^{-k/(k+1)}$.

6. (i) The trace of the graph in the plane $x = c$ is the parabola $z = y^2$ for all values of $c \in \mathbb{R}$. Thus, the graph of f consists of infinitely many copies of the standard parabola $z = y^2$ placed along the x axis.

(ii) The trace of the graph in the plane $x = c$ is the parabola $z = y^2 - (2c + 3)y$. The graph of f is depicted in Figure 12.2.

7. Let (x, y, z) be a point on the surface. Then $x \in [0, a]$, $y = g(x)$ and the distance from (x, y, z) to $(x, 0, 0)$ is $|g(x)|$. Hence,

$$\sqrt{y^2 + z^2} = |g(x)| \quad \text{which implies} \quad y^2 + z^2 = g(x)^2.$$

By the vertical line test (that is, each vertical line must intersect the surface in at most one point), we deduce that this surface is not the graph of any function $f : D \subset \mathbb{R}^2 \to \mathbb{R}$.

8. The distance from P to the y axis is $\sqrt{x^2 + z^2}$ and the distance from P to the origin is $\sqrt{x^2 + y^2 + z^2}$. Thus,

$$\sqrt{x^2 + z^2} = \frac{\sqrt{x^2 + y^2 + z^2}}{2} \quad \text{which yields} \quad y^2 = 3(x^2 + z^2).$$

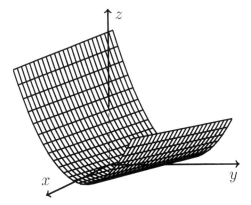

Figure 12.2
The graph of $f(x,y) = y^2 - 2xy - 3y$.

9. The distance from P to the x axis is $\sqrt{y^2 + z^2}$ and the distance from P to the plane $x = -1$ is $|x+1|$. Thus, $\sqrt{y^2 + z^2} = |x+1|$ which implies $y^2 + z^2 = (x+1)^2$.

10. (i) We solve $f(x,y) = c$, $c \in \mathbb{R}$.

    If $c = 0$, then $xy = 0$, so $L(f,0) = \{(x,0),(0,y) : x,y \in \mathbb{R}\}$.

    If $c \neq 0$, then $xy \neq 0$ and $f(x,y) = c$ yields $xy = cx^2 + cy^2$. We can divide by y^2 and derive that $t = \frac{x}{y}$ satisfies $ct^2 - t + c = 0$. If $|c| > 1/2$, then the quadratic equation has no real solutions for the unknown t. If $|c| < 1/2$ we find

    $$x = t_{1,2}y \quad \text{where} \quad t_{1,2} = \frac{1 \pm \sqrt{1 - 4c^2}}{2c}.$$

    Note that if $|c| = 1/2$, then $t_1 = t_2 = 2c$. Thus,

    $$L(f,c) = \begin{cases} \emptyset & \text{if } |c| > 1/2 \\ \text{line } x = 2cy & \text{if } |c| = 1/2 \\ \text{lines } x = t_1 y \text{ and } x = t_2 y & \text{if } 0 < |c| < 1/2 \\ \text{lines } x = 0 \text{ and } y = 0 & \text{if } c = 0. \end{cases}$$

    (ii) $L(f,c) = \emptyset$ if $c \neq -3$ and $L(f,c) = \mathbb{R}^3$, if $c = -3$.

    (iii) $L(f,c)$ is the plane of equation $x + 2y - 7z = c$.

    (iv) $L(f,c) = \emptyset$ if $c \leq 0$ and $L(f,c)$ is the line of equation $x - 2y = \ln c$ if $c > 0$.

(v) If $|c| > 1$, then $L(f, c) = \emptyset$.

If $|c| \leq 1$, then one can solve the trigonometric equation $\cos(xy) = c$ which yields $L(f, c) = \{(x, y) \in \mathbb{R}^2 : xy = \pm \arccos c + 2k\pi, k \in \mathbb{Z}\}$. The level curve is a collection of hyperbolas.

(vi) If $c < 0$, then $L(f, c) = \emptyset$ while for $c = 0$ the level surface $L(f, 0)$ is the y axis. If $c > 0$, then $L(f, c) = \{(x, y, z) \in \mathbb{R}^3 : x^2 + z^2 = c\}$. Geometrically, this represents an infinite cylinder along y axis.

11. (i) We impose $2 - \dfrac{\sin^2 x}{y} \geq 0$ and $y \neq 0$, that is,

$$y < 0 \quad \text{OR} \quad \begin{cases} y \geq \dfrac{\sin^2 x}{2} \\ y > 0 \end{cases}.$$

The domain is depicted in Figure 12.3.

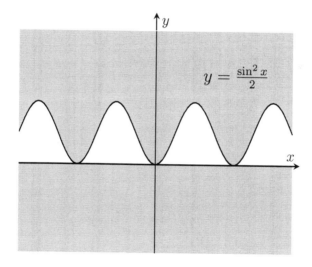

$$y = \frac{\sin^2 x}{2}$$

Figure 12.3
The domain D of f.

(ii) Solving $f(x, y) = 1$ we find $\dfrac{\sin^2 x}{y} = 1$ so $y = \sin^2 x$. The level curve is the graph of $y = \sin^2 x$ from which we remove the point $(0, 0)$ (because this point does not belong to the domain D of f).

12. Solving $f(x, y) = c$ we have $\sqrt{2x^2 + y^2} = x + c$ and by squaring both sides one has

$$2x^2 + y^2 = x^2 + 2xc + c^2 \Longrightarrow (x^2 - 2xc + c^2) + y^2 = 2c^2$$

$$\Longrightarrow (x - c)^2 + y^2 = (c\sqrt{2})^2.$$

This is the equation of a circle centred at $(c,0)$ and having radius $c\sqrt{2}$. We impose $c\sqrt{2} = 1$, so $c = \frac{1}{\sqrt{2}}$.

13. $L(f,17)$ contains all points $(x,y,z) \in \mathbb{R}^3$ that satisfy $f(x,y,z) = 17$, that is,

$$x^2 + ay^2 + z^2 - 2by - 4z = 10. \tag{12.3}$$

Since the equation of a sphere is $(x-x_0)^2+(y-y_0)^2+(z-z_0)^2 = r^2$, identifying the coefficients we must have $a = 1$. Thus, using the completion of squares method we write (12.3) as

$$x^2 + (y - b)^2 + (z - 2)^2 = 14 + b^2.$$

Since $(-1,2,3)$ is a solution of this equation we deduce $1 + (2 - b)^2 + 1 = 14 + b^2$ from which $b = -2$.

14. (i) The equation of the circle centred at $(-2,3)$ and having radius $2c$ is $(x + 2)^2 + (y - 3)^2 = (2c)^2$. This also represents the equation $f(x,y) = c$. From the former equation we find $c = \frac{1}{2}\sqrt{(x + 2)^2 + (y - 3)^2}$. Hence, $f(x,y) = \frac{1}{2}\sqrt{(x + 2)^2 + (y - 3)^2}$.

(ii) The equation of the sphere centred at $(1,0,-1)$ and having radius $2\sqrt{c}$ is $(x - 1)^2 + y^2 + (z + 1)^2 = (2\sqrt{c})^2$, that is, $(x - 1)^2 + y^2 + (z + 1)^2 = 4c$. This represents the equation $f(x,y) = c$, so $f(x,y) = \frac{1}{4}[(x - 1)^2 + y^2 + (z + 1)^2]$.

15. (i) No! Take for instance $f : \mathbb{R}^2 \to \mathbb{R}$, $f(x,y) = \sqrt[3]{x^2 + y^2}$ whose all level curves are circles.

(ii) The equation $g(f(x,y)) = c$ may have either no solution or yield $f(x,y) = C$ where C is a constant taking one or several distinct values. Thus, the level curves of $g \circ f$ are either empty sets or circles or union of disjoint circles.

16. (i) Using the completion of squares, we have $f(x,y,z) = (x - z)^2 + (2y + 1)^2 + (z + y)^2 - 1$. Thus, solving $f(x,y,z) = c$ one has

$$(x - z)^2 + (2y + 1)^2 + (z + y)^2 = c + 1. \tag{12.4}$$

As in Example 2.10, if $c < -1$ the above equation has no solutions while for $c \geq -1$ the level curve $L(f,c)$ is nonempty. We may simply take the points

$$(x,y,z) = \left(-\frac{1}{2} \pm \sqrt{c + 1}, -\frac{1}{2}, \frac{1}{2}\right) \tag{12.5}$$

which satisfy (12.4) and thus belong to $L(f,c)$. Hence $L(f,c)$ is nonempty if and only if $c \geq -1$.

(ii) We have seen that for all $c \geq -1$, the level curve $L(f,c)$ contains the points given by (12.5). Thus, if $L(f,c)$ reduces to a single element we must have $-\sqrt{c+1} = \sqrt{c+1}$, so $c = -1$ and then (12.4) yields $(x - z)^2 + (2y + 1)^2 + (z + y)^2 = 0$. Hence $L(f,-1)$ reduces to the single element $(-1/2, -1/2, 1/2)$.

17. (i) By completion of squares, we have $f(x, y, z) = (2x - 1)^2 + (y^2 + yz)^2 + (x + y^2)^2 + 1$.

If $(x, y, z) \in L(f, 0)$, then $f(x, y, z) = 0$ which implies

$$(2x - 1)^2 + (y^2 + yz)^2 + (x + y^2)^2 = -1.$$

The above equation has no real solutions since the left-hand side is a sum of perfect squares, hence a nonnegative quantity, while the right-hand side is negative. Thus, $L(f, 0) = \emptyset$.

Let now $(x, y, z) \in L(f, 1)$. This entails that x, y, z satisfy

$$(2x - 1)^2 + (y^2 + yz)^2 + (x + y^2)^2 = 0.$$

This means that all perfect squares must be zero, so $2x - 1 = 0$, $y^2 + yz = 0$ and $x + y^2 = 0$. From the first equality we deduce $x = 1/2$ and from the third equality, we have now $y^2 + 1/2 = 0$ which has no real solutions. Hence $L(f, 1) = \emptyset$.

(ii) $L(f, 3)$ contains all triples (x, y, z) which satisfy $(2x - 1)^2 + (y^2 + yz)^2 + (x + y^2)^2 = 2$. By taking $y = 0$ we make the z variable disappear from the equation and we are left with $(2x - 1)^2 + x^2 = 2$. This has solution $x = 1$ so $(1, 0, z) \in L(f, 3)$ for all $z \in \mathbb{R}$. This shows that $L(f, 3)$ is nonempty and also unbounded.

18. Let $f(x, y) = x^3$. Then $L(f, c) = \{x, y) \in \mathbb{R}^2 : x = \sqrt[3]{c}, \ y \in \mathbb{R}\}$.

Another example is $g : \mathbb{R}^2 \to \mathbb{R}$, $g(x, y) = x^2 - y^2$. Then

$$L(g, c) = \begin{cases} \{(x, y) \in \mathbb{R}^2 : x \in \mathbb{R}, \ y = \pm x\} & \text{if } c = 0 \\ \{(x, y) \in \mathbb{R}^2 : x = \pm\sqrt{y^2 + c}, \ y \in \mathbb{R}\} & \text{if } c > 0 \\ \{(x, y) \in \mathbb{R}^2 : x \in \mathbb{R}, \ y = \pm\sqrt{x^2 - c}\} & \text{if } c < 0. \end{cases}$$

19. We can use the vertical line test to check that the output is the graph of the function $g : \overline{B}_1(0, 0) \to \mathbb{R}$, $g(x, y) = f(|y|)$, where $\overline{B}_1(0, 0)$ is the closed unit disc in the plane. The level curves of g are $L(g, c) = \{(x, y) \in \overline{B}_1(0, 0) : f(|y|) = c\}$.

20. (i) $(y + z)^2 + (2x - y)^2 - (x - 1)^2 = 4$, hyperboloid of one sheet.

(ii) $(x - y)^2 + 2(y - 1)^2 = z$, elliptic paraboloid.

(iii) $(x + 2z)^2 + 3(y + 1)^2 = 2x^2$, double cone.

(iv) $(2x + z)^2 - (x + y)^2 - (y - 1)^2 = 5$, hyperboloid of two sheets.

(v) $2x = (y - z)^2 - (x + 2z)^2$, hyperbolic paraboloid.

(vi) We multiply the whole equation by 2 to use the completion of squares. $(x - y)^2 + (y - z)^2 + (z - x)^2 = 14$, ellipsoid.

21. (i) $f(x,y,z) = c$ implies $x^2 + y^2 - cz^2 = c$. If $c < 0$ the equation has no solutions. If $c = 0$, then $L(f,0) = \{(0,0,z) : z \in \mathbb{R}\}$. This is the equation of the z axis. If $c > 0$ we obtain that $L(f,c)$ is geometrically a hyperboloid of one sheet.

    (ii) $f(x,y,z) = c$ implies $x^2 - y^2 - cz^2 = c$. If $c = 0$ we find $L(f,0) = \{(x,y,z) \in \mathbb{R}^3 : |x| = |y|\}$. Geometrically, $L(f,0)$ is the union of two planes of equation $y = x$ and $y = -x$. If $c \neq 0$, $L(f,c)$ is a hyperboloid of two sheets.

    (iii) The domain of the function is $D = \mathbb{R}^3 \setminus \{(x,0,0) : x \in \mathbb{R}\}$, that is, the whole space $\mathbb{R}^3$ except the x axis. Solving $f(x,y,z) = c$ we are led to $x = c(y^2 + z^2)$. If $c = 0$, then $x = 0$ and $L(f,0)$ is the whole yz plane except the origin. If $c \neq 0$ the level surface $L(f,c)$ is an elliptic paraboloid.

    (iv) The domain is $D = \mathbb{R}^3 \setminus \{y\,\text{axis}\}$. Solving $f(x,y,z) = c$ we find $1 - y^2 = c(x^2 + z^2)$. If $c = 0$, then $L(f,0) = \{(x,y,z) \in \mathbb{R}^3 : y = \pm 1\}$ which is the union of two parallel planes. If $c > 0$, then $L(f,c)$ is an ellipsoid from which we remove the points $(0,\pm 1,0)$ while if $c < 0$ the level surface is a hyperboloid of two sheets from which we again remove the same points $(0,\pm 1,0)$.

22. (i) We use $x = y - 2$ in the equation of the hyperboloid $x^2 - y^2 - z^2 = 2$ to derive $(y-2)^2 - y^2 - z^2 = 2$, so $y = \frac{2-z^2}{4}$ and then $x = -\frac{z^2+6}{4}$. Letting $z = t$ we find $\gamma : \mathbb{R} \to \mathbb{R}^3$, $\gamma(t) = \left(-\frac{t^2+6}{4}, \frac{2-t^2}{4}, t\right)$.

    (ii) We parametrize the cylinder $x^2 + z^2 = 4$ as

    $$x = 2\cos t, \quad z = 2\sin t, \quad t \in [0, 2\pi].$$

    Then $y = 4(\cos^2 t - \sin^2 t) = 4\cos(2t)$. Thus, $\gamma : [0, 2\pi] \to \mathbb{R}^3$, $\gamma(t) = (2\cos t, 4\cos(2t), 2\sin t)$.

    (iii) $x = -\sqrt{y^2 + z^2}$ implies $x^2 = y^2 + z^2$. Use this equality in the equation of the ellipsoid to deduce $2y^2 + 3z^2 = 6$ which is the equation of an ellipse. The parametrization of this ellipse is $y = \sqrt{3}\cos t$, $z = \sqrt{2}\sin t$, $t \in [0, 2\pi]$. Then

    $$x = -\sqrt{(\sqrt{3}\cos t)^2 + (\sqrt{2}\sin t)^2} = -\sqrt{2 + \cos^2 t}.$$

    Hence $\gamma : [0, 2\pi] \to \mathbb{R}^3$, $\gamma(t) = \left(-\sqrt{2 + \cos^2 t}, \sqrt{3}\cos t, \sqrt{2}\sin t\right)$.

23. Let $t \in \mathbb{R}$ be be such that the tangent line at $\gamma(t)$ is parallel to z axis. The direction of the tangent line is $\gamma'(t) = (-\sin t, 2t, e^t)$ and the direction of the z axis is $(0,0,1)$. From $\gamma'(t) = \lambda(0,0,1)$, $\lambda \in \mathbb{R}$, we deduce $t = 0$ and thus, $\gamma(0) = (1,2,0)$ is the only point on the curve with the required property.

24. We impose the tangent vectors $\gamma'(t)$ and $\eta'(t)$ to be orthogonal, so $\gamma'(t) \bullet \eta'(t) = 0$. This yields

$$\frac{2t(2t-1)}{t^2+1} + t - 2 = 0 \Longrightarrow t = -2, -1, 1.$$

25. (i) From $\gamma(t) = \eta(s)$ we find $t = s = 1$, so the intersection point is $\gamma(1) = \eta(1) = (1, 1, 1)$.

    (ii) The tangent line to γ at $(1, 1, 1)$ has the direction $\mathbf{n}_1 = \gamma'(1) = (1, 2, 2)$ and the tangent line to η at $(1, 1, 1)$ has the direction $\mathbf{n}_2 = \eta'(1) = (2, -2, 1)$. By formula (1.2) we find $\cos\theta = 0$ so the angle is $\pi/2$.

26. (i) $c = 1/2$. (ii) There are no points with the required property. This shows that the Mean Value Theorem does not hold for vector valued functions.

27. Let $\gamma = (\gamma_1, \gamma_2, \ldots, \gamma_p)$ and $\eta = (\eta_1, \eta_2, \ldots, \eta_p)$. Then, all four requirements follow from the properties of differentiation of single variable calculus if one applies them to γ_j and η_j, $1 \le j \le p$.

28. Since the trace lies on the unit sphere, one has $\|\gamma(t)\| = 1$ for all $t \in I$. Thus, $1 = \|\gamma(t)\|^2 = \gamma(t) \bullet \gamma(t)$ for all $t \in I$. We differentiable this last equality using Exercise 27(iv) and deduce $2\gamma'(t) \bullet \gamma(t) = 0$ which yields $\gamma(t) \perp \gamma'(t)$ for each $t \in I$.

29. The parametric equation of the tangent line to the trace of γ is

$$\big(\gamma_1(\tau) + t\gamma_1'(\tau), \gamma_2(\tau) + t\gamma_2'(\tau)\big) \qquad t \in \mathbb{R}, \tau \in I.$$

    According to our hypothesis, for all $\tau \in I$, there exists $t \in \mathbb{R}$ such that

$$\begin{cases} \gamma_1(\tau) + t\gamma_1'(\tau) = 0 \\ \gamma_2(\tau) + t\gamma_2'(\tau) = 0 \end{cases} \Longrightarrow \begin{cases} \gamma_1(\tau) = -t\gamma_1'(\tau) \\ \gamma_2(\tau) + t\gamma_2'(\tau) = 0. \end{cases}$$

    From the above one has

$$\gamma_1(\tau)\gamma_2'(\tau) - \gamma_2(\tau)\gamma_1'(\tau) = 0 \quad \text{for all } \tau \in I.$$

    This further implies

$$\frac{d}{d\tau}\left(\frac{\gamma_1}{\gamma_2}\right) = 0 \quad \text{for all } \tau \in I.$$

    This means that $\dfrac{\gamma_1}{\gamma_2}$ is constant, so $\gamma_1 = c\gamma_2$ for some $c \in \mathbb{R}$.

12.3   Solutions Chapter 3

1. Recall the following standard limits from one-variable Calculus:

$$\lim_{t\to 0}\frac{\sin t}{t}=1,\quad \lim_{t\to 0}\frac{e^t-1}{t}=1,$$

$$\lim_{t\to 0}\frac{\ln(1+t)}{t}=1,\quad \lim_{t\to 0}\frac{\cos t-1}{t^2}=-\frac{1}{2}.$$

Using the above limits we compute:

(i) $\displaystyle\lim_{(x,y)\to(0,0)}\frac{e^{xy}-1}{xy}=\lim_{t=xy\to 0}\frac{e^t-1}{t}=1;$

(ii) $\displaystyle\lim_{(x,y)\to(0,0)}\frac{\ln(x^2+y^2+1)}{x^2+y^2}=\lim_{t=x^2+y^2\to 0}\frac{\ln(t+1)}{t}=1;$

(iii) $\displaystyle\lim_{(x,y)\to(0,0)}\frac{\cos(xe^y)-1}{x^2e^{2y}}=\lim_{t=xe^y\to 0}\frac{\cos(t)-1}{t^2}=-\frac{1}{2};$

(iv) $\displaystyle\lim_{(x,y)\to(0,0)}\frac{\sin(x^2+y^2)}{x^2+y^2}=\lim_{t=x^2+y^2\to 0}\frac{\sin t}{t}=1.$

2. Let $f:\mathbb{R}^2\to\mathbb{R}$, $f(x,y)=\begin{cases}\dfrac{x^3y}{x^6+y^2} & \text{if }(x,y)\neq(0,0)\\[2mm] 0 & \text{if }(x,y)=(0,0).\end{cases}$

It is easy to check that f satisfies the conditions (i)–(iii). Since f approaches distinct values along the paths $y=x$ and $y=x^3$, we conclude by Theorem 3.3 that f has no limit at $(0,0)$.

3. Along the line $x=0$, we have

$$f(x,y)=f(0,y)=0\to 0\quad\text{as }(x,y)\to(0,0).$$

Along the parabola $x=y^2$, we have

$$f(x,y)=f(y^2,y)=\frac{3\cos y}{2}\to\frac{3}{2}\quad\text{as}\quad (x,y)\to(0,0).$$

Hence, f has no limit at $(0,0)$.

4. Along the line $x=-1$, we have $f(x,y)=f(-1,y)=0\to 0$ as $(x,y)\to(-1,2)$. To find the second path, we look for curves that pass through $(-1,2)$ so that $y-2$ and $x+1$ have the same contribution to the denominator. One example is $y=2+(x+1)^2$. Along this parabola we find

$$f(x,y)=f(x,2+(x+1)^2)=\frac{1}{2}\cdot\frac{\sin(x+1)^2}{(x+1)^2}=\frac{1}{2}\cdot\frac{\sin z}{z}\to\frac{1}{2}$$

as $z=(x+1)^2\to 0$.

(ii) By Theorem 3.3, we deduce that f has no limit at $(-1, 2)$.

5. (i) Let $\mathbf{a} \in \partial A$. Then, in any ball $B_r(\mathbf{a})$ there are points in A and points in $\mathbb{R}^m \setminus A$. Taking $r = 1/n$, we have that for any $n \geq 1$, there exists $\mathbf{a}_n \in B_{1/n}(\mathbf{a}) \cap A$ and $\mathbf{b}_n \in B_{1/n}(\mathbf{a}) \setminus A$. Because A is open, it follows that $B_{1/n}(\mathbf{a}) \cap A$, as intersection of open sets, is also open. Thus, it contain infinitely many elements. In particular, one may assume that $\mathbf{a}_n \in B_{1/n}(\mathbf{a}) \cap A \setminus \{\mathbf{a}\}$. We have thus found $\mathbf{a}_n, \mathbf{b}_n \neq \mathbf{a}$ such that $\|\mathbf{a}_n - \mathbf{a}\| \leq 1/n$ and $\|\mathbf{b}_n - \mathbf{a}\| \leq 1/n$ for all $n \geq 1$. This shows that $\mathbf{a}_n, \mathbf{b}_n \to \mathbf{a}$ as $n \to \infty$.

Since $\mathbf{a}_n \in A$ we now have $f(\mathbf{a}_n) = 1 \to 1$ as $n \to \infty$. Since $\mathbf{b}_n \notin A$ we have $f(\mathbf{b}_n) = 0 \to 0$ as $n \to \infty$. By the Definition 3.1 it follows that f has no limit at $\mathbf{a} \in \partial A$.

Let now $\mathbf{a} \in A^\circ$. Then, because A is open, there exists a ball $B_r(\mathbf{a}) \subset A$. Hence, $f(\mathbf{x}) = 1$ for all $\mathbf{x} \in B_r(\mathbf{a})$. In particular, this shows that $\lim_{\mathbf{x} \to \mathbf{a}} f(\mathbf{x}) = 1$.

Finally, assume now $\mathbf{a} \in \mathbb{R}^m \setminus \overline{A}$. Then, since $\mathbb{R}^m \setminus \overline{A}$ is open, there exists $r > 0$ such that $B_r(\mathbf{a}) \subset \mathbb{R}^m \setminus \overline{A}$. Thus, $f(\mathbf{x}) = 0$ for all $x \in B_r(\mathbf{a})$ so $\lim_{\mathbf{x} \to \mathbf{a}} f(\mathbf{x}) = 0$.

(ii) Let $A = B_1(\mathbf{0}) \cup \{\mathbf{x}_0\} \subset \mathbb{R}^m$, where $x_0 \in \mathbb{R}^m$, $\|\mathbf{x}_0\| = 2$. Then $\mathbf{x}_0 \in \partial A$ and observe that $f = 0$ on $B_1(\mathbf{x}_0) \setminus \{\mathbf{x}_0\}$ which implies $\lim_{\mathbf{x} \to \mathbf{x}_0} f(\mathbf{x}) = 0$. Thus, $\mathbf{x}_0$ is a boundary point of A at which f has limit. This comes from the fact that A is not open and $\mathbf{x}_0$ is an isolated point of A.

6. (i) Using (3.3) and the inequality $|\sin u| \leq |u|$ for all $u \in \mathbb{R}$ we find

$$|f(x, y)| \leq \frac{|\sin(xy^2)|}{x^2 + y^2} \leq \frac{|xy^2|}{x^2 + y^2}$$

$$\leq |x| \cdot \underbrace{\frac{y^2}{x^2 + y^2}}_{\leq 1} \leq |x| \to 0 \quad \text{as } (x, y) \to (0, 0).$$

Thus, by Corollary 3.12 we deduce $\lim_{(x,y) \to (0,0)} f(x, y) = 0$.

(ii) Using (3.3) we estimate

$$|f(x, y)| \leq \frac{|x|^3 + |y|^3}{x^2 + y^2} \leq \underbrace{\frac{x^2}{x^2 + y^2}}_{\leq 1} \cdot |x| + \underbrace{\frac{y^2}{x^2 + y^2}}_{\leq 1} \cdot |y|$$

$$\leq |x| + |y| \to 0 \quad \text{as } (x, y) \to (0, 0).$$

Hence, Corollary 3.12 implies $\lim_{(x,y) \to (0,0)} f(x, y) = 0$.

(iii) Along the x axis, we have

$$f(x,y) = f(x,0) = 0 \to 0 \quad \text{as } (x,y) \to 0.$$

Along the parabola $y = x^2$, we have

$$f(x,y) = f(x,x^2) = \frac{\sin(x^2)}{3x^2} \to \frac{1}{3} \quad \text{as } (x,y) \to (0,0).$$

Hence, f has no limit at $(0,0)$.

(iv) Along the x axis, we have

$$f(x,y) = f(x,0) = \frac{\ln(x^2+1)}{x^2} \to 1 \quad \text{as } (x,y) \to 0.$$

Along y axis, we have

$$f(x,y) = f(0,y) = \frac{\ln(y^2+1)}{2y^2} \to \frac{1}{2} \quad \text{as } (x,y) \to 0.$$

Hence, f has no limit at $(0,0)$.

(v) Using the inequality $|\sin u| \leq |u|$ for all $u \in \mathbb{R}$ we find

$$|f(x,y)| = \frac{|\sin(xy)|}{\sqrt{x^2+y^2}}$$

$$\leq \frac{|xy|}{\sqrt{x^2+y^2}} = |x| \cdot \underbrace{\sqrt{\frac{y^2}{x^2+y^2}}}_{\leq 1}$$

$$\leq |x| \to 0 \quad \text{as } (x,y) \to 0.$$

Hence, by Corollary 3.12 we deduce $\lim\limits_{(x,y)\to(0,0)} f(x,y) = 0.$

(vi) Observe first that $f(x,y) = 1 + \dfrac{x^2y + xy^2}{x^2+y^2}$. Thus, by (3.3) we may estimate

$$|f(x,y) - 1| \leq \frac{x^2}{x^2+y^2} \cdot |y| + \frac{y^2}{x^2+y^2} \cdot |x|$$

$$\leq |x| + |y| \to 0 \quad \text{as } (x,y) \to 0.$$

Hence, by Corollary 3.12 we derive $\lim\limits_{(x,y)\to(0,0)} f(x,y) = 0.$

7. (i) Let first $k = 1$. Along the x axis, we have

$$f(x,y) = f(x,0) = 0 \to 0 \quad \text{as } (x,y) \to (0,0).$$

Along the line $y = x$, we have

$$f(x,y) = f(x,x) = \frac{x^4}{x^4 + 3x^6} = \frac{1}{1 + 3x^2} \to 1 \quad \text{as } (x,y) \to (0,0).$$

Thus, f has no limit at $(0,0)$ if $k = 1$.

Assume now $k = 2$. As above, along the x axis we obtain that $f(x,y) \to 0$. Along the line $y = x^{2/3}$ we find

$$f(x,y) = f(x,x^{2/3}) = \frac{x^4}{x^4 + 3x^4} = \frac{1}{4} \to \frac{1}{4} \quad \text{as } (x,y) \to (0,0).$$

Hence, f has no limit at $(0,0)$ if $k = 2$.

(ii) In light of inequality (3.4) in Remark 3.13 in which we take $A = x^2$, $B = y^3$, we have

$$x^2 |y|^3 \le \frac{1}{2}(x^4 + y^6) \quad \text{for all } x, y \in \mathbb{R}.$$

Thus, since $k > 2$ we may estimate

$$|f(x,y)| = \frac{x^2 |y|^3}{x^4 + 3y^6} \cdot |x|^{k-2} \le \underbrace{\frac{x^2 |y|^3}{x^4 + y^6}}_{\le 1/2} \cdot |x|^{k-2}$$

$$\le \frac{|x|^{k-2}}{2} \to 0 \quad \text{as } (x,y) \to (0,0).$$

Thus, by Corollary 3.12 we conclude $\displaystyle\lim_{(x,y)\to(0,0)} f(x,y) = 0$.

8. (i) By direct calculations, we have

$$2x^6 + y^{12} - 3x^4 y^4 = (2x^6 - 2x^4 y^4) + (y^{12} - x^4 y^4)$$
$$= (x^2 - y^4)^2 (2x^2 + y^4) \ge 0.$$

(ii) Using the above inequality, we have

$$|f(x,y)| = \underbrace{\frac{x^4 y^4}{2x^6 + y^{12}}}_{\le 1/3} \cdot |y| \le \frac{|y|}{3} \to 0 \text{ as } (x,y) \to (0,0).$$

Now Corollary 3.12 yields the conclusion. Due to the specific expression of the numerator, we cannot apply the inequalities in Remark 3.13. However, thanks to the inequality (i) we were able to estimate $|f(x,y)|$ and prove that f tends to zero at $(0,0)$.

9. (i) Along the line $x = y = z$, we have $f(x, y, z) = f(x, x, x) \to 3$ as $(x, y, z) \to (0, 0, 0)$. Along the x axis (where we have $y = z = 0$) we find $f(x, y, z) = f(x, 0, 0) \to 1$ as $(x, y, z) \to (0, 0, 0)$. Thus, f has no limit at $(0, 0, 0)$.

(ii) Note that $(x + y + z)^2 = x^2 + y^2 + z^2 + 2xy + 2yz + 2zx$. Using inequality (3.4) of Remark 3.13, we have

$$2xy \le x^2 + y^2, \quad 2yz \le y^2 + z^2, \quad 2zx \le z^2 + x^2.$$

Thus,

$$(x + y + z)^2 \le 3(x^2 + y^2 + z^2).$$

This shows that $|f(x, y, z)| \le 3|z| \to 0$ as $(x, y, z) \to (0, 0, 0)$, and in light of Corollary 3.12, it follows that f has limit zero at $(0, 0, 0)$.

10. Using the Two-path test for the nonexistence of a limit (see Theorem 3.3) we deduce that if

$$(a, b) = (0, 0), (1, 0), (0, 1), (1, 1), (2, 0), (0, 2), (2, 1), (3, 0), (4, 0),$$

then the function f has no limit at $(0, 0)$. Assume now that (a, b) is none of the above pairs.

Case 1: $b = 0$. Then, we must have $a \ge 5$ and then, using inequality (3.3) in the Remark 3.13, one has

$$|f(x, y)| = \underbrace{\frac{x^4}{x^4 + y^2}}_{\le 1} \cdot |x|^{a-4} \le |x|^{a-4} \to 0 \quad \text{as } (x, y) \to (0, 0).$$

Case 2: $b = 1$. Then $a \ge 3$ and by inequality (3.4) in the Remark 3.13 one has $x^2|y| \le \frac{1}{2}(x^4 + y^2)$ which yields

$$|f(x, y)| = \underbrace{\frac{x^2|y|}{x^4 + y^2}}_{\le 1/2} \cdot |x|^{a-2} \le \frac{|x|^{a-2}}{2} \to 0 \quad \text{as } (x, y) \to (0, 0).$$

Case 3: $b = 2$. Then $a \ge 1$ and by inequality (3.3) in the Remark 3.13 one has

$$|f(x, y)| = \underbrace{\frac{y^2}{x^4 + y^2}}_{\le 1} \cdot |x|^a \le |x|^a \to 0 \quad \text{as } (x, y) \to (0, 0).$$

Case 4: $b \ge 3$. By inequality (3.3) in the Remark 3.13 one has

$$|f(x, y)| = \underbrace{\frac{y^2}{x^4 + y^2}}_{\le 1} \cdot |x|^a |y|^{b-2} \le |x|^a |y|^{b-2} \to 0 \quad \text{as } (x, y) \to (0, 0).$$

The above four cases and Corollary 3.12 show that f has limit at $(0,0)$.

11. Along $x = 0$, we have $f(x,y) = f(0,y) = 0 \to 0$ as $(x,y) \to (0,-1)$.
    Along $y = x - 1$, we have $f(x,y) = f(x,x-1) = \frac{(x+c-1)^2}{2x}$. Thus, if f has limit as $(x,y) \to (0,-1)$, then $f(x,x-1)$ has limit as $x \to 0$. This implies $c - 1 = 0$, otherwise the limit does not exist. Hence $c = 1$. If $c = 1$ we can check by (3.4) that

$$|f(x,y)| = \underbrace{\frac{(y+1)^2}{x^2 + (y+1)^2}}_{\leq 1} \cdot |x| \leq |x| \to 0 \qquad \text{as } (x,y) \to (0,-1).$$

By Corollary 3.12 it follows that $f(x,y) \to 0$ as $(x,y) \to (0,-1)$.

12. We apply the Squeeze Principle by noting that

$$|(fg)(\mathbf{x})| \leq M|f(\mathbf{x})| \to 0 \qquad \text{as } \mathbf{x} \to \mathbf{a}.$$

13. Let $(a,b) \in \mathbb{R}^2$. If (a,b) lies outside of x and y axis (that is, $a \neq 0$ and $b \neq 0$), then $f(x,y) = 0$ for all $(x,y) \in B_R(a,b)$, where $R = \min\{|a|,|b|\} > 0$. Thus, taking a sequence $\{(x_n,y_n)\} \subset \mathbb{R}^2 \setminus \{(a,b)\}$ that converges to (a,b), we have for $n \geq N$ (with $N \geq 1$ large) that $(x_n,y_n) \in B_R(a,b)$. Hence, $f(x_n,y_n) = 0 \to 0 = f(a,b)$ as $n \to \infty$. This shows that f is continuous at all points in the plane that lie outside of x and y axis.

    Assume now that (a,b) lies on the x axis, so $b = 0$. Along the x axis (so $y = 0$), we have $f(x,y) = f(x,0) = 1 \to 1$ as $(x,y) \to (a,0)$. Along the line $y = x - a$, we have $f(x,y) = f(x,x-a) = 0 \to 0$ as $(x,y) \to (a,0)$. This shows that f has no limit at $(a,b) = (a,0)$, and thus, f is not continuous at this point.

    If (a,b) lies on the y axis, that is $a = 0$, a similar approach as above shows that f has no limit at (a,b), so, it is discontinuous at this point.

14. (i) Observe that $|f(x,y)| \leq |x|$ for all $(x,y) \in \mathbb{R}^2$. Using Corollary 3.12 it follows that $\displaystyle\lim_{(x,y)\to(0,0)} f(x,y) = 0 = f(0,0)$, so f is continuous at $(0,0)$. Also $\displaystyle\lim_{t\to 0} h(x) = \lim_{t\to 0} 1 = 1$.

    (ii) We have

$$(h \circ f)(x,y) = h(f(x,y))$$

$$= \begin{cases} 0 & \text{if } f(x,y) = (0,0) \\ 1 & \text{if } f(x,y) \neq (0,0) \end{cases}$$

$$= \begin{cases} 0 & \text{if } x \leq 0 \\ 1 & \text{if } x > 0. \end{cases}$$

Along the semiline $y = x$, $x > 0$, we have $(h \circ f)(x, y) = 1 \to 1$ as $(x, y) \to (0, 0)$ and along the semiline $y = x$, $x < 0$, we have $(h \circ f)(x, y) = 0 \to 0$ as $(x, y) \to (0, 0)$. Thus, $h \circ f$ has no limit at $(0, 0)$.

This fact does not contradict Theorem 3.22(iii) since we do not fulfil all the hypotheses in this result: observe that f is continuous at $(0, 0)$ while the function h only has limit at zero but it is not continuous at this point.

15. (i) Observe that $x^2 - xy + y^2 = \frac{1}{2}\left(x^2 + (x - y)^2 + y^2\right) \geq 0$ so f is defined on the whole plane $\mathbb{R}^2$. The function f is the composition of two elementary functions, namely $(x, y) \mapsto x^2 - xy + y^2$ and $t \mapsto \sqrt{t}$. By Theorem 3.22, it follows that f is continuous on $\mathbb{R}^2$.

(ii) Since $-1 \leq \cos(xy) \leq 1$, we have $2 + \cos(xy) > 0$ for all $(x, y) \in \mathbb{R}^2$, so f is defined on the whole $\mathbb{R}^2$. f is the quotient of composition of elementary functions, so f is continuous on $\mathbb{R}^2$.

(iii) The domain of f is $D = \{(x, y) \in \mathbb{R}^2 : |x| > |y|\}$. On D the function f is continuous as the composition of two elementary functions (logarithmic and polynomial).

16. Let $c \in \mathbb{R}$ and $S = \{\mathbf{x} \in D : f(\mathbf{x}) = c\}$ be the level set of f at level c. We want to show $S = \bar{S}$. If S is empty this is clearly true. Assume in the following that S is nonempty and let $\mathbf{a} \in \bar{S}$. By Theorem 1.19, there exists a sequence $\{\mathbf{a}_n\} \subset S$ that converges to $\mathbf{a}$. Since $\mathbf{a}_n \in S$, one has $f(\mathbf{a}_n) = c$. Now, using the fact that f is continuous one has $f(\mathbf{a}) = \lim_{n \to \infty} f(\mathbf{a}_n) = c$, so $\mathbf{a} \in S$, and thus $\bar{S} \subset S$. Since we also have $S \subset \bar{S}$, it follows that $S = \bar{S}$, that is, S is closed.

17. (i) Let $h : (0, \infty) \to \mathbb{R}$, $h(t) = \ln t$. Then, by Theorem 3.22, it follows that $h \circ e^f = f$ is continuous at (a, b).

(ii) The above argument does not work in this case since $\cos : \mathbb{R} \to [-1, 1]$ has no inverse. As a counterexample let us take

$$f : \mathbb{R}^2 \to \mathbb{R}, \quad f(x, y) = \begin{cases} 0 & \text{if } (x, y) \neq (0, 0) \\ 2\pi & \text{if } (x, y) = (0, 0). \end{cases}$$

Then f is not continuous at $(0, 0)$ but $\cos(f)$ is.

18. We follow verbatim the proof of Theorem 3.25 by using the $\varepsilon - \delta$ characterization of the limit in Theorem 3.2.

19. Using the inequality $|\sin y| \leq |y|$ and (3.4) of Remark 3.13, we have

$$|f(x, y)| \leq \frac{|x - 1||\sin^3 y|}{(x - 1)^2 + y^4} \leq \underbrace{\frac{|x - 1|y^2}{(x - 1)^2 + y^4}}_{\leq 1/2} \cdot |y| \leq \frac{|y|}{2}.$$

This yields $\lim\limits_{(x,y)\to(1,0)} f(x,y) = 0$ and the extension of f is

$$\tilde{f} : \mathbb{R}^2 \to \mathbb{R}, \quad \tilde{f}(x,y) = \begin{cases} \dfrac{(x-1)\sin^3 y}{(x-1)^2 + y^4} & \text{if } (x,y) \ne (1,0) \\ 0 & \text{if } (x,y) = (1,0). \end{cases}$$

20. (i) f is a quotient of continuous functions, so it is continuous on its domain of definition.

(ii) f has a continuous extension at $(2,0)$ if and only if $\lim\limits_{(x,y)\to(2,0)} f(x,y)$ exists as a real number.

Along the line $x = 2$, we have

$$f(x,y) = f(2,y) = \frac{(2+b)y + (2a+c)}{|y|}$$

which must have a limit as $y \to 0$. This yields $2 + b = 0$ and $2a + c = 0$. We can convince ourselves about this fact by taking sided limits as $y \to 0^+$ and $y \to 0^-$. In particular, $c = -2a$.

Along the line $y = 0$, we have

$$f(x,y) = f(x,0) = \frac{ax + c}{|x-2|} = \frac{a(x-2)}{|x-2|}$$

which has a limit as $x \to 2$ if and only if $a = 0$. Hence, $a = c = 0$ and $b = -2$. For these values, we have

$$f(x,y) = \frac{y(x-2)}{\sqrt{(x-2)^2 + y^2}}$$

and then

$$|f(x,y)| = \underbrace{\frac{|x-2|}{\sqrt{(x-2)^2 + y^2}}}_{\le 1} \cdot |y| \le |y| \to 0 \qquad \text{as } (x,y) \to (2,0).$$

By Corollary 3.12 one has $f(x,y) \to 0$ as $(x,y) \to (2,0)$.

21. (i) Along the line $y = 0$, we have

$$f(x,y) = f(x,0) = \frac{\sin(x^2)}{x^2} \to 1 \qquad \text{as } (x,y) \to (0,0).$$

Along the parabola $y = x^2$, we have

$$f(x,x^2) = \frac{\sin(x^2 + x^8)}{x^2\sqrt{2}} = \frac{\sin(x^2 + x^8)}{x^2 + x^8} \cdot \frac{x^2 + x^8}{x^2\sqrt{2}} \to \frac{1}{\sqrt{2}}$$

as $(x, y) \to (0, 0)$. Hence, f has no limit at the origin, and thus, has no continuous extension at this point.

(ii) Using inequality (3.4), we have

$$|f(x, y, z)| = \frac{|xy|}{x^2 + y^2 + z^2} \cdot |z|$$

$$\leq \underbrace{\frac{|xy|}{x^2 + y^2}}_{\leq 1/2} \cdot |z| \leq \frac{|z|}{2} \to 0 \quad \text{as } (x, y, z) \to (0, 0, 0).$$

Corollary 3.12 yields f has limit 0 at the origin and this gives the value of the continuous extension of f at $(0, 0, 0)$.

(iii) $\lim\limits_{\mathbf{x} \to \mathbf{0}} f(\mathbf{x}) = \lim\limits_{t = \|\mathbf{x}\| \to 0^+} \frac{e^{-1/t}}{t} = 0$, so f has a continuous extension at $\mathbf{0}$.

22. We need to impose $\lim_{(x,y) \to (0,0)} f(x, y)$ exists and is a real number. If $a \leq 1$, then along the semiline $y = x$, $x > 0$, we have

$$f(x, y) = f(x, x) = \frac{\sin\left(x^a \sin x\right)}{2x^2}$$

$$= \frac{1}{2} \cdot \frac{\sin\left(x^a \sin x\right)}{x^a \sin x} \cdot \frac{\sin x}{x} \cdot x^{a-1}.$$

Hence,

$$\lim_{x \to 0, x > 0} f(x, x) = \begin{cases} +\infty & \text{if } a < 1 \\ \dfrac{1}{2} & \text{if } a = 1. \end{cases}$$

Also, along the line $y = 0$ one has $f(x, y) \to 0$ as $(x, y) \to (0, 0)$. In conclusion, if $0 < a \leq 1$, then the limit of $f(x, y)$ as $(x, y) \to (0, 0)$ does not exist. On the other hand, if $a > 1$, then using $|\sin z| \leq |z|$ for all $z \in \mathbb{R}$, one has

$$|f(x, y)| \leq \frac{|x|^a |\sin y|}{x^2 + y^2} \leq \frac{|x|^a |y|}{x^2 + y^2}$$

$$= \underbrace{\frac{|xy|}{x^2 + y^2}}_{\leq 1/2} \cdot |x|^{a-1} \leq \frac{|x|^{a-1}}{2} \to 0 \quad \text{as } (x, y) \to (0, 0).$$

Hence, for $a > 1$ the function f has a continuous extension at $(0, 0)$.

23. (i) By (3.4) we deduce

$$|f(x, y) - 2| \leq \frac{x^2 y^4}{x^4 + y^4} = \underbrace{\frac{x^2 y^2}{x^4 + y^4}}_{\leq 1/2} \cdot y^2 \leq \frac{y^2}{2} \to 0 \quad \text{as } (x, y) \to (0, 0).$$

Using Corollary 3.12 we find $\lim\limits_{(x,y)\to(0,0)} f(x,y) = 2$ and thus f has a continuous extension at the origin.

(ii) We take $f_1, f_2 : D \to \mathbb{R}$ by

$$f_1(x,y) = 2 + \begin{cases} \dfrac{x^2 y^4}{x^4 + y^4} & \text{if } (x,y) \neq (0,0) \\ 0 & \text{if } (x,y) = (0,0), \end{cases}$$

$$f_2(x,y) = 2 + \begin{cases} \dfrac{x^2 y^2}{x^4 + y^4} & \text{if } (x,y) \neq (0,0) \\ 0 & \text{if } (x,y) = (0,0). \end{cases}$$

Using the above method it can be shown that f_1 has limit 2 at $(0,0)$ and thus admits a continuous extension at the origin. On the other hand, along the lines $y = 0$ and $y = x$ the function f_2 approaches different values as $(x,y) \to (0,0)$. Thus, by Theorem 3.3, f_2 has no limit at $(0,0)$ and thus no continuous extension of f_2 at this point exists.

12.4   Solutions Chapter 4

1. (i) $\nabla f(x,y) = \left(e^x \cos(x^2 y) - 2xy e^x \sin(x^2 y), -x^2 e^x \sin(x^2 y) \right)$;

(ii) $\nabla f(x,y,z) = \left(-\dfrac{2xz^4 e^{-3y}}{(1 + x^2 z^4)^2}, -\dfrac{3e^{-3y}}{1 + x^2 z^4}, -\dfrac{4x^2 z^3 e^{-3y}}{(1 + x^2 z^4)^2} \right)$;

(iii) We have

$$\frac{\partial f}{\partial x}(x,y,z) = \frac{2xyz}{x^2 + z^2 + 1}$$

$$\frac{\partial f}{\partial y}(x,y,z) = z\ln(x^2 + z^2 + 1)$$

$$\frac{\partial f}{\partial z}(x,y,z) = y\ln(x^2 + z^2 + 1) + \frac{2yz^2}{x^2 + z^2 + 1}.$$

(iv) We have

$$\frac{\partial f}{\partial x_1}(x_1, x_2, x_2, x_4) = -\frac{2x_1 x_2 x_3}{(1 + x_1^2 + x_4^2)^2}$$

$$\frac{\partial f}{\partial x_2}(x_1, x_2, x_2, x_4) = \frac{x_3}{1 + x_1^2 + x_4^2}$$

$$\frac{\partial f}{\partial x_3}(x_1, x_2, x_2, x_4) = \frac{x_2}{1 + x_1^2 + x_4^2}$$

$$\frac{\partial f}{\partial x_4}(x_1, x_2, x_2, x_4) = -\frac{2x_2 x_3 x_4}{(1 + x_1^2 + x_4^2)^2}.$$

2. (i) We impose $\frac{2x}{x^2+y^2} = \frac{2}{5}$ and $\frac{2y}{x^2+y^2} = -\frac{6}{5}$. Dividing these equations we derive $\frac{x}{y} = -\frac{1}{3}$, so $y = -3x$. Using this last equation in $\frac{2x}{x^2+y^2} = \frac{2}{5}$ we are led to $5x = 10x^2$, so $x = 0$ or $x = \frac{1}{2}$. If $x = 0$ we find $y = 0$ but $(x, y) = (0,0)$ does not belong to the domain of definition of f so we discard this solution. We finally obtain $(x, y) = \left(\frac{1}{2}, -\frac{3}{2}\right)$.

(ii) $\nabla f(x, y, z) = (2, -1, -1)$ yields $\left(2xy, x^2 - \frac{z}{y^2}, \frac{1}{y}\right) = (2, -1, -1)$ and then $(x, y, z) = (-1, -1, 2)$.

3. Let $\mathbf{a} = (a_1, a_2, \ldots, a_m)$. Assume first $\mathbf{a} \neq \mathbf{0}$. Then

$$\lim_{t \to 0} \frac{f(\mathbf{0} + t\mathbf{e}_j) - f(\mathbf{0})}{t} = \lim_{t \to 0} \frac{\|\mathbf{a} + t\mathbf{e}_j\| - \|\mathbf{a}\|}{t}$$
$$= \lim_{t \to 0} \frac{\|\mathbf{a} + t\mathbf{e}_j\|^2 - \|\mathbf{a}\|^2}{t(\|\mathbf{a} + t\mathbf{e}_j\| + \|\mathbf{a}\|)}$$
$$= \lim_{t \to 0} \frac{(a_j + t)^2 - a_j^2}{t(\|\mathbf{a} + t\mathbf{e}_j\| + \|\mathbf{a}\|)}$$
$$= \lim_{t \to 0} \frac{2a_j + t}{\|\mathbf{a} + t\mathbf{e}_j\| + \|\mathbf{a}\|} = \frac{2a_j}{2\|\mathbf{a}\|} = \frac{a_j}{\|\mathbf{a}\|}.$$

Hence, if $\mathbf{a} \neq \mathbf{0}$, then $\frac{\partial f}{\partial x_j}(\mathbf{a}) = \frac{a_j}{\|\mathbf{a}\|}$. If $\mathbf{a} = \mathbf{0}$, we have

$$\lim_{t \to 0} \frac{f(\mathbf{0} + t\mathbf{e}_j) - f(\mathbf{0})}{t} = \lim_{t \to 0} \frac{\sqrt{t^2}}{t} = \lim_{t \to 0} \frac{|t|}{t}$$

which, by taking sided limits, does not exist.

4. By Definition 4.1, we have

$$\frac{\partial f}{\partial x}(0,0,0) = \lim_{t \to 0} \frac{f(0 + t, 0, 0) - f(0,0,0)}{t} = \lim_{t \to 0} \frac{t}{t} = 1,$$

$$\frac{\partial f}{\partial y}(0,0,0) = \lim_{t \to 0} \frac{f(0, 0 + t, 0) - f(0,0,0)}{t} = \lim_{t \to 0} t = 0,$$

$$\frac{\partial f}{\partial z}(0,0,0) = \lim_{t \to 0} \frac{f(0, 0, 0 + t) - f(0,0,0)}{t} = \lim_{t \to 0} \frac{1}{t} \quad \text{does not exist.}$$

5. (i) The argument is the same as in the proof of Theorem 4.7. The crucial part is the estimate (4.6) which in case of a function of $m \geq 2$ variables reads

$$|f(\mathbf{x}) - f(\mathbf{a})| \leq \sum_{j=1}^{m} |f(\mathbf{x}_j) - f(\mathbf{x}_{j-1})|$$

where $\mathbf{x}_j = (x_1, x_2, \ldots, x_j, a_{j+1}, \ldots, a_m)$ for all $1 \leq j \leq m$.

(ii) Consider $f : \mathbb{R} \to \mathbb{R}$, $f(t) = \begin{cases} t & \text{if } t \le 0 \\ t+1 & \text{if } t > 0 \end{cases}$, which satisfies the hypotheses in Theorem 4.7 and yet f is not continuous at $t = 0$.

6. To fix the ideas, assume $m = 3$ and that $\nabla f = 0$ on the ball $B_r(a,b,c)$, $r > 0$. As in the proof of Theorem 4.7, we have

$$|f(x,y,z) - f(a,b,c)| \le |f(x,y,z) - f(a,y,z)| \\ + |f(a,y,z) - f(a,b,z)| + |f(a,b,z) - f(a,b,c)|. \quad (12.6)$$

By the Mean Value Theorem there exists α, β, γ such that

$$f(x,y,z) - f(a,y,z) = \frac{\partial f}{\partial x}(\alpha, y, z)(x - a) = 0,$$

$$f(a,y,z) - f(a,b,z) = \frac{\partial f}{\partial y}(a, \beta, z)(y - b) = 0,$$

$$f(a,b,z) - f(a,b,c) = \frac{\partial f}{\partial z}(a, b, \gamma)(z - c) = 0.$$

Using these estimates in (12.6) we find $f(x,y,z) = f(a,b,c)$.

7. Let $g(x,y) = f(x,y) - 4x - 5y$. Then, $\nabla g = 0$ on $\mathbb{R}^2$ which by Exercise 6 above yields $g(x,y) =$ constant. Hence $f(x,y) = 4x + 5y + c$ and from $f(0,1) = 3$ we find $c = -2$, so $f(x,y) = 4x + 5y - 2$.

8. Use the same argument as in the proof of Theorem 4.7. From the estimate (4.6) and the Mean Value Theorem we find

$$|f(x,y) - f(a,b)| \le |f(x,y) - f(a,y)| + |f(a,y) - f(a,b)|$$

$$\le \left| \frac{\partial f}{\partial x}(c_{x,y}, y) \right| |x - a| + \left| \frac{\partial f}{\partial y}(a, d_y) \right| |y - b|,$$

where $c_{x,y} \in \mathbb{R}$ lies between a and x, and $d_y \in \mathbb{R}$ lies between y and b. We may now consider $g : \mathbb{R}^2 \to \mathbb{R}$,

$$g(x,y) = \begin{cases} \left| \frac{\partial f}{\partial x}(c_{x,y}) \right| |x - a| + \left| \frac{\partial f}{\partial y}(d_y) \right| |y - b| & \text{if } (x,y) \ne (a,b) \\ 0 & \text{if } (x,y) = (a,b). \end{cases}$$

As $(x,y) \to (a,b)$ we have $c_{x,y} \to a$ and $d_y \to b$, so that $g(x,y) \to 0$. By Corollary 3.12 we find $f(x,y) \to f(a,b)$ as $(x,y) \to (a,b)$ which shows that f has limit at (a,b) and thus, it admits a continuous extension at this point.

9. (i) $z = x - y$, $(x,y,z) = (1 + t, 1 - t, -t)$, $t \in \mathbb{R}$.

(ii) $z + x = 0$, $(x,y,z) = (\pi - t, 0, -\pi - t)$, $t \in \mathbb{R}$.

(iii) $z = 4e(x + y)$, $(x,y,z) = (2 + 4et, -1 + 4et, 4e - t)$, $t \in \mathbb{R}$.

(iv) $z + \frac{2}{3} = \frac{5}{27}(x+2) + \frac{4}{27}(y-2)$, $(x,y,z) = (-2 + \frac{5t}{27}, 2 + \frac{4t}{27}, -\frac{2}{3} - t)$, $t \in \mathbb{R}$.

10. Let σ be a plane that passes through M, N and is tangent to $z = f(x, y)$ at a point $P(a, b, c)$ where $c = f(a, b) = 3a^2 - b^2$. The equation of the tangent plane is $\sigma : z - 3a^2 + b^2 = 6a(x-a) - 2b(y-b)$. Imposing M, $N \in \sigma$ we find $(a, b) = (1, -2)$, $(-1, 4)$ so the planes are $\sigma_1 : 6x + 4y - z = -1$, $\sigma_2 : 6x + 8y + z = 13$.

11. Let $P(a, b, c)$ be such a point on the surface. Then, the normal line through P to the surface has the direction $\mathbf{n} = (3a^2 - 2b, -2a, -1)$. The normal vectors to the two planes are $\mathbf{n_1} = (2, 1, 0)$ and $\mathbf{n_2} = (1, 0, 2)$ respectively. Since $\mathbf{n}$ is orthogonal to both $\mathbf{n_1}$ and $\mathbf{n_2}$ we impose next $\mathbf{n} \bullet \mathbf{n_1} = 0$ and $\mathbf{n} \bullet \mathbf{n_2} = 0$ to obtain $a = 2$, $b = 5$. Finally, $c = a^3 - 2ab$, so the point is $P(2, 5, -12)$.

12. We know that the equation of the tangent plane to $z = f(x, y)$ at $(-1, 3)$ is

$$z - f(-1, 3) = \frac{\partial f}{\partial x}(-1, 3)(x + 1) + \frac{\partial f}{\partial y}(-1, 3)(y - 3)$$

and on the other hand this is given by $z = -2x + \frac{3}{2}y + \frac{7}{2}$. Identifying the coefficients, one has $\frac{\partial f}{\partial x}(-1, 3) = -2$, $\frac{\partial f}{\partial y}(-1, 3) = \frac{3}{2}$ and then $f(-1, 3) = 10$.

13. Let $P(a, b, c)$ be a point on the surface $z = f(x, y)$, so $c = f(a, b)$. The normal vector to the tangent plane at P is

$$\mathbf{n} = (e^{b^2}, 2abe^{b^2}, -1)$$

while a normal vector to the plane $z = 0$ is $\mathbf{u} = (0, 0, 1)$. Thus, the angle θ between the tangent plane and the plane $z = 0$ satisfies

$$\cos\theta = \frac{|\mathbf{n} \bullet \mathbf{u}|}{\|\mathbf{n}\| \cdot \|\mathbf{u}\|} = \frac{1}{\|\mathbf{n}\| \cdot \|\mathbf{u}\|}.$$

Note that $\|\mathbf{u}\| = 1$ and

$$\|\mathbf{n}\| = \sqrt{e^{2b^2} + (2ab)^2 e^{2b^2} + 1} \geq \sqrt{e^{2b^2} + 1} \geq \sqrt{2}.$$

Hence, $\cos\theta \leq 1/\sqrt{2}$, that is $\pi/4 \leq \theta \leq \pi/2$.

14. (i) Let $f(x, y) = x^2 + 4y^2$. Then $f(1, -2) = 17 > 8$, so P is exterior to the paraboloid.

(ii) Let $R(a, b, c)$ be a point on the paraboloid at which the tangent plane passes through $P(1, -2, 8)$. Then $c = f(a, b) = a^2 + 4b^2$ and the tangent plane at $R(a, b, c)$ has the equation

$$\sigma : z - a^2 - 4b^2 = 2a(x - a) + 8b(y - b).$$

Since $P \in \sigma$ we deduce $8 - a^2 - 4b^2 = 2a(1 - a) + 8b(-2 - b)$, that is,

$$(a - 1)^2 + 4(b + 2)^2 = 9. \tag{12.7}$$

Let now $b \in \mathbb{R}$ be such that $-\frac{7}{2} < b < -\frac{1}{2}$. Then $-\frac{3}{2} < b + 2 < \frac{3}{2}$ and thus $4(b+2)^2 \in [0,9)$. Take now $a = -1 + \sqrt{9 - 4(b+2)^2}$ and it is easy to see that (a, b) satisfies (12.7), that is, P lies on the tangent plane to the paraboloid at $R(a, b, c)$.

(iii) If $Q(-2, 1, 14) \in \sigma$, then $14 - a^2 - 4b^2 = 2a(-2 - a) + 8b(1 - b)$ which combined with (12.7) yields $a = 4b - 1$. Using this last equality in (12.7) we deduce $20b^2 = -11$ with no real solutions. Thus, none of the tangent planes that pass through P contain also Q.

15. Let $P(a, b, c)$ be a required point. The directions of the normal line to the tangent plane at P and the plane $x - z = 5$ are given by

$$\mathbf{n}_1 = \left(b^2 e^{ab}, e^{ab}(1 + ab), -1 \right) \quad \text{and} \quad \mathbf{n}_2 = (1, 0, -1).$$

Thus, the two planes are parallel if there exists $\lambda \in \mathbb{R}$ such that $\mathbf{n}_1 = \lambda \mathbf{n}_2$, which amounts to $\lambda = 1$ and

$$\begin{cases} b^2 e^{ab} = 1 \\ (1 + ab)e^{ab} = 0 \end{cases} \implies \begin{cases} ab = -1 \\ b^2 e^{ab} = 1. \end{cases}$$

We find

$$(a, b, c) = \left(\frac{1}{\sqrt{e}}, -\sqrt{e}, -\frac{1}{\sqrt{e}} \right), \quad \left(-\frac{1}{\sqrt{e}}, \sqrt{e}, \frac{1}{\sqrt{e}} \right).$$

16. (i) By Theorem 4.10(i), the vector $\mathbf{n}_1 = (\nabla f(a, b), -1)$ is orthogonal to the tangent plane to the surface $z = f(x, y)$ at (a, b, c). Similarly, the vector $\mathbf{n}_2 = (\nabla g(a, b), -1)$ is orthogonal to the tangent plane to the surface $z = g(x, y)$ at (a, b, c). If the two tangent planes at (a, b, c) are orthogonal, so are the vectors $\mathbf{n}_1$ and $\mathbf{n}_2$. From $\mathbf{n}_1 \bullet \mathbf{n}_2 = 0$ we reach the conclusion.

(ii) The equation of the tangent plane to $z = f(x, y)$ and $z = g(x, y)$ at (a, b, c) are given by

$$z - c = \frac{\partial f}{\partial x}(a, b)(x - a) + \frac{\partial f}{\partial y}(a, b)(y - b),$$

$$z - c = \frac{\partial g}{\partial x}(a, b)(x - a) + \frac{\partial g}{\partial y}(a, b)(y - b).$$

Since the tangent plane coincide, we identify the coefficients of x and y in the above equalities to deduce $\nabla f(a, b) = \nabla g(a, b)$.

17. (i) By Definition 4.1, we have $\frac{\partial f}{\partial x}(0, 0) = \frac{\partial f}{\partial y}(0, 0) = 0$. Thus,

$$\frac{f(x, y) - \{ f(0, 0) + \nabla f(0, 0) \bullet (x, y) \}}{\| (x, y) \|} = x \to 0 \quad \text{as } (x, y) \to (0, 0),$$

so f is differentiable at $(0,0)$.

(ii) Assume there are points $(a, b, f(a, b))$ on the graph of f at which the tangent plane is orthogonal to the plane $x - 2y - mz = 6$ for some $m > 0$. Then, the corresponding normal vectors, respectively

$$\mathbf{n}_1 = \left(\frac{\partial f}{\partial x}(a, b), \frac{\partial f}{\partial y}(a, b), -1 \right) \quad \text{and} \quad \mathbf{n}_2 = (1, -2, -m)$$

are orthogonal. This yields

$$\frac{\partial f}{\partial x}(a, b) - 2\frac{\partial f}{\partial y}(a, b) + m = 0. \tag{12.8}$$

If $(a, b) = (0, 0)$, then (12.8) cannot hold since

$$\frac{\partial f}{\partial x}(0, 0) = \frac{\partial f}{\partial y}(0, 0) = 0 \quad \text{and} \quad m > 0.$$

If $(a, b) \neq (0, 0)$ we find from (12.8) that

$$\frac{2a^2 + b^2}{\sqrt{a^2 + b^2}} - \frac{2ab}{\sqrt{a^2 + b^2}} + m = 0,$$

and finally

$$2a^2 + b^2 + m\sqrt{a^2 + b^2} = 2ab. \tag{12.9}$$

This is an equation with the unknowns $a, b \in \mathbb{R}$. Observe that by inequality (3.4) in Remark 3.13 one has $a^2 + b^2 \geq 2ab$ which from (12.9) forces $(a, b) = (0, 0)$, impossible. Thus, there are no points on the graph of f at which the tangent plane is orthogonal to the plane $x - 2y - mz = 6$.

18. (i) Using Definition 4.1 we compute $\frac{\partial f}{\partial x}(0, 0) = \frac{\partial f}{\partial x}(0, 0) = 1$. For $(x, y) \in \mathbb{R}^2 \setminus \{(0, 0)\}$ let

$$g(x, y) = \frac{f(x, y) - \{f(0, 0) + \nabla f(0, 0) \bullet (x, y)\}}{\|(x, y)\|}.$$

Then, f is differentiable at $(0, 0)$ if and only if $\lim_{(x,y) \to (0,0)} g(x, y) = 0$.

We have

$$g(x, y) = \frac{2xy(x + y)}{(x^2 + y^2)\sqrt{x^2 + y^2}}$$

and $g(x, y) \to \sqrt{2}$ as $(x, y) \to 0$ along the semiline $y = x$, $x > 0$ while $g(x, y) \to -\sqrt{2}$ as $(x, y) \to 0$ along the semiline $y = x$, $x < 0$. Hence, g has no limit at $(0, 0)$ which implies that f is not differentiable at this point.

(ii) By Definition 4.1, we have $\frac{\partial f}{\partial x}(0,0) = \frac{\partial f}{\partial x}(0,0) = 0$. Then, f is differentiable at $(0,0)$ if and only if $\lim\limits_{(x,y)\to(0,0)} g(x,y) = 0$, where for $(x,y) \in \mathbb{R}^2 \setminus \{(0,0)\}$ we set

$$g(x,y) = \frac{f(x,y) - \{f(0,0) + \nabla f(0,0) \bullet (x,y)\}}{\|(x,y)\|}.$$

Note that

$$g(x,y) = \frac{x^2 y^2}{(x^4 + y^4)\sqrt{x^2 + y^2}} \sin x \sin y$$

and by the standard estimates in Remark 3.13, we have

$$|g(x,y)| = \underbrace{\frac{x^2 y^2}{x^4 + y^4}}_{\leq 1/2} \cdot \underbrace{\frac{|x|}{\sqrt{x^2 + y^2}}}_{\leq 1} \cdot \underbrace{\left|\frac{\sin x}{x}\right|}_{\leq 1} \cdot |\sin y|$$

$$\leq \frac{|\sin y|}{2} \to 0 \quad \text{as } (x,y) \to (0,0).$$

This implies that $g(x,y) \to 0$ as $(x,y) \to (0,0)$, so f is differentiable at $(0,0)$.

19. (i) Let $(a,b) \in \mathbb{R}^2$. If $|a| \neq |b|$, then (a,b) lies outside the lines $y = x$ and $y = -x$ (which we call the first and second diagonal of the plane). Then, for $t \in \mathbb{R}$ close to zero, we have $(a+t,b)$ and $(a,b+t)$ lie outside of these diagonals as well. We thus have

$$\frac{\partial f}{\partial x}(a,b) = \lim_{t\to 0} \frac{f(a+t,b) - f(a,b)}{t} = \lim_{t\to 0} \frac{(a+t)b - ab}{t} = b$$

and in the same manner $\frac{\partial f}{\partial y}(a,b) = a$.

Suppose now that (a,b) lies on either the line $y = x$ or $y = -x$. Assume first that $a = b$. Then, for $t \in \mathbb{R}$ close to zero, we have

$$\frac{f(a+t,b) - f(a,b)}{t} = \frac{(a+t)a - 0}{t} = \frac{a^2}{t} + a.$$

Thus, the above quotient has no limit as $t \to 0$, unless $a = b = 0$. Thus, $\frac{\partial f}{\partial x}(a,a)$ doe not exist for any $a \in \mathbb{R}^2 \setminus \{0\}$. A similar argument applies for $\frac{\partial f}{\partial y}(a,a)$. If $b = -a$ we proceed similarly to reach the same conclusion. Thus

$$\frac{\partial f}{\partial x}(x,y) = \begin{cases} y & \text{if } |x| \neq |y| \text{ or } (x,y) = (0,0) \\ \text{does not exist} & \text{if } |x| = |y| \neq 0, \end{cases}$$

and similarly

$$\frac{\partial f}{\partial y}(x,y) = \begin{cases} x & \text{if } |x| \neq |y| \text{ or } (x,y) = (0,0) \\ \text{does not exist} & \text{if } |x| = |y| \neq 0. \end{cases}$$

(ii) Observe that

$$\left| \frac{f(x,y) - \{f(0,0) + \nabla f(0,0) \bullet (x,y)\}}{\|(x,y)\|} \right| \leq \frac{|xy|}{\sqrt{x^2 + y^2}} \leq |y| \to 0$$

as $(x,y) \to (0,0)$. By Corollary 3.12 we conclude that f is differentiable at $(0,0)$.

(iii) f cannot be differentiable at $(-1,1)$ because the partial derivatives do not exist at this point.

20. (i) f is continuous on $\mathbb{R}^2 \setminus \{(0,b) : b \in \mathbb{R}\}$. Also,

$$|f(x,y)| \leq |x|^{4/3} \to 0 \qquad \text{as } (x,y) \to (0,b),$$

so f is continuous on $\mathbb{R}^2$.

(ii) If $(x,y) \neq (0,b)$, we have

$$\frac{\partial f}{\partial x}(x,y) = \frac{4}{3}x^{1/3} \sin \frac{y}{x} - x^{-2/3} y \cos \frac{y}{x}$$

$$\frac{\partial f}{\partial y}(x,y) = x^{1/3} \cos \frac{y}{x}.$$

For $(x,y) = (0,b)$, $b \in \mathbb{R}$, we use the Definition 4.12 to obtain

$$\frac{\partial f}{\partial x}(0,b) = \lim_{t \to 0} \frac{f(t,b) - f(0,b)}{t} = \lim_{t \to 0} t^{1/3} \sin \frac{b}{t} = 0$$

$$\frac{\partial f}{\partial y}(0,b) = \lim_{t \to 0} \frac{f(0,b+t) - f(0,b)}{t} = 0.$$

(iii) Let $(a_n, b_n) = \left(\frac{1}{2\pi n}, 1 \right)$ and $(\bar{a}_n, \bar{b}_n) = \left(\frac{1}{2\pi n + \frac{\pi}{2}}, 1 \right)$, $n \geq 1$. Then $(a_n, b_n) \to (0,1)$ and $(\bar{a}_n, \bar{b}_n) \to (0,1)$ as $n \to \infty$. Let us note that as $n \to \infty$, we have

$$\frac{\partial f}{\partial y}(a_n, b_n) = a_n^{1/3} \to \infty \qquad \text{and} \qquad \frac{\partial f}{\partial y}(\bar{a}_n, \bar{b}_n) = \bar{a}_n^{1/3} \cdot 0 \to 0.$$

By Definition 3.1, we conclude that $\dfrac{\partial f}{\partial y}$ has no limit at $(0,1)$.

(iv) Since f is differentiable on $\mathbb{R}^2 \setminus \{(0,b) : b \in \mathbb{R}\}$, it remains to discuss its differentiability at $(x,y) = (0,b)$, $b \in \mathbb{R}$. Let

$$g(x,y) = \frac{f(x,y) - \{f(0,b) + \nabla f(0,b) \bullet (x - 0, y - b)\}}{\|(x,y) - (0,b)\|}$$

$$= \frac{x^{4/3} \sin \frac{y}{x}}{\sqrt{x^2 + (y-b)^2}} \qquad \text{for all } (x,y) \in \mathbb{R}^2 \setminus \{(0,b)\}.$$

Then

$$|g(x,y)| = \underbrace{\frac{|x|}{\sqrt{x^2 + (y-b)^2}}}_{\leq 1} \cdot \left| x^{1/3} \sin \frac{y}{x} \right|$$

$$\leq |x|^{1/3} \to 0 \quad \text{as } (x,y) \to (0,b).$$

Hence, g has limit 0 at $(0,b)$ and thus f is differentiable at this point. In conclusion, f is differentiable on $\mathbb{R}^2$.

21. (i) Letting $(x,y) = (0,0)$ in the main inequality we find $f(0,0) = 0$.

(ii) Letting $y = 0$ in the main inequality we find $|f(x,0)| \leq x^2$ for all $|x| < 1$. Thus,

$$\left| \frac{f(t,0) - f(0,0)}{t} \right| \leq |t| \quad \text{for all } |t| < 1.$$

Hence

$$\frac{\partial f}{\partial x}(0,0) = \lim_{t \to 0} \frac{f(t,0) - f(0,0)}{t} = 0,$$

and in the same way, we have $\dfrac{\partial f}{\partial y}(0,0) = 0$.

(iii) Observe first that for all $(x,y) \in B_1(0,0)$, we have

$$|f(x,y)| \leq |f(x,y) - \sin(xy)| + |\sin(xy)| \\ \leq x^2 + y^2 + |\sin(xy)|. \tag{12.10}$$

For $(x,y) \in B_1(0,0) \setminus \{(0,0)\}$ denote

$$g(x,y) = \frac{f(x,y) - \{f(0,0) + \nabla f(0,0) \bullet (x,y)\}}{\|(x,y)\|}.$$

Using (12.10) we estimate

$$|g(x,y)| = \frac{|f(x,y)|}{\|(x,y)\|}$$

$$\leq \frac{x^2 + y^2 + |\sin(xy)|}{\|(x,y)\|}$$

$$= \sqrt{x^2 + y^2} + \frac{|\sin(xy)|}{\|(x,y)\|}$$

$$\leq \sqrt{x^2 + y^2} + \frac{|xy|}{\|(x,y)\|} \quad (\text{use } |\sin u| \leq |u|)$$

$$\leq \sqrt{x^2 + y^2} + \underbrace{\sqrt{\frac{x^2}{x^2 + y^2}}}_{\leq 1} \cdot |y|$$

$$\leq \sqrt{x^2 + y^2} + |y| \to 0 \quad \text{as } (x,y) \to (0,0).$$

By Corollary 3.12 it follows that $g(x,y) \to 0$ as $(x,y) \to (0,0)$, hence f is differentiable at $(0,0)$.

22. By the differentiability of f at $\mathbf{a}$, we have

$$\lim_{\mathbf{x} \to \mathbf{a}} \frac{f(\mathbf{x}) - \{f(\mathbf{a}) + \nabla f(\mathbf{a}) \bullet (\mathbf{x} - \mathbf{a})\}}{\|\mathbf{x} - \mathbf{a}\|} = 0.$$

Letting $\mathbf{x} = a + t\mathbf{h}$, we deduce

$$\lim_{t \to 0} \frac{f(\mathbf{a} + t\mathbf{h}) - \{f(\mathbf{a}) + \nabla f(\mathbf{a}) \bullet (t\mathbf{h})\}}{\|t\mathbf{h}\|} = 0.$$

This yields

$$\lim_{t \to 0} \frac{g(t) - g(0) - t\nabla f(\mathbf{a}) \bullet \mathbf{h}}{|t|} = 0.$$

In particular

$$\lim_{t \to 0} \left| \frac{g(t) - g(0) - t\nabla f(\mathbf{a}) \bullet \mathbf{h}}{t} \right| = 0$$

which further implies

$$\lim_{t \to 0} \frac{g(t) - g(0)}{t} = \nabla f(\mathbf{a}) \bullet \mathbf{h},$$

so g is differentiable at 0 and $g'(0) = \nabla f(\mathbf{a}) \bullet \mathbf{h}$.

23. Let $r > 0$ be such that $B_r(\mathbf{a}) \subset D$. Define $\eta : B_r(\mathbf{0}) \to \mathbb{R}$, $\eta(\mathbf{h}) = f(\mathbf{a} + \mathbf{h}) - f(\mathbf{a})$. Then, $f(\mathbf{a} + \mathbf{h}) = f(\mathbf{a}) + \eta(\mathbf{h})$ for all $\mathbf{h} \in B_r(\mathbf{0})$ and $\frac{\eta(\mathbf{h})}{\|\mathbf{h}\|} \to 0$ as $\mathbf{h} \to \mathbf{0}$. By Theorem 4.16, it follows that f is differentiable at $\mathbf{a}$ and $\nabla f(\mathbf{a}) = \mathbf{0}$.

24. (i) From the main inequality we deduce (by letting $\mathbf{y} = \mathbf{h} \in \mathbb{R}^m$):

$$|f(\mathbf{x} + \mathbf{h}) - f(\mathbf{x}) - \mathbf{x} \bullet \mathbf{h}| \le \|\mathbf{h}\|^2 \quad \text{for all } \mathbf{x}, \mathbf{h} \in \mathbb{R}^m.$$

Thus, if we define $\eta(\mathbf{h}) = f(\mathbf{x} + \mathbf{h}) - f(\mathbf{x}) - \mathbf{x} \bullet \mathbf{h}$, we have that

$$f(\mathbf{x} + \mathbf{h}) = f(\mathbf{x}) + \mathbf{x} \bullet \mathbf{h} + \eta(\mathbf{h}) \quad \text{for all } \mathbf{x}, \mathbf{h} \in \mathbb{R}^m,$$

and $\eta : \mathbb{R}^m \to \mathbb{R}$ satisfies

$$\frac{\eta(\mathbf{h})}{\|\mathbf{h}\|} \to 0 \quad \text{as } \|\mathbf{h}\| \to 0.$$

We thus fulfil the hypotheses in Theorem 4.16, so f is differentiable at any $\mathbf{x} \in \mathbb{R}^m$ and $\nabla f(\mathbf{x}) = \mathbf{x}$.

(ii) Let $g(\mathbf{x}) = f(\mathbf{x}) - \frac{\|\mathbf{x}\|^2}{2}$. Since $\nabla f(\mathbf{x}) = \mathbf{x}$ we deduce $\nabla g(\mathbf{x}) = 0$ for all $\mathbf{x} \in \mathbb{R}^m$.

(iii) Since $\nabla g(\mathbf{x}) = 0$ for all $\mathbf{x} \in \mathbb{R}^m$, we may use Exercise 6 to derive $g = c$ is a constant function, so $f(x) = \frac{\|\mathbf{x}\|^2}{2} + c$ for all $\mathbf{x} \in \mathbb{R}^m$.

25. Let $L = \lim\limits_{(x,y)\to(a,b)} \dfrac{\partial f}{\partial x}(x,y)$ and $M = \lim\limits_{(x,y)\to(a,b)} \dfrac{\partial f}{\partial y}(x,y)$. As in the proof of Theorem 4.18 we estimate

$$\begin{aligned}
\Big|f(x,y) &- f(a,b) - (L,M) \bullet (x-a, y-b)\Big| \\
&\leq \Big|f(x,y) - f(a,y) - L(x-a)\Big| \\
&\quad + \Big|f(a,y) - f(a,b) - M(y-b)\Big|.
\end{aligned} \tag{12.11}$$

Apply next the Mean Value Theorem to $g(t) = f(t,y) - Lt$ over the interval with endpoints at a and x to obtain

$$\Big|f(x,y) - f(a,y) - L(x-a)\Big| = \left|\dfrac{\partial f}{\partial x}(c_{x,y}, y) - L\right| |x-a|,$$

where $c_{x,y} \in \mathbb{R}$ lies between a and x. Similarly,

$$\Big|f(a,y) - f(a,b) - M(y-b)\Big| = \left|\dfrac{\partial f}{\partial y}(a, d_y) - M\right| |y-b|,$$

where $d_y \in \mathbb{R}$ lies between b and y. Thus, from (12.11) we derive

$$\begin{aligned}
\Big|f(x,y) &- f(a,b) - (L,M) \bullet (x-a, y-b)\Big| \\
&\leq \left|\dfrac{\partial f}{\partial x}(c_{x,y}, y) - L\right| |x-a| + \left|\dfrac{\partial f}{\partial y}(a, d_y) - M\right| |y-b| \\
&\leq \left(\left|\dfrac{\partial f}{\partial x}(c_{x,y}, y) - L\right| + \left|\dfrac{\partial f}{\partial y}(a, d_y) - M\right|\right) \|(x,y) - (a,b)\|.
\end{aligned}$$

Let $\eta : \mathbb{R}^2 \to \mathbb{R}$ be given by

$$\eta(x,y) = \dfrac{f(x,y) - f(a,b) - (L,M) \bullet (x-a, y-b)}{\|(x,y) - (a,b)\|}$$

if $(x,y) \neq (a,b)$, and $\eta(a,b) = 0$. By the above estimates, we have

$$|\eta(x,y)| \leq \left|\dfrac{\partial f}{\partial x}(c_{x,y}, y) - L\right| + \left|\dfrac{\partial f}{\partial y}(a, d_y) - M\right|.$$

Since $c_{x,y} \to a$, $d_y \to b$ as $(x,y) \to (a,b)$, by property (ii), we have $\eta(x,y) \to 0$ as $(x,y) \to (a,b)$. Note also that

$$f(x,y) = f(a,b) + (L,M) \bullet (x-a, y-b) + \eta(x,y)\|(x,y) - (a,b)\|.$$

This means that f satisfies the conditions in Theorem 4.16 around the point (a,b). By this result f is differentiable at (a,b).

26.   (i) If $k < 3$, then

$$\lim_{t \to 0} \frac{f(t,0) - f(0,0)}{t} = \lim_{t \to 0} t^{k-3} \quad \text{does not exist.}$$

This means that if $k < 3$, then $\dfrac{\partial f}{\partial x}(0,0)$ does not exist, so f is not differentiable at $(0,0)$. Assume $k = 3$. Then, $\dfrac{\partial f}{\partial x}(0,0) = 1$, $\dfrac{\partial f}{\partial x}(0,0) = 0$ and f is differentiable at $(0,0)$ if and only if the function

$$g(x,y) = \frac{f(x,y) - \{f(0,0) + \nabla f(0,0) \bullet (x - 0, y - 0)\}}{\|(x,y)\|}$$

$$= -\frac{x^2 y}{(x^2 + y^2)\sqrt{x^2 + y^2}}$$

has limit at $(0,0)$. Along the semilines $y = x$, $x > 0$ and $y = x$, $x < 0$ we see that g approaches different values, so f is not differentiable at the origin.

(ii) Assume $k > 3$. Then, for $(x,y) \neq (0,0)$, we compute

$$\frac{\partial f}{\partial x}(x,y) = \frac{(k-2)x^{k+1} + kx^{k-1}y^2}{(x^2 + y^2)^2}, \quad \frac{\partial f}{\partial y}(x,y) = -\frac{2x^k y}{(x^2 + y^2)^2}.$$

This shows that the partial derivatives are continuous on $\mathbb{R}^2 \setminus \{(0,0)\}$. By direct calculation $\dfrac{\partial f}{\partial x}(0,0) = \dfrac{\partial f}{\partial x}(0,0) = 0$. Furthermore,

$$\left| \frac{\partial f}{\partial x}(x,y) \right| \leq k \frac{|x|^{k+1} + |x|^{k-1}y^2}{(x^2 + y^2)^2} = k \frac{|x|^{k-1}}{x^2 + y^2}$$

$$= k \underbrace{\frac{x^2}{x^2 + y^2}}_{\leq 1} \cdot |x|^{k-3} \to 0,$$

$$\left| \frac{\partial f}{\partial y}(x,y) \right| \leq \underbrace{\frac{2|xy|}{x^2 + y^2}}_{\leq 1} \cdot \underbrace{\frac{x^2}{x^2 + y^2}}_{\leq 1} \cdot |x|^{k-3} \to 0,$$

as $(x,y) \to (0,0)$. Hence $\dfrac{\partial f}{\partial x}$ and $\dfrac{\partial f}{\partial y}$ are continuous on $\mathbb{R}^2$. By Theorem 4.18, it follows that f is differentiable at $(0,0)$.

27.   (i) $J_F(x,y) = \begin{pmatrix} 2x \cos y & -x^2 \sin y \\ -ye^{-x} & e^{-x} \\ yx^{y-1} & x^y \ln x \end{pmatrix}$

(ii) $J_F(x, y, z) = \begin{pmatrix} 0 & \dfrac{1}{2\sqrt{y}} & -\dfrac{1}{2\sqrt{z}} \\ \dfrac{1}{yz} & -\dfrac{x}{y^2 z} & -\dfrac{x}{yz^2} \end{pmatrix}$

(iii) $J_F(x, y, z) = \begin{pmatrix} -\dfrac{xy}{\sqrt{(x^2+z^2)^3}} & \dfrac{1}{\sqrt{x^2+z^2}} & -\dfrac{yz}{\sqrt{(x^2+z^2)^3}} \\[2ex] y^z x^{y^z-1} & zy^{z-1}x^{y^z}\ln x & y^z x^{y^z}(\ln x)(\ln y) \\[2ex] \dfrac{1}{x} & \dfrac{1}{y\ln(y\ln z)} & \dfrac{1}{z\ln z \cdot \ln(y\ln z)} \end{pmatrix}$

28. Let $F = (f_1, f_2, f_3)$. We have

$$\frac{\partial f_1}{\partial x}(x, y) = 2xy^3 - 2 \qquad \frac{\partial f_1}{\partial y}(x, y) = 3x^2 y^2 - 6y$$

$$\frac{\partial f_2}{\partial x}(x, y) = \sin y \qquad \frac{\partial f_2}{\partial y}(x, y) = x\cos y - \sin y$$

$$\frac{\partial f_3}{\partial x}(x, y) = -e^{x-y-1} \qquad \frac{\partial f_3}{\partial y}(x, y) = e^{x-y-1}.$$

We first integrate $\dfrac{\partial f_1}{\partial x}(x, y) = 2xy^3 - 2$ with respect to x variable to deduce $f_1(x, y) = x^2 y^3 - 2x + g(y)$. Then, $\dfrac{\partial f_1}{\partial y}(x, y) = 3x^2 y^2 - 6y$ yields $g'(y) = -6y$ so $g(y) = -3y^2 + C$ and then $f_1(x, y) = x^2 y^3 - 2x - 3y^2 + C$. From $f_1(1, 0) = 1$ we find $C = 3$ and $f_1(x, y) = x^2 y^3 - 2x - 3y^2 + 3$. In the same way we find $f_2(x, y) = x\sin y + \cos y$ and $f_3(x, y) = -e^{x-y-1}$.

29. Let $F = (f_1, f_2)$. We have

$$\frac{\partial f_1}{\partial x}(x, y, z) = y \qquad \frac{\partial f_2}{\partial x}(x, y, z) = z\cos(xz)$$

$$\frac{\partial f_1}{\partial y}(x, y, z) = x + 2yz \qquad \frac{\partial f_2}{\partial y}(x, y, z) = 1$$

$$\frac{\partial f_1}{\partial z}(x, y, z) = y^2 - 6z^2 \qquad \frac{\partial f_2}{\partial z}(x, y, z) = x\cos(xz).$$

We integrate $\dfrac{\partial f_1}{\partial x}(x, y, z) = y$ with respect to x variable to find

$$f_1(x, y, z) = xy + g(y, z).$$

Then,

$$\begin{cases} \dfrac{\partial f_1}{\partial y}(x,y,z) = x + 2yz \\[2mm] \dfrac{\partial f_1}{\partial z}(x,y,z) = y^2 - 6z^2 \end{cases} \implies \begin{cases} \dfrac{\partial g}{\partial y}(y,z) = 2yz \\[2mm] \dfrac{\partial g}{\partial z}(y,z) = y^2 - 6z^2. \end{cases}$$

From here we deduce

$$g(y,z) = y^2 z + h(z) \quad \text{and} \quad h'(z) = -6z^2.$$

Thus, $h(z) = -2z^3 + C$ and $f_1(x,y,z) = xy + y^2 z - 2z^3 + C$. From $f_1(0,0,1) = -2$, we have $C = 0$, so $f_1(x,y,z) = xy + y^2 z - 2z^3$. Similarly we find $f_2(x,y,z) = y + \sin(xz)$.

30. $J_F(x,y) = \begin{pmatrix} 1 & 0 \\ 0 & 1 \\ \dfrac{\partial \phi}{\partial x} & \dfrac{\partial \phi}{\partial y} \end{pmatrix}$ and $J_G(x,y,z) = \begin{pmatrix} 1 & 0 & 0 \\ 0 & 1 & 0 \\ z\dfrac{\partial \phi}{\partial x} & z\dfrac{\partial \phi}{\partial y} & \phi \end{pmatrix}.$

31. $F = (f_1, \ldots, f_m)$ where

$$f : \mathbb{R}^m \to \mathbb{R}, \quad f(\mathbf{x}) = \begin{cases} x_i \|\mathbf{x}\|^\alpha \ln \|\mathbf{x}\| & \text{if } \mathbf{x} \neq \mathbf{0} \\ 0 & \text{if } \mathbf{x} = \mathbf{0}. \end{cases}$$

Then

$$\frac{\partial f_i}{\partial x_j}(\mathbf{0}) = \lim_{t \to 0} \frac{f_i(t\mathbf{e}_j) - f(\mathbf{0})}{t} = \lim_{t \to 0} \delta_{ij}|t|^\alpha \ln |t| = 0,$$

where δ_{ij} denotes the Kronecker symbol.

For $\mathbf{x} \neq \mathbf{0}$, we have

$$\frac{\partial f_i}{\partial x_j}(\mathbf{x}) = \delta_{ij}\|\mathbf{x}\|^\alpha \ln \|\mathbf{x}\| + (1 + \alpha \ln \|\mathbf{x}\|)x_i x_j \|\mathbf{x}\|^{\alpha-2}.$$

Then, the functions $\dfrac{\partial f_i}{\partial x_j}(\mathbf{x})$ are continuous on $\mathbb{R}^m \setminus \{\mathbf{0}\}$. Furthermore,

$$\left| \frac{\partial f_i}{\partial x_j}(\mathbf{x}) \right| \leq \|\mathbf{x}\|^\alpha |\ln \|\mathbf{x}\|| + \left(1 + \alpha \big| \ln \|\mathbf{x}\| \big| \right) |x_i||x_j| \|\mathbf{x}\|^{\alpha-2}$$

$$\leq \|\mathbf{x}\|^\alpha \ln \|\mathbf{x}\| + \left(1 + \alpha \big| \ln \|\mathbf{x}\| \big| \right) \|\mathbf{x}\|^\alpha \to 0 \quad \text{as } \mathbf{x} \to \mathbf{0}.$$

This implies that the partial derivatives are continuous on $\mathbb{R}^m$ and by Theorem 4.18 we deduce f_i, and thus F, are differentiable at $\mathbf{0}$. The Jacobian of F is given by

$$J_F(\mathbf{x})_{ij} = \begin{cases} \delta_{ij}\|\mathbf{x}\|^\alpha \ln \|\mathbf{x}\| + (1 + \alpha \ln \|\mathbf{x}\|)x_i x_j \|\mathbf{x}\|^{\alpha-2} & \text{if } \mathbf{x} \neq \mathbf{0} \\ 0 & \text{if } \mathbf{x} = \mathbf{0}. \end{cases}$$

32. Let $F = (f_1, \ldots, f_p)$ and $G = (g_1, \ldots, g_p)$. Then $h(\mathbf{x}, \mathbf{y}) = \sum_{j=1}^{p} f_j(\mathbf{x}) g_j(\mathbf{y})$ which shows that h is the sum of differentiable functions, so h is also differentiable. Also

$$\frac{\partial h}{\partial x_i}(\mathbf{x}, \mathbf{y}) = \sum_{j=1}^{p} \frac{\partial f_j}{\partial x_i}(\mathbf{x}) g_j(\mathbf{y}) = \left(G(\mathbf{y}) \cdot J_F(\mathbf{x}) \right)_i,$$

$$\frac{\partial h}{\partial y_i}(\mathbf{x}, \mathbf{y}) = \sum_{j=1}^{p} \frac{\partial g_j}{\partial y_i}(\mathbf{y}) f_j(\mathbf{x}) = \left(F(\mathbf{y}) \cdot J_G(\mathbf{y}) \right)_i,$$

which yields the desired formula for the gradient of h.

33. (i) Let $F = (f_1, \ldots, f_p)$. For any $\mathbf{a}, \mathbf{x} \in \mathbb{R}^m$, we have

$$|f_i(\mathbf{x}) - f_i(\mathbf{a})| \leq \|F(\mathbf{x}) - F(\mathbf{a})\| \leq \|\mathbf{x} - \mathbf{a}\|^{\alpha}$$

so that $f_i(\mathbf{x}) \to f_i(\mathbf{a})$ as $\mathbf{x} \to \mathbf{a}$. Hence f_i, $1 \leq i \leq p$ are continuous, which is equivalent to the continuity of F.

(ii) For $1 \leq j \leq m$ and $t \in \mathbb{R}$, we have $|f_i(\mathbf{a} + t\mathbf{e}_j) - f_i(\mathbf{a})| \leq C|t|^{\alpha}$. Then, since $\alpha > 1$ we deduce

$$\left| \frac{f_i(\mathbf{a} + t\mathbf{e}_j) - f_i(\mathbf{a})}{t} \right| \leq C|t|^{\alpha - 1} \to 0 \qquad \text{as } t \to 0.$$

This yields $\dfrac{\partial f_i}{\partial x_j}(\mathbf{a}) = 0$ for all $1 \leq i \leq p$, $1 \leq j \leq m$. From Exercise 6 we deduce f_i is constant for all $1 \leq i \leq p$, so F is constant.

(iii) Take $F(\mathbf{x}) = \begin{cases} (x_1, x_2, \ldots, x_p) & \text{if } p \leq m \\ (x_1, x_2, \ldots, x_m, 0, \ldots, 0) & \text{if } p > m. \end{cases}$

(iv) The fact that $F(x) = \|\mathbf{x}\|$ is Lipschitz follows from triangle inequality (see Exercise 3(i) in Chapter 1). F is not differentiable at the origin by Exercise 3 of this chapter.

12.5   Solutions Chapter 5

1. $V = \frac{\pi r^2 h}{3}$, $\frac{dr}{dt} = 2$ and $\frac{dh}{dt} = -3$. By the Chain Rule, we have

$$\frac{dV}{dt} = \frac{\partial V}{\partial r}\frac{dr}{dt} + \frac{\partial V}{\partial h}\frac{dh}{dt} = \frac{2\pi}{3}rh\frac{dr}{dt} + \frac{\pi}{3}r^2\frac{dh}{dt} = \frac{4\pi}{3}rh - \pi r^2.$$

At $r = 5$ and $h = 12$ we find $\frac{dV}{dt} = 55\pi$ cm^3/s.

2. By Chain Rule, we have

$$\frac{dT}{dt} = \frac{\partial T}{\partial x}\frac{dx}{dt} + \frac{\partial T}{\partial y}\frac{dy}{dt} = e^{-\cos t}(\sin^2 t + \cos t) + \frac{4\cos t \sin t}{(1 + \sin^2 t)^2}.$$

3. $g : \mathbb{R} \setminus \{0\} \to \mathbb{R}^2$, $g(t) = (t, 6/t)$ is a parametrization of the curve $xy = 6$. Let $\phi = f \circ g$. Then by the Chain Rule one has

$$\frac{d\phi}{dt}(t) = \frac{\partial f}{\partial x}\left(t, \frac{6}{t}\right) - \frac{6}{t^2}\frac{\partial f}{\partial y}\left(t, \frac{6}{t}\right).$$

At $t = 3$, we have

$$\frac{d\phi}{dt}(3) = \frac{\partial f}{\partial x}(3, 2) - \frac{2}{3}\frac{\partial f}{\partial y}(3, 2) = 3.$$

4. We are required to find $\nabla g(x, y)$ at $(x, y) = (2, -2)$ where $g(x, y) = f\left(x, y, \frac{8}{xy}\right)$. Using the Chain Rule one has

$$\frac{\partial g}{\partial x} = \frac{\partial f}{\partial x}\left(x, y, \frac{8}{xy}\right) - \frac{8}{x^2y}\frac{\partial f}{\partial z}\left(x, y, \frac{8}{xy}\right)$$

$$\frac{\partial g}{\partial y} = \frac{\partial f}{\partial y}\left(x, y, \frac{8}{xy}\right) - \frac{8}{xy^2}\frac{\partial f}{\partial z}\left(x, y, \frac{8}{xy}\right).$$

At $(x, y) = (2, -2)$ we find $\nabla g(2, -2) = (-3, 4)$.

5. To find a parametric equation of the line ℓ let $z = t$. Then $x = 1 - t$ and $y = 2t - 3$. Let $g : \mathbb{R} \to \mathbb{R}^3$, be given by $g(t) = (-t + 1, 2t - 3, t)$ and $\phi = f \circ g$. By the Chain Rule one finds

$$\frac{d\phi}{dt}(t) = -\frac{\partial f}{\partial x}(-t + 1, 2t - 3, t) + 2\frac{\partial f}{\partial y}(-t + 1, 2t - 3, t)$$

$$+ \frac{\partial f}{\partial z}(-t + 1, 2t - 3, t).$$

At $t = 0$, we have $\dfrac{d\phi}{dt}(0) = 0$.

6. We have

$$\frac{\partial x}{\partial s} = 8te^{2s}, \quad \frac{\partial x}{\partial t} = 4e^{2s},$$

and

$$\frac{\partial y}{\partial s} = 4s, \quad \frac{\partial y}{\partial t} = -2t.$$

By Chain Rule we deduce

$$\begin{cases} \dfrac{\partial g}{\partial s} = \dfrac{\partial g}{\partial x}\dfrac{\partial x}{\partial s} + \dfrac{\partial g}{\partial y}\dfrac{\partial y}{\partial s} = 8te^{2s}\dfrac{\partial g}{\partial x} + 4s\dfrac{\partial g}{\partial y} \\[2mm] \dfrac{\partial g}{\partial t} = \dfrac{\partial g}{\partial x}\dfrac{\partial x}{\partial t} + \dfrac{\partial g}{\partial y}\dfrac{\partial y}{\partial t} = 4e^{2s}\dfrac{\partial g}{\partial x} - 2t\dfrac{\partial g}{\partial y}. \end{cases}$$

Hence,

$$\frac{\partial g}{\partial s} - 2t\frac{\partial g}{\partial t} = (4s + 4t^2)\frac{\partial g}{\partial y}.$$

7. By Chain Rule, we have $\nabla h(x,y) = (2xy f'(x^2 y), x^2 f'(x^2 y))$.
    Letting $(x,y) = (-1,1)$ we find

    $$\frac{\partial h}{\partial x}(-1,1) = -2f'(1) \quad \text{and} \quad \frac{\partial h}{\partial y}(-1,1) = f'(1)$$

    and the conclusion follows.

8. (i) By Chain Rule, we have

    $$\frac{\partial f}{\partial r} = \frac{\partial f}{\partial x} \cdot \frac{\partial x}{\partial r} + \frac{\partial f}{\partial y} \cdot \frac{\partial y}{\partial r} = 3(r^2 - s^2)\frac{\partial f}{\partial x} + 6rs\frac{\partial f}{\partial y}$$
    $$\frac{\partial f}{\partial s} = \frac{\partial f}{\partial x} \cdot \frac{\partial x}{\partial s} + \frac{\partial f}{\partial y} \cdot \frac{\partial y}{\partial s} = -6rs\frac{\partial f}{\partial x} + 3(r^2 - s^2)\frac{\partial f}{\partial y}.$$

    (ii) The conclusion follows by squaring the above two equalities.

9. We have

    $$\frac{\partial f}{\partial r} = \frac{\partial f}{\partial x}\frac{\partial x}{\partial r} + \frac{\partial f}{\partial y}\frac{\partial y}{\partial r} + \frac{\partial f}{\partial z}\frac{\partial z}{\partial r} = e^r \cos t \frac{\partial f}{\partial x} + e^r \sin t \frac{\partial f}{\partial y} + 2r\frac{\partial f}{\partial z}$$

    $$\frac{\partial f}{\partial t} = \frac{\partial f}{\partial x}\frac{\partial x}{\partial t} + \frac{\partial f}{\partial y}\frac{\partial y}{\partial t} + \frac{\partial f}{\partial z}\frac{\partial z}{\partial t} = -e^r \sin t \frac{\partial f}{\partial x} + e^r \cos t \frac{\partial f}{\partial y}.$$

10. By Chain Rule and Remark 5.5 we find

    $$\frac{\partial f}{\partial r} = \frac{\partial f}{\partial x}\cos\varphi\sin\theta + \frac{\partial f}{\partial y}\sin\varphi\sin\theta + \frac{\partial f}{\partial z}\cos\theta$$

    $$\frac{\partial f}{\partial \varphi} = -\frac{\partial f}{\partial x}r\sin\varphi\sin\theta + \frac{\partial f}{\partial y}r\cos\varphi\sin\theta$$

    $$\frac{\partial f}{\partial r} = \frac{\partial f}{\partial x}r\cos\varphi\cos\theta + \frac{\partial f}{\partial y}r\sin\varphi\cos\theta - \frac{\partial f}{\partial z}r\sin\theta.$$

11. (i) We impose $x + y \neq 0$, so $D = \{(x,y) \in \mathbb{R}^2 : y \neq -x\}$.

    (ii) Direct application of Chain Rule yields

    $$\frac{\partial \phi}{\partial x}(x,y) = \frac{y^2}{(x+y)^2}f'\left(\frac{xy}{x+y}\right), \quad \frac{\partial \phi}{\partial y}(x,y) = \frac{x^2}{(x+y)^2}f'\left(\frac{xy}{x+y}\right).$$

    From here the conclusion is immediate.

12. We have

    $$\frac{\partial \phi}{\partial x}(x,y) = f(x^3 - y^2) + 3x^3 f'(x^3 - y^2), \quad \frac{\partial \phi}{\partial y}(x,y) = -2yf'(x^3 - y^2).$$

    The conclusion follows by direct calculation.

13. Let $g(x, y, t) = (tx, ty)$. Then $\phi = f \circ g$ and by Chain Rule one has

$$\nabla\phi(x, y, t) = \nabla f(g(x, y, t)) \bullet J_g(x, y, t).$$

This yields

$$\frac{\partial \phi}{\partial x}(x, y, t) = t\frac{\partial f}{\partial x}(tx, ty)$$

$$\frac{\partial \phi}{\partial y}(x, y, t) = t\frac{\partial f}{\partial y}(tx, ty)$$

$$\frac{\partial \phi}{\partial t}(x, y, t) = x\frac{\partial f}{\partial x}(tx, ty) + y\frac{\partial f}{\partial y}(tx, ty).$$

Part (i) follows by letting $(x, y, t) = (0, 0, 0)$ while for part (ii) we let $(x, y, t) = (1, 1, 1)$ in the above equalities.

14. (i) Direct calculus!

(ii) Let $\phi(x, y, t) = f(tx, ty)$. By Chain Rule one has

$$\frac{\partial \phi}{\partial x}(x, y, t) = t\frac{\partial f}{\partial x}(tx, ty). \tag{12.12}$$

On the other hand $\phi(x, y, t) = t^n f(x, y)$ so

$$\frac{\partial \phi}{\partial x}(x, y, t) = t^n \frac{\partial f}{\partial x}(x, y). \tag{12.13}$$

Combining (12.12) and (12.13) above we are led to

$$\frac{\partial f}{\partial x}(tx, ty) = t^{n-1}\frac{\partial f}{\partial x}(x, y) \quad \text{for all } t, x, y \in \mathbb{R}, t \neq 0. \tag{12.14}$$

To show that (12.14) also holds for $t = 0$ we let $(x, y) = (0, 0)$ and $t = 2$ in (12.14) and have

$$\frac{\partial f}{\partial x}(0, 0) = 2^{n-1}\frac{\partial f}{\partial x}(0, 0) \implies \frac{\partial f}{\partial x}(0, 0) = 0.$$

Thus, (12.14) holds for all $x, y, t \in \mathbb{R}$, which shows that $\dfrac{\partial f}{\partial x}$ is homogeneous of degree $n - 1$. In a similar manner we find that $\dfrac{\partial f}{\partial y}$ is homogeneous of degree $n - 1$.

(iii) With the function ϕ defined above, we have

$$\frac{\partial \phi}{\partial t}(x, y, t) = x\frac{\partial f}{\partial x}(tx, ty) + y\frac{\partial f}{\partial y}(tx, ty). \tag{12.15}$$

On the other hand $\phi(x, y, t) = t^n f(x, y)$ so

$$\frac{\partial \phi}{\partial t}(x, y, t) = nt^{n-1}f(x, y). \tag{12.16}$$

Now, (12.15) and (12.16) yield

$$t\frac{\partial f}{\partial x}(tx, ty) + t\frac{\partial f}{\partial y}(tx, ty) = nt^{n-1}f(x, y) \quad \text{for all } t, x, y \in \mathbb{R}.$$

The conclusion follows by letting $t = 1$ in the above equality.

15. Let $g(\mathbf{x}) = \|x\|$. Then, by Theorem 4.18, one has that g is differentiable on $\mathbb{R}^m \setminus \{\mathbf{0}\}$ and g is not differentiable at $\mathbf{0}$ (see Exercise 3 in Chapter 4). By the Chain Rule one has that $f = \phi \circ g$ is differentiable on $\mathbb{R}^m \setminus \{\mathbf{0}\}$ and

$$\nabla f(\mathbf{x}) = \phi'(\|\mathbf{x}\|)\nabla g(\mathbf{x}) = \phi'(\|\mathbf{x}\|)\frac{\mathbf{x}}{\|\mathbf{x}\|} \quad \text{for all } \mathbf{x} \in \mathbb{R}^m \setminus \{\mathbf{0}\}.$$

16. Taking $(r, \theta) = \left(1, \frac{\pi}{2}\right)$ in the implicit equation we find $z\left(1, \frac{\pi}{2}\right) = 1$. Implicit differentiation yields

$$\begin{cases} \dfrac{\partial z}{\partial r}\cos\theta + 2z\dfrac{\partial z}{\partial r} = \dfrac{\sin\theta}{z} - \dfrac{r\sin\theta}{z^2}\dfrac{\partial z}{\partial r} \\[4mm] \dfrac{\partial z}{\partial \theta}\cos\theta - z\sin\theta + 2z\dfrac{\partial z}{\partial \theta} = \dfrac{r\cos\theta}{z} - \dfrac{r\sin\theta}{z^2}\dfrac{\partial z}{\partial \theta}. \end{cases}$$

Letting $(r, \theta) = \left(1, \frac{\pi}{2}\right)$ we find $\dfrac{\partial z}{\partial r}\left(1, \frac{\pi}{2}\right) = \dfrac{\partial z}{\partial \theta}\left(1, \frac{\pi}{2}\right) = \dfrac{1}{3}$.

17. We want to find $\dfrac{\partial z}{\partial x}(-1, 2)$ and $\dfrac{\partial z}{\partial y}(-1, 2)$. We differentiate in the implicit equation to find

$$\begin{cases} y - y\dfrac{\partial z}{\partial x} + 2z + 2x\dfrac{\partial z}{\partial x} = 0 \\[4mm] x - z - y\dfrac{\partial z}{\partial y} + 2x\dfrac{\partial z}{\partial y} = 0. \end{cases}$$

At $(x, y, z) = (-1, 2, -2)$ we find

$$\frac{\partial z}{\partial x}(-1, 2) = -\frac{1}{2} \quad \text{and} \quad \frac{\partial z}{\partial y}(-1, 2) = \frac{1}{4}.$$

18. (i) Let $(x, y) \in (-\infty, 0) \times \mathbb{R}$ and consider

$$\Phi : \mathbb{R} \to \mathbb{R}, \quad \Phi(z) = e^{x+y-z} - x^2 z + y^3.$$

Then $\Phi'(z) = -e^{x+y-z} - x^2 < 0$ for all $z \in \mathbb{R}$. This shows that Φ is a decreasing function. Also, $\lim_{z \to \pm\infty} \Phi(z) = \mp\infty$, so Φ is bijective. Hence, for all $(x, y) \in (-\infty, 0) \times \mathbb{R}$ there exists a unique $z = z(x, y)$ such that $\Phi(z) = 8$, and so, the function $z = z(x, y)$ is unique.

(ii) Letting $(x, y) = (-1, 2)$ in the implicit equation we find $e^{1-z} - z = 0$. Since $z = 1$ satisfies the equation and we know from part (i) that this equation has a unique solution, it follows that $z(-1, 2) = 1$.

Differentiate with respect to x and y in the implicit equation to find

$$\begin{cases} e^{x+y-z}\left(1 - \dfrac{\partial z}{\partial x}\right) - 2xz - x^2\dfrac{\partial z}{\partial x} = 0 \\[2mm] e^{x+y-z}\left(1 - \dfrac{\partial z}{\partial y}\right) - x^2\dfrac{\partial z}{\partial y} + 3y^2 = 0. \end{cases}$$

Letting $(x, y) = (-1, 2)$ we find

$$\frac{\partial z}{\partial x}(-1, 2) = \frac{3}{2} \quad \text{and} \quad \frac{\partial z}{\partial y}(-1, 2) = \frac{13}{2}.$$

19. Implicit differentiation yields

$$\begin{cases} e^u \dfrac{\partial u}{\partial x} = 3\left(\dfrac{\partial v}{\partial x} + 1\right) \\[2mm] e^v \dfrac{\partial v}{\partial x} = 2\dfrac{\partial u}{\partial x} \end{cases} \quad \text{and} \quad \begin{cases} e^u \dfrac{\partial u}{\partial y} = 3\dfrac{\partial v}{\partial y} \\[2mm] e^v \dfrac{\partial v}{\partial y} = 2\left(\dfrac{\partial u}{\partial y} + 1\right). \end{cases}$$

Since $u(0, 0) = v(0, 0) = 0$ we easily deduce $\nabla u(0, 0) = \left(-\frac{3}{5}, -\frac{6}{5}\right)$ and $\nabla v(0, 0) = \left(-\frac{6}{5}, -\frac{2}{5}\right)$.

20. Let $\mathbf{x} \in B_r(\mathbf{a})$. By the Mean Value Theorem 5.10 there exists $\mathbf{c} \in (\mathbf{a}, \mathbf{x})$ such that $f(\mathbf{x}) - f(\mathbf{a}) = \nabla f(\mathbf{c}) \bullet (\mathbf{x} - \mathbf{a}) = 0$. Thus, $f(\mathbf{x}) = f(\mathbf{a})$ for all $\mathbf{x} \in B_r(\mathbf{a})$, which shows that f is constant. Compare this short and elegant solution with the solution of Exercise 6 in Chapter 4.

21. $f(\mathbf{a}) = 1$, $f(\mathbf{b}) = -1$ and by the Mean Value Theorem there exists $\mathbf{c} = (c_1, c_2) \in (\mathbf{a}, \mathbf{b})$ such that $f(\mathbf{b}) - f(\mathbf{a}) = \nabla f(\mathbf{c}) \bullet (\mathbf{b} - \mathbf{a})$. This means that

$$-2 = (2c_1 - c_2, -c_1) \bullet (2, 2) \implies c_2 = c_1 + 1.$$

This is exactly the equation of segment $(\mathbf{a}, \mathbf{b})$ for $c_1 \in (-1, 1)$.

22. We have $f(\mathbf{a}) = 0$, $f(\mathbf{b}) = e$ and by the Mean Value Theorem there exists $\mathbf{c} \in (\mathbf{a}, \mathbf{b})$ such that $f(\mathbf{b}) - f(\mathbf{a}) = \nabla f(\mathbf{c}) \bullet (\mathbf{b} - \mathbf{a})$. Thus, letting $\mathbf{c} = (c_1, c_2)$, we solve

$$(2c_1 e^{c_2}, c_1^2 e^{c_2}) \bullet (-1, 2) = e \implies 2c_1^2 - 2c_1 = e^{1-c_2}. \qquad (12.17)$$

From $\mathbf{c} \in (\mathbf{a}, \mathbf{b})$, we have $c_1 \in (-1, 0)$ and $c_2 = -2c_1 - 1$. Thus, equation (12.17) reads $2c_1^2 - 2c_1 = e^{2+2c_1}$ where $c_1 \in (-1, 0)$. The function $g : (-1, 0) \to \mathbb{R}$, $g(t) = 2t^2 - 2t - e^{2+2t}$ is decreasing since $g'(t) < 0$ on $(-1, 0)$. Thus, the equation $g(t) = 0$ has at most one solution, which follows that there exists exactly one point $\mathbf{c} \in (\mathbf{a}, \mathbf{b})$ that satisfies the Mean Value Theorem.

23. Let $g : [a, b] \to \mathbb{R}$ be given by $g(t) = f(t) \bullet \mathbf{u}$. Then g satisfies the conditions in the one dimensional Mean Value Theorem and $g(b) = g(a)$. It follows that there exists $c \in (a, b)$ such that $g'(c) = 0$, that is, $f'(c) \bullet \mathbf{u} = 0$. Hence, for some $c \in (a, b)$ the vector $f'(c)$ is orthogonal to $\mathbf{u}$.

24. Applying the Mean Value Theorem to $g(t) = f(t^2, t^2)$ on the interval $[0, 1] \subset \mathbb{R}$, there exists $c \in (0, 1)$ such that $g(1) - g(0) = g'(c)$. By the Chain Rule one has

$$g'(c) = 2c\frac{\partial f}{\partial x}(c^2, c^2) + 2c\frac{\partial f}{\partial y}(c^2, c^2).$$

25. Let $g : [0, 1] \to \mathbb{R}$, $g(t) = u \bullet F(\mathbf{a} + t(\mathbf{b} - \mathbf{a}))$. By the standard Mean Value Theorem, we have $g(1) - g(0) = g'(\tau)$, for some $\tau \in (0, 1)$. Using the Chain Rule (see Theorem 5.1), one has

$$g'(\tau) = \mathbf{u}\, J_F\big(\mathbf{a} + \tau(\mathbf{b} - \mathbf{a})\big) \bullet (\mathbf{b} - \mathbf{a}).$$

Now, $c = \mathbf{a} + \tau(\mathbf{b} - \mathbf{a}) \in (\mathbf{a}, \mathbf{b})$ and the conclusion follows.

26. The result is false! Take for instance $f : \mathbb{R}^2 \to \mathbb{R}$, $f(x, y) = x + y$ and $\mathbf{a} = (0, 1)$, $\mathbf{b} = (1, 0)$. Then $f(\mathbf{a}) = f(\mathbf{b}) = 1$ and $\nabla f(\mathbf{c}) = (1, 1) \neq (0, 0)$ for all $\mathbf{c} \in [\mathbf{a}, \mathbf{b}]$. For a correct interpretation of Rolle's Theorem in the context of functions of several variables, see Exercise 7 in the next chapter.

27. (i) $f(x, y) = e^{x+y}$.

(ii) Let $(x, y) \in \mathbb{R}^2$. Using the Mean Value Theorem with $\mathbf{a} = (x, y)$ and $\mathbf{b} = (x + y, 0)$, we have for some $\mathbf{c} \in [\mathbf{a}, \mathbf{b}]$ that

$$f(x + y, 0) - f(x, y) = \nabla f(\mathbf{c}) \bullet (y, -y) = \frac{\partial f}{\partial x}(\mathbf{c})y - \frac{\partial f}{\partial y}(\mathbf{c})y = 0.$$

Hence, $f(x + y, 0) = f(x, y)$.

(iii) If $g : \mathbb{R} \to \mathbb{R}$ is given by $g(t) = f(t, 0)$, it follows from (ii) above that $f(x, y) = g(x + y)$.

(iv) If $\dfrac{\partial f}{\partial x} = -\dfrac{\partial f}{\partial y}$ we proceed as before by proving first that $f(x, y) = f(x - y, 0)$ for all $(x, y) \in \mathbb{R}^2$. Thus, letting $h(t) = f(t, 0)$ one has $f(x, y) = h(x - y)$ for all $(x, y) \in \mathbb{R}^2$.

12.6 Solutions Chapter 6

1. All functions are differentiable, so one can employ Theorem 6.5.

(i) A unit vector in the direction $\mathbf{i} - \mathbf{j}$ is $\mathbf{u} = \frac{1}{\sqrt{2}}\mathbf{i} - \frac{1}{\sqrt{2}}\mathbf{j}$.

We have $\nabla f(x,y) = \left(2yx^{2y-1}, 2x^{2y}\ln x\right)$ and thus

$$D_{\mathbf{u}}f(e,1) = \nabla f(e,1) \bullet \mathbf{u} = (2e, 2e^2) \bullet \left(\frac{1}{\sqrt{2}}, -\frac{1}{\sqrt{2}}\right) = e(1-e)\sqrt{2}.$$

(ii) A unit vector in the direction $-3\mathbf{i} + 4\mathbf{j}$ is $\mathbf{u} = -\frac{3}{5}\mathbf{i} + \frac{4}{5}\mathbf{j}$.

We have $\nabla f(x,y) = \left(-y\sin(xy), -x\sin(xy)\right)$ and thus

$$D_{\mathbf{u}}f\left(1, \frac{\pi}{2}\right) = \nabla f\left(1, \frac{\pi}{2}\right) \bullet \mathbf{u} = \left(-\frac{\pi}{2}, -1\right) \bullet \left(-\frac{3}{5}, \frac{4}{5}\right) = \frac{3\pi - 8}{10}.$$

(iii) A unit vector in the direction $\mathbf{i} + 2\mathbf{j} - 2\mathbf{k}$ is $\mathbf{u} = \frac{1}{3}\mathbf{i} + \frac{2}{3}\mathbf{j} - \frac{2}{3}\mathbf{k}$.

We have $\nabla f(x,y) = e^{-x+y+z}\left(2xz - x^2z, x^2z, x^2 + x^2z\right)$ and thus

$$D_{\mathbf{u}}f(1,0,1) = \nabla f(1,0,1) \bullet \mathbf{u} = -\frac{1}{3}.$$

(iv) A unit vector in the direction $-2\mathbf{i} - 6\mathbf{j} + 3\mathbf{k}$ is $\mathbf{u} = -\frac{2}{7}\mathbf{i} - \frac{6}{7}\mathbf{j} + \frac{3}{7}\mathbf{k}$.

We have $\nabla f(x,y) = \left(\cos y, -x\sin y + \sin z, y\cos z\right)$ and thus

$$D_{\mathbf{u}}f\left(\frac{\pi}{2}, \frac{\pi}{2}, \frac{\pi}{2}\right) = \nabla f\left(\frac{\pi}{2}, \frac{\pi}{2}, \frac{\pi}{2}\right) \bullet \mathbf{u} = \frac{6}{7}\left(\frac{\pi}{2} - 1\right).$$

2. (i) We have $|f(x,y)| = \dfrac{|x|y^2}{x^2 + 3y^2} \le |x| \to 0$ as $(x,y) \to (0,0)$.

By Corollary 3.12 it follows that $\lim\limits_{(x,y)\to(0,0)} f(x,y) = 0 = f(0,0)$,

hence f is continuous at $(0,0)$.

(ii) $\dfrac{\partial f}{\partial x}(0,0) = \lim\limits_{t\to 0} \dfrac{f(t,0) - f(0,0)}{t} = 0$ and similarly $\dfrac{\partial f}{\partial x}(0,0) = 0$.

(iii) For $(x,y) \in \mathbb{R}^2 \setminus \{(0,0)\}$ let

$$g(x,y) = \frac{f(x,y) - \{f(0,0) + \nabla f(0,0) \bullet (x,y)\}}{\|(x,y)\|}.$$

Then, f is differentiable at $(0,0)$ if and only if $\lim\limits_{(x,y)\to(0,0)} g(x,y) = 0$.

We have

$$g(x,y) = \frac{xy^2}{(x^2 + 3y^2)\sqrt{x^2 + y^2}} \qquad \text{for all } (x,y) \in \mathbb{R}^2 \setminus \{(0,0)\}.$$

Along the line $y = 0$ one has and $g(x,y) \to 0$ as $(x,y) \to 0$. Along the semiline $y = x$, $x > 0$, we have

$$g(x,y) \to \frac{1}{4\sqrt{2}} \qquad \text{as } (x,y) \to 0.$$

This implies that g has no limit as $(x, y) \to (0, 0)$ so f is not differentiable at $(0, 0)$.

(iv) Let $\mathbf{u} = (u_1, u_2)$ be a unit vector. By Definition 6.1 we find

$$D_{\mathbf{u}}f(0, 0) = \lim_{t \to 0} \frac{f(tu_1, tu_2) - f(0, 0)}{t} = \frac{u_1 u_2^2}{u_1^2 + 3u_2^2}.$$

3. (i) Observe first that

$$f(x, y) = \frac{x^5 + y^5 + x^3 y^2 + x^2 y^3}{x^4 + y^4} \qquad \text{for all } (x, y) \in \mathbb{R}^2 \setminus \{(0, 0)\}.$$

Using the standard inequalities (3.3)–(3.4) for all $(x, y) \in \mathbb{R}^2 \setminus \{(0, 0)\}$ one has

$$\frac{x^4}{x^4 + y^4} \le 1, \qquad \frac{y^4}{x^4 + y^4} \le 1, \qquad \frac{x^2 y^2}{x^4 + y^4} \le \frac{1}{2}.$$

Thus, for all $(x, y) \in \mathbb{R}^2 \setminus \{(0, 0)\}$ we estimate

$$|f(x, y)| \le \frac{x^4}{x^4 + y^4}|x| + \frac{y^4}{x^4 + y^4}|y| + \frac{x^2 y^2}{x^4 + y^4}(|x| + |y|)$$
$$\le \frac{3}{2}(|x| + |y|) \to 0 \qquad \text{as } (x, y) \to (0, 0)$$

and now by Corollary 3.12 we deduce $\lim_{(x, y) \to (0, 0)} f(x, y) = 0 = f(0, 0)$. Hence f is continuous at $(0, 0)$.

(ii) By Definition 4.1, we have $\dfrac{\partial f}{\partial x}(0, 0) = \lim_{t \to 0} \dfrac{f(t, 0) - f(0, 0)}{t} = 1$ and similarly $\dfrac{\partial f}{\partial x}(0, 0) = 1$. As in the previous exercise, for $(x, y) \in \mathbb{R}^2 \setminus \{(0, 0)\}$ we set

$$g(x, y) = \frac{f(x, y) - \{f(0, 0) + \nabla f(0, 0) \bullet (x, y)\}}{\|(x, y)\|}$$
$$= \frac{x^3 y^2 + x^2 y^3 - x^4 y - xy^4}{(x^4 + y^4)\sqrt{x^2 + y^2}}.$$

Along the line $y = 0$, we have $g(x, y) \to 0$ as $(x, y) \to (0, 0)$ while along the line $y = 2x$, $x > 0$, we have

$$g(x, y) = g(x, 2x) \to -\frac{6}{17\sqrt{5}} \qquad \text{as } (x, y) \to (0, 0).$$

This shows that g has no limit at $(0, 0)$ and thus f is not differentiable at $(0, 0)$.

(iii) By Definition 6.1 and using the fact that $u_1^2 + u_2^2 = 1$, we have

$$D_{\mathbf{u}}f(0, 0) = \lim_{t \to 0} \frac{f(tu_1, tu_2) - f(0, 0)}{t} = \frac{u_1^3 + u_2^3}{u_1^4 + u_2^4}.$$

4. Let $\mathbf{u} = (u_1, u_2) \in \mathbb{R}^2$ be a unit vector. Then

$$D_{\mathbf{u}}f(1,0) = \lim_{t \to 0} \frac{f(1+tu_1, tu_2) - f(1,0)}{t} = \lim_{t \to 0} \frac{f(1+tu_1, tu_2)}{t}.$$

For $|t| > 0$ small, we have $1 + tu_1 > 0$.

If $u_2 \neq 0$, then $f(1+tu_1, tu_2) = 1$ and the above limit does not exist.

If $u_2 = 0$, then $\|\mathbf{u}\| = 1$ implies $u_1 = \pm 1$. Also,

$$f(1+tu_1, tu_2) = f(1 \pm t, 0) = 0 \quad \text{and thus} \quad D_{\mathbf{u}}f(1,0) = 0.$$

Hence, f has directional derivative at $(1,0)$ only in the directions $(1,0)$ and $(-1,0)$.

5. (i) The implication $(D) \Longrightarrow (A)$, $(D) \Longrightarrow (C)$ and $(C) \Longrightarrow (B)$ follow from Corollary 4.17, Theorem 6.5 and Remark 6.2 respectively.

(ii) The norm function $f : \mathbb{R}^m \to \mathbb{R}$ is continuous but has no directional derivative in any direction $\mathbf{u} \in \mathbb{R}^m$ (see Example 6.4). This shows that the implication $(A) \Longrightarrow (B)$ and thus $(A) \Longrightarrow (C)$, do not always hold.

Example 6.3 provides a function which has directional derivatives (in any direction) but is not differentiable. This shows that the implication $(C) \Longrightarrow (D)$ is not always true.

Finally, consider the function f defined in Exercise 4. It can be shown that $\dfrac{\partial f}{\partial x}(0,0) = \dfrac{\partial f}{\partial y}(0,0) = 0$ but $D_{\mathbf{u}}f(0,0)$ does not exist if both components of $\mathbf{u}$ are nonzero. Thus, the implication $(B) \Longrightarrow (C)$ does not always hold.

6. Let $\nabla f(a,b) = (\alpha, \beta)$. A unit vector in the direction $\mathbf{i} - \mathbf{j}$ is $\mathbf{u} = \frac{1}{\sqrt{2}}\mathbf{i} - \frac{1}{\sqrt{2}}\mathbf{j}$ and a unit vector in the direction $2\mathbf{i} + \mathbf{j}$ is $\mathbf{v} = \frac{2}{\sqrt{5}}\mathbf{i} + \frac{1}{\sqrt{5}}\mathbf{j}$. Since f is differentiable at (a,b) we may use Theorem 6.5 to obtain

$$\begin{cases} D_{\mathbf{u}}f(a,b) = \nabla f(a,b) \bullet \mathbf{u} = \sqrt{5} \\ D_{\mathbf{v}}f(a,b) = \nabla f(a,b) \bullet \mathbf{v} = 2\sqrt{2} \end{cases} \Longrightarrow \begin{cases} \alpha - \beta = \sqrt{10} \\ 2\alpha + \beta = 2\sqrt{10}. \end{cases}$$

We find $\nabla f(a,b) = (\sqrt{10}, 0)$.

A unit vector in the direction $3\mathbf{i} + \mathbf{j}$ is $\mathbf{w} = \frac{3}{\sqrt{10}}\mathbf{i} + \frac{1}{\sqrt{10}}\mathbf{j}$. Hence, $D_{\mathbf{w}}f(a,b) = \nabla f(a,b) \bullet \mathbf{w} = 3$.

7. By the Mean Value Theorem 5.10 for several variables there exists $\mathbf{c} \in (\mathbf{a}, \mathbf{b})$ such that $f(\mathbf{b}) - f(\mathbf{a}) = \nabla f(\mathbf{c}) \bullet (\mathbf{b} - \mathbf{a})$. Note that $\mathbf{u} = (\mathbf{b} - \mathbf{a})/\|\mathbf{b} - \mathbf{a}\|$. Thus,

$$\frac{f(\mathbf{b}) - f(\mathbf{a})}{\|\mathbf{b} - \mathbf{a}\|} = \nabla f(\mathbf{c})\frac{\mathbf{b} - \mathbf{a}}{\|\mathbf{b} - \mathbf{a}\|} = \nabla f(\mathbf{c}) \cdot \mathbf{u} = D_{\mathbf{u}}f(\mathbf{c}).$$

8. (i) Assume for instance that f is continuous at $\mathbf{a}$. We have

$$\frac{(fg)(\mathbf{a}+t\mathbf{u}) - (fg)(\mathbf{a})}{t}$$

$$= f(\mathbf{a}+t\mathbf{u}) \cdot \frac{g(\mathbf{a}+t\mathbf{u}) - g(\mathbf{a})}{t} + g(\mathbf{a}) \cdot \frac{f(\mathbf{a}+t\mathbf{u}) - f(\mathbf{a})}{t}.$$

Since f is continuous at $\mathbf{a}$ one has $f(\mathbf{a}+t\mathbf{u}) \to g(\mathbf{a})$ as $t \to 0$. Thus, passing to the limit as $t \to 0$ in the above equality we deduce

$$D_{\mathbf{u}}(fg)(\mathbf{a}) = \lim_{t \to 0} \frac{(fg)(\mathbf{a}+t\mathbf{u}) - (fg)(\mathbf{a})}{t}$$

$$= f(\mathbf{a})D_{\mathbf{u}}g(\mathbf{a}) + g(\mathbf{a})D_{\mathbf{u}}f(\mathbf{a}).$$

(ii) Let $f = g : \mathbb{R}^2 \to \mathbb{R}$ be given by

$$f(x,y) = \begin{cases} \dfrac{xy}{x^2+y^2} & \text{if } (x,y) \neq (0,0) \\ 0 & \text{if } (x,y) = (0,0). \end{cases}$$

Then f and g are not continuous at $(0,0)$. Let $\mathbf{u} = (u_1, u_2)$ be a unit vector with $u_1, u_2 \neq 0$. Then

$$D_{\mathbf{u}}(fg)(0,0) = \left(\frac{u_1 u_2}{u_1^2 + u_2^2}\right)^2 \neq 0$$

while

$$f(0,0)D_{\mathbf{u}}g(0,0) + g(0,0)D_{\mathbf{u}}f(0,0) = 0.$$

9. By Exercise 16 in Chapter 5 one has that f is differentiable on $\mathbb{R}^m \setminus \{\mathbf{0}\}$ and

$$\nabla f(\mathbf{x}) = \phi'(\|\mathbf{x}\|)\frac{\mathbf{x}}{\|\mathbf{x}\|} \quad \text{for all} \quad \mathbf{x} \in \mathbb{R}^m \setminus \{\mathbf{0}\}.$$

(i) From the above equality and the fact that $\|\mathbf{u}\| = \|\mathbf{v}\| = 1$ we deduce $D_{\mathbf{u}}f(\mathbf{v}) = \phi'(1)\mathbf{u} \bullet \mathbf{v} = D_{\mathbf{v}}f(\mathbf{u})$.

(ii) $D_{\mathbf{u}}f(\mathbf{a}) = D_{\mathbf{v}}f(\mathbf{a})$ implies

$$\phi'(\|\mathbf{a}\|)\frac{\mathbf{a} \bullet \mathbf{u}}{\|\mathbf{a}\|} = \phi'(\|\mathbf{a}\|)\frac{\mathbf{a} \bullet \mathbf{v}}{\|\mathbf{a}\|}$$

which yields

$$\phi'(\|\mathbf{a}\|)\mathbf{a} \bullet (\mathbf{u} - \mathbf{v}) = 0 \Longrightarrow \mathbf{u} - \mathbf{v} \perp \phi'(\|\mathbf{a}\|)\mathbf{a}.$$

10. We need to find all unit vectors $\mathbf{u} = (u_1, u_2)$ such that $D_{\mathbf{u}}f(-1,1) = \frac{3}{2}$. We have $\nabla f(-1,1) = \left(-\frac{1}{2}, \frac{3}{2}\right)$ and u_1, u_2 satisfy

$$u_1^2 + u_2^2 = 1 \quad \text{and} \quad \left(-\frac{1}{2}, \frac{3}{2}\right) \bullet (u_1, u_2) = \frac{3}{2}.$$

We find $\mathbf{u} = (0,1)$ and $\mathbf{u} = \left(-\frac{3}{5}, \frac{4}{5}\right)$.

11. The vector $\overrightarrow{PQ}$ has the components $(-1, 3)$ and the unit vector in this direction is $\mathbf{u} = \left(-\frac{1}{\sqrt{10}}, \frac{3}{\sqrt{10}}\right)$. We want to maximize/minimize the function

$$D_{\mathbf{u}}g(x, y) = \nabla g(x, y) \bullet \mathbf{u} = \frac{4}{\sqrt{10}}xe^{-2x^2} - \frac{48}{\sqrt{10}}ye^{-8y^2}.$$

The function $h : \mathbb{R} \to \mathbb{R}$, $h(t) = te^{-\alpha t^2}$, $\alpha > 0$, has a minimum at $t = -1/\sqrt{2\alpha}$ and a maximum at $t = 1/\sqrt{2\alpha}$. Thus, the maximum of $D_{\mathbf{u}}g$ occurs at $(x, y) = \left(\frac{1}{2}, -\frac{1}{4}\right)$ and this equals $\frac{14}{\sqrt{10e}}$. Similarly, the minimum of $D_{\mathbf{u}}g$ occurs at $(x, y) = \left(-\frac{1}{2}, \frac{1}{4}\right)$ and this equals $-\frac{14}{\sqrt{10e}}$.

12. (i) $T(-1, 3, 1) = 2$ and the temperature remains constant on the level surface

$$T(x, y, z) = 2, \quad \text{that is} \quad 2 + z^2 = (x + y)^2.$$

(ii) We have

$$\nabla T(x, y, z) = \left(\frac{2(x + y)}{1 + z^2}, \frac{2(x + y)}{1 + z^2}, -\frac{2z(x + y)^2}{(1 + z^2)^2}\right)$$

so $\nabla T(-1, 3, 1) = (2, 2, -2)$. The temperature decreases the most rapidly at $(-1, 3, 1)$ in the direction

$$-\nabla T(-1, 3, 1) = (-2, -2, 2).$$

The unit vector in this direction is

$$-\frac{\nabla T(-1, 3, 1)}{\|\nabla T(-1, 3, 1)\|} = \left(-\frac{1}{\sqrt{3}}, -\frac{1}{\sqrt{3}}, \frac{1}{\sqrt{3}}\right).$$

13. $\nabla z(x, y) = \left(-\frac{x^3}{180}, -\frac{y + 1}{20}\right)$.

(i) $\nabla z(6, -5) = \left(-\frac{6}{5}, \frac{1}{5}\right)$ and the unit vector pointing north is $\mathbf{u} = (0, 1)$. Thus,

$$D_{\mathbf{u}}z(6, -5) = \nabla z(6, -5) \bullet (0, 1) = \frac{1}{5} > 0$$

which shows that you ascend when move north.

(ii) $\nabla z(-6, 3) = \left(\frac{6}{5}, -\frac{1}{5}\right)$ and the unit vector pointing south-west is $\mathbf{v} = \left(-\frac{1}{\sqrt{2}}, -\frac{1}{\sqrt{2}}\right)$. Thus,

$$D_{\mathbf{v}}z(-6, 3) = \nabla z(-6, 3) \bullet \left(-\frac{1}{\sqrt{2}}, -\frac{1}{\sqrt{2}}\right) = -\frac{1}{\sqrt{2}} < 0,$$

so you descend when move south-west.

14. (i) A unit vector in direction $2\mathbf{i}+\mathbf{j}$ is $\mathbf{u} = \frac{1}{\|(2,1)\|}(2,1) = \left(\frac{2}{\sqrt{5}}, \frac{1}{\sqrt{5}}\right)$. Hence

$$D_{\mathbf{u}}f(1,-2) = \nabla f(1,-2) \bullet u = (3,-4) \bullet \left(\frac{2}{\sqrt{5}}, \frac{1}{\sqrt{5}}\right) = \frac{2}{\sqrt{5}}.$$

(ii) f decreases the most rapidly in the direction $-\nabla f(1,-2) = (-3,4)$. The unit vector in this direction is $\left(-\frac{3}{5}, \frac{4}{5}\right)$.

(iii) We know that $D_{\mathbf{u}}f(1,-2) = \|\nabla f(1,-2)\| \cos\theta$, where θ is the angle between $\mathbf{u}$ and the vector $\nabla f(2,1)$. Thus,

$$D_{\mathbf{u}}f(1,-2) \le \|\nabla f(1,-2)\| = 5$$

and equality holds if and only if $\theta = 0$, that is, $\mathbf{u}$ is the unit vector in the direction given by $\nabla f(1,-2) = (3,-4)$. Similarly, $D_{\mathbf{v}}f(1,-2) \le 5$. Thus,

$$D_{\mathbf{u}}f(1,-) + D_{\mathbf{v}}f(1,-2) \le 10,$$

and equality holds if and only if $\mathbf{u}$ and $\mathbf{v}$ are the unit vectors in the direction given by $\nabla f(1,-2)$. Hence,

$$\mathbf{u} = \mathbf{v} = \frac{1}{\|\nabla f(1,-2)\|} \nabla f(1,-2).$$

15. (i) Using the Chain Rule in the definition of an even function one finds

$$-\nabla f(-x,-y) = \nabla f(x,y) \quad \text{for all} \quad (x,y) \in \mathbb{R}^2.$$

In particular, for $(x,y) = 0$ one gets $\nabla f(0,0) = (0,0)$.

(ii) From the above equality we find $\nabla f(-2,3) = -\nabla f(2,-3)$ which shows that f increases the most rapidly at $(-2,3)$ in the same direction along which f decreases the most rapidly at $(2,-3)$.

16. Using the Chain Rule in the definition of an odd function we find

$$-\nabla g(-x,-y) = -\nabla g(x,y) \quad \text{for all} \quad (x,y) \in \mathbb{R}^2.$$

In particular, $\nabla g(-2,3) = \nabla g(2,-3)$ which shows that g increases the most rapidly at $(-2,3)$ in the same direction along which g increases the most rapidly at $(2,-3)$.

17. $\nabla f(x,y) = (2x+ay, ax+3by^2)$, so $\nabla f(2,-1) = (4-a, 2a+3b) = \lambda(3,1)$. Since $\nabla f(2,-1)$ has the same direction as $3\mathbf{i}+\mathbf{j}$ we impose $\lambda > 0$. Thus,

$$\begin{cases} 4 - a = 3\lambda \\ 2a + 3b = \lambda \end{cases} \implies \begin{cases} a = 4 - 3\lambda \\ b = \dfrac{7\lambda - 8}{3}. \end{cases}$$

Since $\lambda > 0$, it follows immediately that $a < 4$ and $b > -8/3$. From the above we also derive $7a + 9b = 4$.

18. Since $\mathbf{u} \neq \pm\mathbf{v}$, we have that at least one of these two vectors is not orthogonal to $\nabla f(\mathbf{a})$ (otherwise, if both $\mathbf{u}$ and $\mathbf{v}$ are orthogonal to $\nabla f(\mathbf{a})$, it follows that either $\mathbf{u} = \mathbf{v}$ or $\mathbf{u} = -\mathbf{v}$). Assuming that $\mathbf{u}$ is not orthogonal to $\nabla f(\mathbf{a})$, we have $D_\mathbf{u} f(\mathbf{a}) = \|\nabla f(\mathbf{a})\| \cos\theta$, where θ is the angle between $\nabla f(\mathbf{a})$ and $\mathbf{u}$. Since $\mathbf{u}$ and $\nabla f(\mathbf{a})$ are not orthogonal, we have $\cos\theta \neq 0$ and $D_\mathbf{u} f(\mathbf{a}) = 0$ implies $\|\nabla f(\mathbf{a})\| = 0$, so $\nabla f(\mathbf{a}) = (0,0)$.

19. We impose

$$\begin{cases} \nabla f(a,b) = \lambda(2,-1) & \lambda > 0 \quad \text{and} \quad \|\nabla f(a,b)\| = 3 \\ \nabla g(a,b) = \mu(-1,2) & \mu > 0 \quad \text{and} \quad \|\nabla g(a,b)\| = 4. \end{cases}$$

This yields $\lambda = \frac{3}{\sqrt{5}}$ and $\mu = \frac{4}{\sqrt{5}}$, so

$$\nabla f(a,b) = \left(\frac{6}{\sqrt{5}}, -\frac{3}{\sqrt{5}} \right) \quad \text{and} \quad \nabla g(a,b) = \left(-\frac{4}{\sqrt{5}}, \frac{8}{\sqrt{5}} \right).$$

The function $h = f - g$ increases the most rapidly at (a,b) in the direction

$$\nabla h(a,b) = \nabla f(a,b) - \nabla g(a,b) = \left(2\sqrt{5}, -\frac{11}{\sqrt{5}} \right)$$

and the value of this increase is $\|\nabla h(a,b)\| = \sqrt{221/5}$.

20. (i) Since $|D_\mathbf{u} f(\mathbf{a})| \leq \|\nabla f(a,b)\|$ and $|D_\mathbf{v} f(\mathbf{a})| \leq \|\nabla f(a,b)\|$, we have

$$|D_\mathbf{u} f(\mathbf{a})||D_\mathbf{u} f(\mathbf{a})| \leq \|\nabla f(a,b)\|^2$$

and equality holds if and only if $D_\mathbf{u} f(\mathbf{a})$ and $D_\mathbf{v} f(\mathbf{a})$ have the same sign and $|D_\mathbf{u} f(\mathbf{a})| = |D_\mathbf{v} f(\mathbf{a})| = \|\nabla f(a,b)\|$. This means that either $\mathbf{u}, \mathbf{v}$ have the same direction as $\nabla f(a,b)$ or both have the direction given by $-\nabla f(a,b)$. Since $\mathbf{u}$ and $\mathbf{v}$ are unit vectors, it follows that they are equal.

(ii) As above, we deduce $|D_\mathbf{u} f(\mathbf{a})| = |D_\mathbf{v} f(\mathbf{a})| = |D_\mathbf{w} f(\mathbf{a})| = \|\nabla f(a,b)\|$. Thus, the unit vectors have the direction $\nabla f(a,b)$ or $-\nabla f(a,b)$. It follows that at least two of the unit vectors are equal.

21. Let θ be the angle between $\mathbf{u}$ and $\nabla f(\mathbf{a})$. Then, the angle ω between $\mathbf{v}$ and $\nabla f(\mathbf{a})$ is computed by one of the equalities (see Figure 12.4):

$$\omega = 90^0 - \theta, \quad \omega = 90^0 + \theta, \quad \omega = \theta - 90^0, \quad \omega = 270^0 - \theta.$$

In all the above cases $\cos\omega = \pm \sin\theta$.

(i) Since $D_\mathbf{u} f(\mathbf{a}) = \|\nabla f(\mathbf{a})\| \cos\theta$ and $D_\mathbf{v} f(\mathbf{a}) = \|\nabla f(\mathbf{a})\| \cos\omega$, we

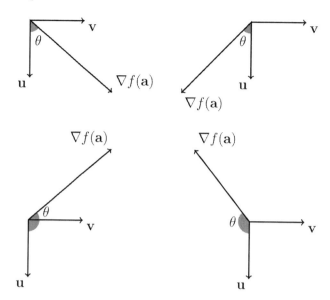

Figure 12.4
The orthogonal vectors **u** and **v**.

have

$$D_{\mathbf{u}}f(\mathbf{a}) + D_{\mathbf{v}}f(\mathbf{a}) = \|\nabla f(\mathbf{a})\| \cos\theta + \|\nabla f(\mathbf{a})\| \cos\omega$$
$$= \|\nabla f(\mathbf{a})\|\big(\cos\theta \pm \sin\theta\big)$$
$$= \sqrt{2}\,\|\nabla f(\mathbf{a})\| \cos(\theta \mp 45^0)$$
$$\leq \sqrt{2}\,\|\nabla f(\mathbf{a})\|.$$

(ii) Similar to part (i) above, we have

$$D_{\mathbf{u}}f(\mathbf{a}) \cdot D_{\mathbf{v}}f(\mathbf{a}) = \|\nabla f(\mathbf{a})\|^2 \cos\theta \cos\omega$$
$$= \pm\|\nabla f(\mathbf{a})\|^2 \cos\theta \sin\theta$$
$$= \pm\frac{1}{2}\,\|\nabla f(\mathbf{a})\|^2 \sin(2\theta)$$
$$\leq \frac{1}{2}\,\|\nabla f(\mathbf{a})\|^2.$$

22. Since $(3, -1) \in L(f, 2)$, we have $f(3, -1) = 2$. Denote $\nabla f(3, -1) = (a, b)$. By Theorem 6.11, we have that $\nabla f(3, -1)$ is orthogonal to the tangent line to the level curve $L(f, 2)$. Since the tangent line is $x - 2y = 5$, we have $\nabla f(3, -1) \perp (2, 1)$, so $2a + b = 0$.

We next exploit the information in part (iii). The unit vector **u** in the direction $\mathbf{i} - 2\mathbf{j}$ is $\mathbf{u} = \left(\frac{1}{\sqrt{5}}, -\frac{2}{\sqrt{5}}\right)$. Thus, $D_{\mathbf{u}}f(3, -1) = \sqrt{5}$

yields $a - 2b = 5$. From the above we deduce $a = 1$, $b = -2$, so $\nabla f(3, -1) = (1, -2)$.

23. Since $(2, -3)$ lies on the parabola $y^2 - 2x = 5$, we deduce $(2, -3) \in L(f, c)$, so $c = f(2, -3)$.

    The level curve $L(f, c)$ coincides with the level curve $L(g, 5)$ where $g(x, y) = y^2 - 2x$. By Theorem 6.11, one has that $\nabla f(2, -3)$ and $\nabla g(2, -3)$ are orthogonal to the tangent line to the curve $L(f, c) = L(g, 5)$ at $(2, -3)$ so $\nabla f(2, -3) = \lambda \nabla g(2, -3) = (-2\lambda, -6\lambda)$, for some $\lambda \in \mathbb{R}$.

    Since the equation of the tangent plane to $z = f(x, y)$ at $(2, -3, c)$ is $z = -x + by - 2$, it follows that

    $$\frac{\partial f}{\partial x}(2, -3) = -1, \quad \frac{\partial f}{\partial y}(2, -3) = b \quad \text{and} \quad c = -3b - 4.$$

    Since $\nabla f(2, -3) = (-2\lambda, -6\lambda)$ we obtain

    $$\begin{cases} -2\lambda = -1 \\ -6\lambda = b \\ c = -3b - 4 \end{cases} \implies \begin{cases} \lambda = \dfrac{1}{2} \\ b = -3 \\ c = 5. \end{cases}$$

    Hence, $f(2, -3) = 5$ and $\nabla f(2, -3) = (-1, -3)$.

24. The ellipsoid is the level set $L(f, 12)$, where $f(x, y, z) = 2(x-1)^2 + (y+1)^2 + z^2$. Let (a, b, c) be a point on the ellipsoid with the desired property. Then, a normal vector to the tangent plane at (a, b, c) is given by $\nabla f(a, b, c) = (4(a-1), 2(b+1), 2c)$ while a normal vector to the plane $2x - 3y + z = 7$ is given by $\mathbf{n} = (2, -3, 1)$. Thus, the tangent plane and the plane $2x - 3y + z = 7$ are parallel if and only if $\nabla f(a, b, c)$ and $\mathbf{n}$ are parallel. This means there exists $t \in \mathbb{R}$ such that $\nabla f(a, b, c) = t\mathbf{n}$, that is,

    $$a - 1 = \frac{t}{2}, \quad b + 1 = -\frac{3t}{2}, \quad c = \frac{t}{2}.$$

    Plug the above equalities in the equation of the ellipsoid to deduce $t = \pm 2$ and then $(a, b, c) = (2, -4, 1), (0, 2, -1)$.

25. Let (x, y, z) be a point on the surface at which the property holds. The normal line to the surface is $\mathbf{n} = \nabla f(x, y, z) = (2x, z, y + 3z^2)$, where $f(x, y, z) = x^2 + yz + z^3$. Then $\mathbf{n} \bullet (2, 5, 4) = 0$ which yields $4x + 5z + 4y + 12z^2 = 0$. We have obtained that the point (x, y, z) satisfies two simultaneous equations, namely,

    $$\begin{cases} x^2 + yz + z^3 = 1 \\ 4x + 5z + 4y + 12z^2 = 0. \end{cases}$$

We multiply the first equation by 12 and the second equation by z. Subtracting the resulting equalities we find

$$4xz + 5z^2 + 4yz + 12(1 - x^2 - yz) = 0.$$

The above equation is equivalent to

$$\left(x + \frac{z}{2}\right)^2 + (z - y)^2 = 4x^2 + y^2 - 3. \qquad (12.18)$$

Note. Letting $X = 3\sqrt{2}x - \frac{z}{\sqrt{2}}$, $Y = 3y - z$ and $Z = 3z - 3y$, we deduce that the quadric in (12.18) defines a hyperboloid of one sheet of equation $X^2 + Y^2 - Z^2 = 18$.

26. The hyperboloid is the level surface $L(g, 4)$ where $g(x, y, z) = x^2 + y^2 - z^2$, so $L(f, c) = L(g, 5)$. By Theorem 6.11, one has that $\nabla f(1, 2, 1)$ and $\nabla g(1, 2, 1)$ are orthogonal to the tangent plane to this surface, so $\nabla f(1, 2, 1) = \lambda \nabla g(1, 2, 1)$, for some $\lambda \in \mathbb{R}$. Thus, $\nabla f(1, 2, 1) = (2\lambda, 4\lambda, -2\lambda)$. From here we easily deduce the conclusion.

27. (i) The sphere centred at $(0, 1, 2)$ that passes through $(-1, 0, 3)$ has the equation

$$x^2 + (y - 1)^2 + (z - 2)^2 = 3.$$

This is the level surface $L(g, 3)$ where $g(x, y, z) = x^2 + (y - 1)^2 + (z - 2)^2$. We know by Theorem 6.11 that $\nabla g(-1, 0, 3) = (-2, -2, 2)$ is orthogonal to the tangent plane at $(-1, 0, 3)$. Thus, the equation of the tangent plane is

$$(x + 1, y, z - 3) \bullet (-2, -2, 2) = 0 \implies x + y - z + 4 = 0.$$

(ii) Since $L(f, 1) = L(g, 3)$, Theorem 6.11 yields

$$\nabla f(-1, 0, 3) = \lambda \nabla g(-1, 0, 3) = (-2\lambda, -2\lambda, 2\lambda),$$

for some $\lambda \in \mathbb{R}$. The conclusion follows now immediately.

28. (i) A normal vector at P to the cone is $\mathbf{n}_1 = (2ax_0, 2by_0, -2z_0)$ and a normal vector at P to the paraboloid is $\mathbf{n}_2 = (2bx_0, 2ay_0, -1)$. Since the two surfaces have the same tangent plane $x + y + z = 0$ at $P(x_0, y_0, z_0)$, it follows that

$$\mathbf{n}_1 = \lambda(1, 1, 1) \quad \text{and} \quad \mathbf{n}_2 = \mu(1, 1, 1).$$

From the last equality we deduce $\mu = -1$ and then $x_0 = -\frac{1}{2b}$, $y_0 = -\frac{1}{2a}$. From the first equality in the above, we find

$$\lambda = 2ax_0 = 2by_0 = -2z_0 \implies 2z_0 = -\lambda = \frac{a}{b} = \frac{b}{a}.$$

Note that the above equalities yield $a^2 = b^2$ so $a = \pm b$. Also, $z_0 > 0$ so $\lambda < 0$ which yields $a = b$.

(ii) Since $a = b$, the above equality yields $z_0 = \frac{1}{2}$ and $\lambda = -1$. Finally, we obtain $a = b = 2$, $x_0 = y_0 = -\frac{1}{4}$ and $c = \frac{1}{4}$.

29. Let (a, b, c) be a point on the cone at which the tangent plane is parallel to the plane $\alpha x + \beta y + \gamma z = d$. The normal line to this plane has the direction $\mathbf{n} = (\alpha, \beta, \gamma)$ while the normal line to the tangent plane at (a, b, c) to the cone has the direction given by $\nabla f(a, b, c)$ where $f(x, y, z) = x^2 + (y-1)^2 - z^2$, hence $\nabla f(a, b, c) = (2a, 2(b-1), -2c)$. The two planes are parallel, if and only if there exists $t \in \mathbb{R}$ such that $\nabla f(a, b, c) = t\mathbf{n}$. This yields:

$$a = \frac{\alpha t}{2}, \quad b - 1 = \frac{\beta t}{2}, \quad c = -\frac{\gamma t}{2}. \qquad (12.19)$$

On the other hand, (a, b, c) lies on the cone, so $a^2 + (b-1)^2 = c^2$. Using (12.19) in this last equality we derive $\alpha^2 + \beta^2 = \gamma^2$.

The converse statement is also true. Indeed, given any triple $(\alpha, \beta, \gamma) \neq (0, 0, 0)$ such that $\alpha^2 + \beta^2 = \gamma^2$, we let $a = \alpha$, $b = \beta + 1$ and $c = -\gamma$. Then (a, b, c) lies on the cone and the normal to the tangent plane at (a, b, c) is given by $\nabla f(a, b, c) = (2\alpha, 2\beta, 2\gamma)$ which is orthogonal to the plane $\alpha x + \beta y + \gamma z = d$.

30. $\mathcal{S}_1$ and $\mathcal{S}_2$ are described as level surfaces $\mathcal{S}_1 = L(f, 1)$ and $\mathcal{S}_2 = L\{g, 6\}$ corresponding to $f(x, y, z) = x^2 + 2xyz + 2y^2$ and $g(x, y, z) = y^3 - x^2 z$ respectively.

(i) The normal vector to $\mathcal{S}_1$ at $(1, 0, 2)$ is

$$\mathbf{n} = \nabla f(1, 0, 2) = (2x + 2yz, 4y + 2xz, 2xy)\Big|_{(1,0,2)} = (2, 4, 0).$$

The normal line has the direction $(2, 4, 0)$ and passes through $(1, 0, 2)$. Its parametric equation is

$$\begin{pmatrix} x \\ y \\ z \end{pmatrix} = \begin{pmatrix} 1 \\ 0 \\ 2 \end{pmatrix} + t \begin{pmatrix} 2 \\ 4 \\ 0 \end{pmatrix} \implies \begin{pmatrix} x \\ y \\ z \end{pmatrix} = \begin{pmatrix} 2t + 1 \\ 4t \\ 2 \end{pmatrix}, t \in \mathbb{R}$$

and the equation of the tangent plane is

$$(2, 4, 0) \bullet \left\{ (x, y, z) - (1, 0, 2) \right\} = 0 \implies x + 2y = 1.$$

(ii) Let (x, y, z) be a point on the surface $\mathcal{S}_2$ at which the normal line is parallel to the vector $(-4, 3, -1)$. This means that there exists $k \in \mathbb{R}$ such that

$$\nabla g(x, y, z) = k(-4, 3, -1) \implies \begin{cases} -2xz = -4k \\ 3y^2 = 3k \\ -x^2 = -k. \end{cases}$$

This implies

$$\begin{cases} xz = 2k \\ y^2 = k \\ x^2 = k. \end{cases} \qquad (12.20)$$

From the first and the third equation of (12.20) we find $2x^2 = xz$. If $x = 0$, then from the last two equations of (12.20) we would have $k = 0$ and $y = 0$ which is impossible since $y^3 - x^2z = 6$. Thus, $x \neq 0$ and $z = 2x$. Also, from the last two equations of (12.20) we find $x^2 = y^2$ so $y = \pm x$.

If $y = x$, then $g(x, y, z) = 6$ implies $-x^3 = 6 \Rightarrow x = -\sqrt[3]{6}$.

Hence $(x, y, z) = \left(-\sqrt[3]{6}, -\sqrt[3]{6}, -2\sqrt[3]{6} \right)$.

If $y = -x$, then $z = 2x$ and $g(x, y, z) = 6$ yield $x^3 = -2$, that is $x = -\sqrt[3]{2}$.

In this case we obtain $(x, y, z) = \left(-\sqrt[3]{2}, \sqrt[3]{2}, -2\sqrt[3]{2} \right)$.

(iii) A normal vector to the level surface $L(f, 1)$ at $(-1, 2, 2)$ is

$$\mathbf{n}_1 = \nabla f(-1, 2, 2) = (2x + 2yz, 4y + 2xz, 2xy)\Big|_{(-1,2,2)} = (6, 4, -4).$$

Similarly, a normal vector to the level surface $L(g, 6)$ at $(-1, 2, 2)$ is

$$\mathbf{n}_2 = \nabla g(-1, 2, 2) = (-2xz, 3y^2, -x^2)\Big|_{(-1,2,2)} = (4, 12, -1).$$

The tangent line is perpendicular to $\mathbf{n}_1$ and $\mathbf{n}_2$, so it has the direction of the cross product

$$\mathbf{n}_1 \times \mathbf{n}_2 = \det \begin{pmatrix} \mathbf{i} & \mathbf{j} & \mathbf{k} \\ 6 & 4 & -4 \\ 4 & 12 & -1 \end{pmatrix} = 44\mathbf{i} - 10\mathbf{j} + 56\mathbf{k}.$$

An equation of the tangent line to $\mathcal{C}$ at the point $(-1, 2, 2)$ is

$$\begin{pmatrix} x \\ y \\ z \end{pmatrix} = \begin{pmatrix} -1 \\ 2 \\ 2 \end{pmatrix} + t \begin{pmatrix} 44 \\ -10 \\ 56 \end{pmatrix} \implies \begin{pmatrix} x \\ y \\ z \end{pmatrix} = \begin{pmatrix} 44t - 1 \\ -10t + 2 \\ 56t + 2 \end{pmatrix}, t \in \mathbb{R}.$$

12.7   Solutions Chapter 7

1.   (i) $\dfrac{\partial^2 f}{\partial x^2} = y^3 e^{xy-1}$, $\dfrac{\partial^2 f}{\partial xy} = \dfrac{\partial^2 f}{\partial yx} = y(2 + xy)e^{xy-1}$ and

$$\frac{\partial^2 f}{\partial y^2} = x(2 + xy)e^{xy-1}.$$

(ii) $\dfrac{\partial^2 f}{\partial x^2} = \dfrac{6xy}{\ln^2 y}, \; \dfrac{\partial^2 f}{\partial xy} = \dfrac{\partial^2 f}{\partial yx} = 3x^2\dfrac{\ln y - 2}{\ln^3 y}, \; \dfrac{\partial^2 f}{\partial y^2} = \dfrac{x^3}{y}\cdot\dfrac{6 - 2\ln y}{\ln^4 y}.$

(iii) $\dfrac{\partial^2 f}{\partial x^2} = (y + z)(y + z - 1)x^{y+z-2}$

$$\frac{\partial^2 f}{\partial xy} = \frac{\partial^2 f}{\partial yx} = \frac{\partial^2 f}{\partial xz} = \frac{\partial^2 f}{\partial zx} = [(y + z)\ln x + 1]x^{y+z-1}$$

$$\frac{\partial^2 f}{\partial y^2} = \frac{\partial^2 f}{\partial z^2} = \frac{\partial^2 f}{\partial yz} = x^{y+z}\ln^2 x.$$

(iv) $\dfrac{\partial^2 f}{\partial x_i \partial x_j} = 4\delta_{ij}\|\mathbf{x}\|^2 + 8x_i x_j, \; 1 \le i, j \le m.$

2. By quotient rule, for all $(x, y) \in \mathbb{R} \setminus \{(0,0)\}$, we have

$$\frac{\partial f}{\partial x}(x, y) = \frac{x^4 + 3x^2y^2 - 2xy^3}{(x^2 + y^2)^2}.$$

Also, by Definition 4.1 we find

$$\frac{\partial f}{\partial x}(0, 0) = \lim_{t \to 0} \frac{f(t, 0) - f(0, 0)}{t} = 1.$$

Hence,

$$\frac{\partial f}{\partial x}(x, y) = \begin{cases} \dfrac{x^4 + 3x^2y^2 - 2xy^3}{(x^2 + y^2)^2} & \text{if } (x, y) \ne (0,0) \\ 1 & \text{if } (x, y) = (0,0). \end{cases}$$

With the same approach, we have

$$\frac{\partial f}{\partial y}(x, y) = \begin{cases} \dfrac{y^4 + 3x^2y^2 - 2x^3y}{(x^2 + y^2)^2} & \text{if } (x, y) \ne (0,0) \\ 1 & \text{if } (x, y) = (0,0). \end{cases}$$

Now, by Definition 4.1 we compute

$$\frac{\partial^2 f}{\partial x^2}(0, 0) = \frac{\partial}{\partial x}\left(\frac{\partial f}{\partial x}\right)(0, 0)$$

$$= \lim_{t \to 0} \frac{\frac{\partial f}{\partial x}(t, 0) - \frac{\partial f}{\partial x}(0, 0)}{t} = \lim_{t \to 0} \frac{1 - 1}{t} = 0$$

and similarly $\dfrac{\partial^2 f}{\partial y^2}(0, 0) = 0$. Further, we have

$$\frac{\partial^2 f}{\partial x \partial y}(0, 0) = \frac{\partial}{\partial x}\left(\frac{\partial f}{\partial y}\right)(0, 0)$$

$$= \lim_{t \to 0} \frac{\frac{\partial f}{\partial y}(t, 0) - \frac{\partial f}{\partial y}(0, 0)}{t} = \lim_{t \to 0} \frac{-1}{t},$$

and the above limit does not exist. A similar approach yields $\dfrac{\partial^2 f}{\partial y \partial x}(0,0)$ does not exist.

3. As above we find

$$\frac{\partial f}{\partial x}(x,y) = \begin{cases} \dfrac{2x(x^2 - y^2)(x^2 + 3y^2)}{(x^2 + y^2)^2} & \text{if } (x,y) \neq (0,0) \\ 0 & \text{if } (x,y) = (0,0) \end{cases}$$

and

$$\frac{\partial f}{\partial y}(x,y) = \begin{cases} \dfrac{2y(y^2 - x^2)(3x^2 + y^2)}{(x^2 + y^2)^2} & \text{if } (x,y) \neq (0,0) \\ 1 & \text{if } (x,y) = (0,0). \end{cases}$$

Next, using Definition 4.1 we find $H_f(0,0) = \begin{pmatrix} 2 & 0 \\ 0 & 2 \end{pmatrix}$.

4. (i) Let $(a,b) \in \mathbb{R}^2$. If $a \neq 0$, then $B_r(a,b)$ with $r = |a| > 0$ is an open disc centred at (a,b) which does not intersect the y axis. Hence, for all $(x,y) \in B_r(a,b)$ one has $x \neq 0$ and thus $f(x,y) = x^2$. This yields $\dfrac{\partial f}{\partial y}(a,b) = 0$.

Assume now $a = 0$. Then, by Definition 4.1 we find

$$\frac{\partial f}{\partial y}(a,b) = \lim_{t \to 0} \frac{f(0, b+t) - f(0,b)}{t} = \lim_{t \to 0} \frac{(b+t)^2 - b^2}{t} = 2b.$$

Hence,

$$\frac{\partial f}{\partial y}(x,y) = \begin{cases} 0 & \text{if } x \neq 0 \\ 2y & \text{if } x = 0. \end{cases}$$

(ii) Using again Definition 4.1 we compute

$$\frac{\partial^2 f}{\partial y^2}(0,0) = \frac{\partial}{\partial y}\left(\frac{\partial f}{\partial y}\right)(0,0) = \lim_{t \to 0} \frac{\frac{\partial f}{\partial y}(0,t) - \frac{\partial f}{\partial y}(0,0)}{t} = 2,$$

$$\frac{\partial^2 f}{\partial x \partial y}(0,0) = \frac{\partial}{\partial x}\left(\frac{\partial f}{\partial y}\right)(0,0) = \lim_{t \to 0} \frac{\frac{\partial f}{\partial y}(t,0) - \frac{\partial f}{\partial y}(0,0)}{t} = 0.$$

5. Let $f_1(x,y) = \frac{\partial f}{\partial x}(x,y)$ and $f_2(x,y) = \frac{\partial f}{\partial y}(x,y)$. Then, f is twice differentiable if and only if f_1 and f_2 are differentiable on $\mathbb{R}^2$.

If $(x,y) \neq (0,0)$, then f is a rational function (quotient of two polynomials) in a neighbourhood of (x,y), so it is differentiable at (x,y) and $f_1(x,y) = -\frac{2xy^a}{(x^2+y^4)^2}$. Also,

$$f_1(0,0) = \frac{\partial f}{\partial x}(0,0) = \lim_{t \to 0} \frac{f(t,0) - f(0,0)}{t} = 0.$$

Hence,

$$f_1(x,y) = \frac{\partial f}{\partial x}(x,y) = \begin{cases} -\dfrac{2xy^a}{(x^2+y^4)^2} & \text{if } (x,y) \neq (0,0) \\ 0 & \text{if } (x,y) = (0,0). \end{cases}$$

Since f is twice differentiable, f_1 must be differentiable. We have

$$\frac{\partial f_1}{\partial x}(0,0) = \lim_{t \to 0} \frac{f_1(t,0) - f_1(0,0)}{t} = 0$$

$$\frac{\partial f_1}{\partial y}(0,0) = \lim_{t \to 0} \frac{f_1(0,t) - f_1(0,0)}{t} = 0.$$

Let us observe that for $(x,y) \neq (0,0)$, the function f_1 is a rational function in a neighbourhood of (x,y), so it is differentiable at (x,y). Therefore, f_1 is differentiable on $\mathbb{R}^2$ if and only if f_1 is differentiable at $(0,0)$. In light of Remark 4.13 this is equivalent to that fact that the function

$$g_1(x,y) := \frac{f_1(x,y) - f_1(0,0) - \nabla f_1(0,0) \bullet (x,y)}{\|(x,y)\|}$$

tends to zero as $(x,y) \to (0,0)$. Note that

$$g_1(x,y) = -\frac{2xy^a}{(x^2+y^4)^2 \sqrt{x^2+y^2}}.$$

Since $\lim\limits_{(x,y)\to(0,0)} g_1(x,y) = 0$, we have in particular that $\lim\limits_{y\to 0} g_1(y^2, y) = 0$ which yields $a > 7$. Since a is an integer, we must have $a \geq 8$.

We show that $a = 8$ is the smallest for which f is twice differentiable. Indeed, for $a = 8$ we see that

$$\begin{aligned} |g_1(x,y)| &= \frac{2|x|y^8}{(x^2+y^4)^2 \sqrt{x^2+y^2}} \\ &= \underbrace{\frac{2|x|y^2}{x^2+y^4}}_{\leq 1} \cdot \underbrace{\frac{y^4}{x^2+y^4}}_{\leq 1} \cdot \underbrace{\frac{|y|}{\sqrt{x^2+y^2}}}_{\leq 1} \cdot |y| \\ &\leq |y| \to 0 \quad \text{as } (x,y) \to (0,0). \end{aligned}$$

Hence $|g_1(x,y)| \to 0$ as $(x,y) \to (0,0)$. By Corollary 3.12 we conclude $\lim\limits_{(x,y)\to(0,0)} g_1(x,y) = 0$. This shows that f_1 is differentiable at

$(0,0)$ and thus, on $\mathbb{R}^2$. It remains to show that for $a = 8$ the function $f_2 = \frac{\partial f}{\partial y}$ is also differentiable. This fact is equivalent to proving the differentiability of f_2 at $(0,0)$. As above, we find

$$f_2(x,y) = \frac{\partial f}{\partial y}(x,y) = \begin{cases} \dfrac{4y^7(2x^2 + y^4)}{(x^2 + y^4)^2} & \text{if } (x,y) \neq (0,0) \\ 0 & \text{if } (x,y) = (0,0). \end{cases}$$

Also, $\dfrac{\partial f_2}{\partial x}(0,0) = \dfrac{\partial f_2}{\partial y}(0,0) = 0$. Hence, f_2 is differentiable at $(0,0)$ if and only if

$$g_2(x,y) := \frac{f_2(x,y) - f_2(0,0) - \nabla f_2(0,0) \bullet (x,y)}{\|(x,y)\|} \to 0$$

as $(x,y) \to (0,0)$. Note that $g_2(x,y) = \dfrac{4y^7(2x^2 + y^4)}{(x^2 + y^4)^2 \sqrt{x^2 + y^2}}$ and then

$$|g_2(x,y)| = 4 \cdot \underbrace{\frac{y^4}{x^2 + y^4}}_{\leq 1} \cdot \underbrace{\frac{2x^2 + y^4}{x^2 + y^4}}_{\leq 2} \cdot \underbrace{\frac{|y|}{\sqrt{x^2 + y^2}}}_{\leq 1} \cdot y^2 \leq 8y^2.$$

This shows that $g_2(x,y) \to 0$ as $(x,y) \to (0,0)$. From here we deduce that f_2 is differentiable at $(0,0)$, thereby on $\mathbb{R}^2$. We have shown that both f_1 and f_2 are differentiable for $a = 8$ and this is equivalent to f is twice differentiable. Hence, $a = 8$ is the smallest nonnegative integer with the required property.

6. Let $g(x,y) = f(x,y) - y^2 - xy$. Then, the Hessian matrix of $g(x,y)$ is the zero matrix. As in Example 7.7, we have

$$\nabla\left(\frac{\partial g}{\partial x}\right) = (0,0) \quad \text{and} \quad \nabla\left(\frac{\partial g}{\partial y}\right) = (0,0).$$

By Exercise 6 in Chapter 4, it follows that $g(x,y) = ax + by + c$ for some constants $a, b, c \in \mathbb{R}$ and thus, $f(x,y) = y^2 + xy + ax + by + c$ for all $(x,y) \in \mathbb{R}^2$.

7. Let $g(x,y) = f(x,y) - \frac{x^2 y^2}{2}$. Then, the Hessian matrix of $g(x,y)$ is the zero matrix and as above we find $g(x,y) = ax + by + c$ for some constants $a, b, c \in \mathbb{R}$. Thus,

$$f(x,y) = \frac{x^2 y^2}{2} + ax + by + c \quad \text{for all } (x,y) \in \mathbb{R}^2.$$

8. By Chain Rule we find

$$\frac{\partial g}{\partial t} = \frac{\partial g}{\partial x}\frac{\partial x}{\partial t} + \frac{\partial g}{\partial y}\frac{\partial y}{\partial t} = -4t\frac{\partial g}{\partial x} + 2st\frac{\partial g}{\partial y}.$$

Hence, by product rule,

$$\frac{\partial^2 g}{\partial s \partial t} = \frac{\partial}{\partial s}\left(\frac{\partial g}{\partial t}\right) = -4t\frac{\partial}{\partial s}\left(\frac{\partial g}{\partial x}\right) + 2t\frac{\partial g}{\partial y} + 2st\frac{\partial}{\partial s}\left(\frac{\partial g}{\partial y}\right). \quad (12.21)$$

Again by Chain Rule we find

$$\begin{aligned}
\frac{\partial}{\partial s}\left(\frac{\partial g}{\partial x}\right) &= \frac{\partial}{\partial x}\left(\frac{\partial g}{\partial x}\right)\frac{\partial x}{\partial s} + \frac{\partial}{\partial y}\left(\frac{\partial g}{\partial x}\right)\frac{\partial y}{\partial s} \\
&= 2s\frac{\partial^2 g}{\partial x^2} + t^2\frac{\partial^2 g}{\partial y \partial x}
\end{aligned} \qquad (12.22)$$

and

$$\begin{aligned}
\frac{\partial}{\partial s}\left(\frac{\partial g}{\partial y}\right) &= \frac{\partial}{\partial x}\left(\frac{\partial g}{\partial y}\right)\frac{\partial x}{\partial s} + \frac{\partial}{\partial y}\left(\frac{\partial g}{\partial y}\right)\frac{\partial y}{\partial s} \\
&= 2s\frac{\partial^2 g}{\partial x \partial y} + t^2\frac{\partial^2 g}{\partial y^2}.
\end{aligned} \qquad (12.23)$$

Finally, use (12.22) and (12.23) in (12.21) to deduce

$$\frac{\partial^2 g}{\partial s \partial t} = 2t\frac{\partial g}{\partial y} - 8st\frac{\partial^2 g}{\partial x^2} + 2st^3\frac{\partial^2 g}{\partial y^2} + 4t(s^2 - t^2)\frac{\partial^2 g}{\partial x \partial y}.$$

9. Using the Chain Rule we find

$$\frac{\partial h}{\partial t} = \frac{\partial h}{\partial x}\frac{\partial x}{\partial t} + \frac{\partial h}{\partial y}\frac{\partial y}{\partial t} = (s+r)\frac{\partial h}{\partial x} + (s-r)\frac{\partial h}{\partial y}.$$

By product rule, we have

$$\begin{aligned}
\frac{\partial^2 h}{\partial r \partial t} &= \frac{\partial}{\partial r}\left(\frac{\partial h}{\partial t}\right) \\
&= \frac{\partial h}{\partial x} + (s+r)\frac{\partial}{\partial r}\left(\frac{\partial h}{\partial x}\right) - \frac{\partial h}{\partial y} + (s-r)\frac{\partial}{\partial r}\left(\frac{\partial h}{\partial y}\right).
\end{aligned}$$

By Chain Rule again, we obtain

$$\begin{cases}
\dfrac{\partial}{\partial r}\left(\dfrac{\partial h}{\partial x}\right) = \dfrac{\partial}{\partial x}\left(\dfrac{\partial h}{\partial x}\right)\dfrac{\partial x}{\partial r} + \dfrac{\partial}{\partial y}\left(\dfrac{\partial h}{\partial x}\right)\dfrac{\partial y}{\partial r} = t\dfrac{\partial^2 h}{\partial x^2} - t\dfrac{\partial^2 h}{\partial y \partial x} \\[2mm]
\dfrac{\partial}{\partial r}\left(\dfrac{\partial h}{\partial y}\right) = \dfrac{\partial}{\partial x}\left(\dfrac{\partial h}{\partial y}\right)\dfrac{\partial x}{\partial r} + \dfrac{\partial}{\partial y}\left(\dfrac{\partial h}{\partial y}\right)\dfrac{\partial y}{\partial r} = t\dfrac{\partial^2 h}{\partial x \partial y} - t\dfrac{\partial^2 h}{\partial y^2}.
\end{cases}$$

Combining the last equalities we find

$$\begin{aligned}
\frac{\partial^2 h}{\partial r \partial t} = {} &\frac{\partial h}{\partial x} - \frac{\partial h}{\partial y} + (s+r)\left\{t\frac{\partial^2 h}{\partial x^2} - t\frac{\partial^2 h}{\partial y \partial x}\right\} \\
&+ (s-r)\left\{t\frac{\partial^2 h}{\partial x \partial y} - t\frac{\partial^2 h}{\partial y^2}\right\}.
\end{aligned}$$

To finalise, it remains to notice that $x = t(s+r)$ and $y = t(s-r)$ which used in the above equality yields the conclusion.

10. (i) Proceeding in a similar way as in Example 7.9 we find

$$\frac{\partial^2 f}{\partial r^2} = e^r \left\{ \cos t \frac{\partial f}{\partial x} + \sin t \frac{\partial f}{\partial y} \right\}$$

$$+ e^{2r} \left\{ \cos^2 t \frac{\partial^2 f}{\partial x^2} + 2 \sin t \frac{\partial^2 f}{\partial x \partial y} + \sin^2 t \frac{\partial^2 f}{\partial y^2} \right\},$$

$$\frac{\partial^2 f}{\partial r \partial t} = e^r \left\{ -\sin t \frac{\partial f}{\partial x} + \cos t \frac{\partial f}{\partial y} \right\} - e^{2r} \sin t \cos t \left\{ \frac{\partial^2 f}{\partial x^2} - \frac{\partial^2 f}{\partial y^2} \right\}$$

$$+ e^{2r} (\cos^2 t - \sin^2 t) \frac{\partial^2 f}{\partial x \partial y},$$

$$\frac{\partial^2 f}{\partial t^2} = - e^r \left\{ \cos t \frac{\partial f}{\partial x} + \sin t \frac{\partial f}{\partial y} \right\}$$

$$+ e^{2r} \left\{ \sin^2 t \frac{\partial^2 f}{\partial x^2} - 2 \sin t \frac{\partial^2 f}{\partial x \partial y} + \cos^2 t \frac{\partial^2 f}{\partial y^2} \right\}.$$

Part (ii) follows by adding the first and the third equality from above.

11. (i) Leting $x = -1$ and $y = 2$ in the implicit equation we find $z^3 - 2z - 4 = 0$, that is, $(z - 2)(z^2 + 2z + 2) = 0$ with the only solution $z = 2$ in the set of real numbers. Thus $z(-1, 2) = 2$. We next differentiate with respect to x and y variables to find

$$\begin{cases} 3z^2 \dfrac{\partial z}{\partial x} + yz + xy \dfrac{\partial z}{\partial x} + y^2 = 0 \\[2mm] 3z^2 \dfrac{\partial z}{\partial y} + xz + xy \dfrac{\partial z}{\partial y} + 2xy = 0 \end{cases}$$

and thus

$$\begin{cases} \dfrac{\partial z}{\partial x} = - \dfrac{y(z + y)}{3z^2 + xy} \\[3mm] \dfrac{\partial z}{\partial y} = - \dfrac{x(z + 2y)}{3z^2 + xy}. \end{cases} \tag{12.24}$$

Taking $x = -1$ and $y = 2$ in the above equalities one has

$$\frac{\partial z}{\partial x}(-1, 2) = -\frac{4}{5} \quad \text{and} \quad \frac{\partial z}{\partial y}(-1, 2) = \frac{3}{5}. \tag{12.25}$$

(ii) From (12.24), we have that $\dfrac{\partial z}{\partial x}$ and $\dfrac{\partial z}{\partial y}$ are differentiable functions as long as $3z^2 + xy \neq 0$. In particular, $\dfrac{\partial z}{\partial x}$ and $\dfrac{\partial z}{\partial y}$ are differentiable at $(x, y) = (-1, 2)$. From the second equation of (12.24), we have

$$(3z^2 + xy)\frac{\partial z}{\partial y} = -x(z + 2y).$$

Differentiate the above equality with respect to x and obtain

$$\left(6z\frac{\partial z}{\partial x}+y\right)\frac{\partial z}{\partial y}+(3z^2+xy)\frac{\partial^2 z}{\partial x\partial y}=-(z+2y)-x\frac{\partial z}{\partial x}.$$

Letting $(x,y,z)=(-1,2,2)$ and using (12.25) we deduce $\dfrac{\partial^2 z}{\partial x\partial y}(-1,2)=-\dfrac{28}{125}.$

12. (i) We have

$$\frac{\partial f}{\partial x}=\frac{2x}{x^2+y^2+z^2}\implies \frac{\partial^2 f}{\partial x^2}=\frac{2}{x^2+y^2+z^2}-\frac{4x^2}{(x^2+y^2+z^2)^2}$$

and similarly

$$\frac{\partial f}{\partial y}=\frac{2y}{x^2+y^2+z^2}\implies \frac{\partial^2 f}{\partial y^2}=\frac{2}{x^2+y^2+z^2}-\frac{4y^2}{(x^2+y^2+z^2)^2},$$

$$\frac{\partial f}{\partial z}=\frac{2z}{x^2+y^2+z^2}\implies \frac{\partial^2 f}{\partial z^2}=\frac{2}{x^2+y^2+z^2}-\frac{4z^2}{(x^2+y^2+z^2)^2}.$$

Hence,

$$\Delta f=\frac{6}{x^2+y^2+z^2}-\frac{4(x^2+y^2+z^2)}{(x^2+y^2+z^2)^2}=\frac{2}{x^2+y^2+z^2}.$$

(ii) Similar to the above calculations, we have

$$\frac{\partial f}{\partial x_i}(\mathbf{x})=k\|\mathbf{x}\|^{k-2}x_i$$

$$\frac{\partial^2 f}{\partial x_i^2}(\mathbf{x})=k\|\mathbf{x}\|^{k-2}+k(k-2)x_i^2\|\mathbf{x}\|^{k-4}.$$

Hence, $\Delta f(\mathbf{x})=k(k+m-2)\|\mathbf{x}\|^{k-2}.$

13. We show $\Delta f=0$ in each case by direct calculation. We may also use the previous exercise. The second order derivatives of f are given below.

(i) $\dfrac{\partial^2 f}{\partial x^2}=\dfrac{y^2-x^2}{(x^2+y^2)^2},\quad \dfrac{\partial^2 f}{\partial y^2}=\dfrac{x^2-y^2}{(x^2+y^2)^2}.$

(ii) $\dfrac{\partial^2 f}{\partial x_i^2}(\mathbf{x})=(2-m)\|\mathbf{x}\|^{-m}+m(m-2)x_i^2\|\mathbf{x}\|^{-m-2}.$

(iii) For $1\le i\le m$, we have

$$\frac{\partial^2 f}{\partial x_i^2}(\mathbf{x})=-2m\delta_{im}x_i\|\mathbf{x}\|^{-m-2}-mx_m\|\mathbf{x}\|^{-m-2}$$

$$+m(m+2)x_mx_i^2\|\mathbf{x}\|^{-m-4}.$$

14. Direct calculations, since $\dfrac{\partial^2(fg)}{\partial x_i^2} = f\dfrac{\partial^2 g}{\partial x_i^2} + 2\dfrac{\partial f}{\partial x_i}\dfrac{\partial g}{\partial x_i} + g\dfrac{\partial^2 f}{\partial x_i^2}$.

15. $\Delta f^3 = 3f^2\Delta f + 6f|\nabla f|^2$. Since f and f^3 are harmonic functions, it follows that $\Delta f = 0$ and then $f|\nabla f|^2 = 0$. This last equality yields $|\nabla f^3| = 0$, so f^3 is constant which further implies that f is constant.

16. Using the Chain Rule, we have

$$\frac{1}{r}\frac{\partial f}{\partial r} = \frac{1}{r}\frac{\partial f}{\partial x}\cos\theta + \frac{1}{r}\frac{\partial f}{\partial y}\sin\theta,$$

$$\frac{\partial^2 f}{\partial r^2} = \frac{\partial^2 f}{\partial x^2}\cos^2\theta + 2\frac{\partial^2 f}{\partial x\partial y}\sin\theta\cos\theta + \frac{\partial^2 f}{\partial y^2}\sin^2\theta,$$

$$\frac{1}{r^2}\frac{\partial^2 f}{\partial \theta^2} = \frac{\partial^2 f}{\partial x^2}\sin^2\theta - 2\frac{\partial^2 f}{\partial x\partial y}\sin\theta\cos\theta + \frac{\partial^2 f}{\partial y^2}\cos^2\theta$$
$$- \frac{1}{r}\frac{\partial f}{\partial x}\cos\theta - \frac{1}{r}\frac{\partial f}{\partial y}\sin\theta.$$

Adding the above three equalities, we find

$$\frac{1}{r}\frac{\partial f}{\partial r} + \frac{\partial^2 f}{\partial r^2} + \frac{1}{r^2}\frac{\partial^2 f}{\partial \theta^2} = \frac{\partial^2 f}{\partial x^2} + \frac{\partial^2 f}{\partial y^2},$$

which we may write

$$\frac{1}{r}\frac{\partial}{\partial r}\left(r\frac{\partial f}{\partial r}\right) + \frac{1}{r^2}\frac{\partial^2 f}{\partial \theta^2} = \frac{\partial^2 f}{\partial x^2} + \frac{\partial^2 f}{\partial y^2},$$

from which the conclusion follows immediately.

17. Let $\phi : \mathbb{R}\setminus\{0\} \to \mathbb{R}$, $\phi(r) = \dfrac{e^r - e^{-r}}{2r}$. Then, by L'Hôpital's rule one has

$$\lim_{r\to 0}\phi(r) = 1, \quad \lim_{r\to 0}\phi'(r) = 0 \quad \text{and} \quad \lim_{r\to 0}\phi''(r) = \frac{1}{3}. \quad (12.26)$$

Another way to deduce the above limits is to use the power series of e^r as follows:

$$e^r = 1 + r + \frac{r^2}{2} + \frac{r^3}{6} + \frac{r^4}{24} + \frac{r^5}{120} + \cdots,$$

$$e^{-r} = 1 - r + \frac{r^2}{2} - \frac{r^3}{6} + \frac{r^4}{24} - \frac{r^5}{120} + \cdots.$$

Hence,

$$\phi(r) = 1 + \frac{r^2}{6} + \frac{r^4}{120} + \cdots,$$

from which we deduce (12.26). Since $f(\mathbf{x}) = \phi(\|\mathbf{x}\|)$, as in Example 7.14, for all $\mathbf{x} \in \mathbb{R}^m \setminus \{\mathbf{0}\}$ we compute

$$\frac{\partial f}{\partial x_i}(\mathbf{x}) = \phi'(\|\mathbf{x}\|)\frac{x_i}{\|\mathbf{x}\|} \tag{12.27}$$

and

$$\frac{\partial^2 f}{\partial x_i \partial x_j}(\mathbf{x}) = \left(\phi''(\|\mathbf{x}\|) - \frac{\phi'(\|\mathbf{x}\|)}{\|\mathbf{x}\|}\right)\frac{x_i x_j}{\|\mathbf{x}\|^2} + \delta_{ij}\frac{\phi'(\|\mathbf{x}\|)}{\|\mathbf{x}\|}. \tag{12.28}$$

First, we note that f is twice continuously differentiable on $\mathbb{R}^m \setminus \{\mathbf{0}\}$. Also, $\lim_{\mathbf{x} \to 0} f(\mathbf{x}) = \lim_{r \to 0} \phi(r) = 1 = f(\mathbf{0})$, so f is continuous at $\mathbf{0}$ and hence on $\mathbb{R}^m$. Since $\lim_{r \to 0} \phi'(r) = 0$, by (12.27) and Corollary 3.12, we have

$$\lim_{\mathbf{x} \to 0} \frac{\partial f}{\partial x_i}(\mathbf{x}) = 0,$$

so f is differentiable on $\mathbb{R}^m$. Since by L'Hôpital's rule, we have

$$\lim_{r \to 0} \frac{\phi'(r)}{r} = \lim_{r \to 0} \phi''(r) = \frac{1}{3},$$

it follows that $\lim_{r \to 0} \phi''(r) - \frac{\phi'(r)}{r} = 0$ and again by Corollary 3.12 one has

$$\lim_{\mathbf{x} \to 0} \left(\phi''(\|\mathbf{x}\|) - \frac{\phi'(\|\mathbf{x}\|)}{\|\mathbf{x}\|}\right)\frac{x_i x_j}{\|\mathbf{x}\|^2} = 0.$$

Thus, from (12.28), we have that f has continuous first and second order derivatives on $\mathbb{R}^m$, thus, f is of class C^2 on $\mathbb{R}^m$.

(ii) Using Example 7.14 one has

$$\Delta f(\mathbf{x}) = \phi''(r) + \frac{m-1}{r}\phi'(r)$$

$$= \begin{cases} \left(1 - \dfrac{m-3}{\|\mathbf{x}\|^2}\right)f(\mathbf{x}) + \dfrac{m-3}{2}\dfrac{e^{\|\mathbf{x}\|} + e^{-\|\mathbf{x}\|}}{2\|\mathbf{x}\|} & \text{if } \mathbf{x} \neq \mathbf{0} \\[2ex] \dfrac{m}{3} & \text{if } \mathbf{x} = \mathbf{0}. \end{cases}$$

12.8   Solutions Chapter 8

1. For any $1 \leq i \leq m$ we compute

$$\frac{\partial g}{\partial x_i} = \frac{\partial f}{\partial x_i} + \sum_{j=1}^{m} x_j \frac{\partial^2 f}{\partial x_i \partial x_j}$$

and then

$$\frac{\partial^2 g}{\partial x_i^2} = 2\frac{\partial^2 f}{\partial x_i^2} + \sum_{j=1}^{m} x_j \frac{\partial^3 f}{\partial x_i^2 \partial x_j}.$$

Hence

$$\Delta g = \sum_{i=1}^{m} \frac{\partial^2 g}{\partial x_i^2} = 2\Delta f + \sum_{j=1}^{m} x_j \frac{\partial}{\partial x_j}(\Delta f) = 2\Delta f(\mathbf{x}) + \mathbf{x} \bullet \nabla(\Delta f(\mathbf{x})).$$

2. (i) $D^\alpha f(x,y) = x^{\alpha_2} y^{\alpha_1} e^{xy}$;

   (ii) $D^\alpha f(x,y,z) = x^{\alpha_2+\alpha_3} y^{\alpha_3+\alpha_1} z^{\alpha_1+\alpha_2} e^{xyz}$.

3. (i)–(ii) We can show by induction that $P_n(x,y) = ny^{n-1} + xy^n$.

4. (i) $Q(x,y) = x + y - 2xy + \dfrac{y^2}{2}$;

   (ii) $Q(x,y) = -4 - 8\ln 2 + 4x + 4(\ln 2)y + (x-2)^2 + 2(1+2\ln 2)(x - 2)(y-2) + 2(\ln^2 2)(y-2)^2$;

   (iii) $Q(x,y) = -2 + (\ln 2)x + 2y + (x-2)(y-1)$;

   (iv) $Q(x,y) = 1 - \dfrac{1}{2}\left(x - \dfrac{\pi}{2}\right)^2 - \dfrac{y^2}{2}$.

5. The linear approximation of f at (2.3) is

$$L_f(x,y) = f(2,3) + \frac{\partial f}{\partial x}(2,3)(x-2) + \frac{\partial f}{\partial y}(2,3)(y-3).$$

Plugging $(x,y) = (2.1,3)$ and then $(x,y) = (2,2.9)$ in the above, we find $\frac{\partial f}{\partial x}(2,3) = 10$ and $\frac{\partial f}{\partial y}(2,3) = 10$. Thus,

$$f(2.1, 3.1) \simeq L_f(2.1, 3.1)$$
$$= f(2,3) + \frac{\partial f}{\partial x}(2,3)(2.1-2) + \frac{\partial f}{\partial y}(2,3)(3.1-3) = 8.$$

6. Use the Chain Rule to find the derivatives of g. The quadratic Taylor approximation of g is $Q_g(x,y) = f(0,0) + \dfrac{\partial f}{\partial x}(0,0)x^2 + \dfrac{\partial f}{\partial y}(0,0)y^2$.

7. $Q_f(x,y) = 3 + 2x - 3y + 9(x-1)^2 - 24(x-1)(y-1) + 15(y-1)^2$.

8. $Q_f(x,y) = -5 + 4x - 2y - 3(x-2)^2 + 2(x-2)(y-2)$.

9. Identifying the coefficients as in Example 8.9, we have $f(1,-1) = 2$, $\frac{\partial f}{\partial x}(1,-1) = 3$, $\frac{\partial f}{\partial y}(1,-1) = 4$ and $g(1,-1) = -1$, $\frac{\partial g}{\partial x}(1,-1) = 2$, $\frac{\partial g}{\partial y}(1,-1) = 1$.

   (i) If $u = fg$, then $u(1,-1) = -2$ and

$$\frac{\partial u}{\partial x}(1,-1) = f(1,-1)\frac{\partial g}{\partial x}(1,-1) + g(1,-1)\frac{\partial f}{\partial x}(1,-1) = 1,$$

$$\frac{\partial u}{\partial y}(1,-1) = f(1,-1)\frac{\partial g}{\partial y}(1,-1) + g(1,-1)\frac{\partial f}{\partial y}(1,-1) = -2.$$

Thus, the linear approximation of fg at $(1, -1)$ is

$$L_{fg}(x, y) = -2 + (x - 1) - 2(y + 1) = x - 2y - 5.$$

(ii) If $v = \dfrac{f}{g}$, then $v(1, -1) = -2$ and

$$\frac{\partial v}{\partial x}(1, -1) = \frac{\frac{\partial f}{\partial x}(1, -1)g(1, -1) - \frac{\partial g}{\partial x}(1, -1)f(1, -1)}{g(1, -1)^2} = -7$$

$$\frac{\partial v}{\partial y}(1, -1) = \frac{\frac{\partial f}{\partial y}(1, -1)g(1, -1) - \frac{\partial g}{\partial y}(1, -1)f(1, -1)}{g(1, -1)^2} = -6.$$

Thus, the linear approximation of $\frac{f}{g}$ at $(1, -1)$ is

$$L_{f/g}(x, y) = -2 - 7(x - 1) - 6(y + 1) = -7x - 6y - 1.$$

(iii) If $w = f^g$, then $w(1, -1) = \frac{1}{2}$ and

$$\frac{\partial u}{\partial x}(1, -1) = \left\{ f^g(\ln f)\frac{\partial g}{\partial x} + gf^{g-1}\frac{\partial f}{\partial x} \right\}(1, -1) = \ln 2 - \frac{3}{4},$$

$$\frac{\partial u}{\partial y}(1, -1) = \left\{ f^g(\ln f)\frac{\partial g}{\partial y} + gf^{g-1}\frac{\partial f}{\partial y} \right\}(1, -1) = \frac{\ln 2}{2} - 1.$$

Thus, the linear approximation of f^g at $(1, -1)$ is

$$L_{f^g}(x, y) = \left(\ln 2 - \frac{3}{4} \right)x + \left(\frac{\ln 2}{2} - 1 \right)y + \frac{1 - 2\ln 2}{4}.$$

10. (i) Let $f(x, y) = \sin(x^2 + xy)$. By Theorem 8.10, we have $f(x, y) = x^2 + xy + g(x, y)$ where $g(x, y) = o(x^2 + y^2)$ as $(x, y) \to (0, 0)$. The limit is 1.

(ii) Let $f(x, y) = \ln(1 + xy - y)$. By Theorem 10.4, we have $f(x, y) = -y + g(x, y)$ where $g(x, y) = o(\|(x, y)\|)$ as $(x, y) \to (0, 0)$. The limit is 0.

(iii) Let $f(x, y) = e^{x^2 + 2y^2}$. By Theorem 8.10, we have $f(x, y) = 1 + x^2 + 2y^2 + g(x, y)$ where $g(x, y) = o(x^2 + y^2)$ as $(x, y) \to (0, 0)$. The limit is 1.

(iv) Let $f(x, y) = \sin(2x + y^3)$. By Theorem 8.10, we have $f(x, y) = 2x + g(x, y)$ where $g(x, y) = o(x^2 + y^2)$ as $(x, y) \to (0, 0)$. The limit is 0.

11. (i) Let $f : \mathbb{R}^m \to \mathbb{R}$, $f(\mathbf{x}) = x_1$, for all $\mathbf{x} = (x_1, x_2, \ldots, x_m)$ and $\mathbf{a} = \mathbf{0}$. Then, f is differentiable (since f is a polynomial function) and

$$\lim_{\mathbf{x} \to \mathbf{a}} \frac{f(\mathbf{x}) - f(\mathbf{a})}{\|\mathbf{x} - \mathbf{a}\|} = \lim_{\mathbf{x} \to 0} \frac{x_1}{\sqrt{x_1^2 + x_2^2 + \ldots + x_m^2}} \quad \text{does not exists,}$$

as we can see by computing the behaviour along the paths $\mathbf{x} = (t, 0, 0, \ldots, 0)$, $t \to 0$, $t > 0$ and then $\mathbf{x} = (t, 0, 0, \ldots, 0)$, with $t \to 0$, $t > 0$.

(ii) By Taylor's Theorem 8.4, we have

$$f(\mathbf{x}) = f(\mathbf{a}) + \nabla f(\mathbf{a}) \bullet (\mathbf{x} - \mathbf{a}) + g(\mathbf{x}), \qquad (12.29)$$

where g is a continuous function at $\mathbf{a}$ and

$$\lim_{\mathbf{x} \to \mathbf{a}} \frac{g(\mathbf{x})}{\|\mathbf{x} - \mathbf{a}\|} = 0. \qquad (12.30)$$

Thus,

$$\lim_{\mathbf{x} \to \mathbf{a}} \frac{f(\mathbf{x}) - f(\mathbf{a})}{\|\mathbf{x} - \mathbf{a}\|} \text{ exists} \iff \lim_{\mathbf{x} \to \mathbf{a}} k(\mathbf{x}) \text{ exists}, \qquad (12.31)$$

where

$$k : D \setminus \{\mathbf{a}\} \to \mathbb{R}, \quad k(\mathbf{x}) = \frac{\nabla f(\mathbf{a}) \bullet (\mathbf{x} - \mathbf{a})}{\|\mathbf{x} - \mathbf{a}\|}.$$

For $1 \le i \le m$, let $\mathbf{e}_i$ be the unit vector in $\mathbb{R}^m$ with 1 on the i-th entry and 0 elsewhere. Thus,

$$k(\mathbf{a} + t\mathbf{e}_i) = \frac{\partial f}{\partial x_i}(\mathbf{a}) \cdot \frac{t}{|t|}$$

and this fact together with (12.31) implies that $\lim_{t \to 0} k(\mathbf{a} + t\mathbf{e}_i)$ exists for all $1 \le i \le m$. This yields $\frac{\partial f}{\partial x_i}(\mathbf{a}) = 0$, so that $\nabla f(\mathbf{a}) = 0$. Conversely, if $\nabla f(\mathbf{a}) = \mathbf{0}$, then by (12.29) and (12.30) we deduce

$$\lim_{\mathbf{x} \to \mathbf{a}} \frac{f(\mathbf{x}) - f(\mathbf{a})}{\|\mathbf{x} - \mathbf{a}\|} = \lim_{\mathbf{x} \to \mathbf{a}} \frac{g(\mathbf{x})}{\|\mathbf{x} - \mathbf{a}\|} = 0.$$

12. Differentiate with respect to x and y in the implicit equations to obtain

$$\begin{cases} z\dfrac{\partial z}{\partial x} + \left(x\dfrac{\partial z}{\partial x} + z\right)e^{xz} = 0 \\[2mm] z\dfrac{\partial z}{\partial y} + x\dfrac{\partial z}{\partial y}e^{xz} + 3y = 0. \end{cases} \qquad (12.32)$$

In the above equations we let $(x, y) = (0, -1)$ and deduce

$$\frac{\partial z}{\partial x}(0, -1) = -1 \quad \text{and} \quad \frac{\partial z}{\partial y}(0, -1) = 3.$$

We further differentiate the first equation of (12.32) with respect to x and then with respect to y. We also differentiate the second equation of (12.32) with respect to y. We find

$$\begin{cases} \left(\dfrac{\partial z}{\partial x}\right)^2 + z\dfrac{\partial^2 z}{\partial x^2} + \left(x\dfrac{\partial z}{\partial x} + z\right)^2 e^{xz} + \left(2\dfrac{\partial z}{\partial x} + x\dfrac{\partial^2 z}{\partial x^2}\right)e^{xz} = 0 \\[2mm] \dfrac{\partial z}{\partial x}\dfrac{\partial z}{\partial y} + z\dfrac{\partial^2 z}{\partial x\partial y} + \left(x\dfrac{\partial^2 z}{\partial x\partial y} + \dfrac{\partial z}{\partial y}\right)e^{xz} + x\dfrac{\partial z}{\partial y}\left(x\dfrac{\partial z}{\partial x} + z\right)e^{xz} = 0 \\[2mm] \left(\dfrac{\partial z}{\partial y}\right)^2 + z\dfrac{\partial^2 z}{\partial y^2} + \left(x\dfrac{\partial z}{\partial y}\right)^2 e^{xz} + x\dfrac{\partial^2 z}{\partial y^2}e^{xz} + 3 = 0. \end{cases}$$

Taking $(x, y) = (0, -1)$ in the above equations we find

$$\frac{\partial^2 z}{\partial x^2}(0, -1) = 0, \quad \frac{\partial^2 z}{\partial x\partial y}(0, -1) = 0 \quad \text{and} \quad \frac{\partial^2 z}{\partial y^2}(0, -1) = -12.$$

Hence, the quadratic Taylor approximation of z at $(0, -1)$ is

$$Q_z(x, y) = 4 - x + 3y - 6(y + 1)^2.$$

13. Differentiating with respect to x in the implicit equations one has

$$\begin{cases} x\dfrac{\partial u}{\partial x} + u + y\dfrac{\partial v}{\partial x} = -y\dfrac{\partial u}{\partial x}\sin(yu) \\[2mm] y\dfrac{\partial u}{\partial x} + x\dfrac{\partial v}{\partial x} + v = \left(v + x\dfrac{\partial v}{\partial x}\right)\cos(xv). \end{cases}$$

Take $(x, y) = (0, 1)$ in the above and use $u(0, 1) = 0$, $v(0, 1) = 1$ to find

$$\begin{cases} \dfrac{\partial v}{\partial x}(0, 1) = 0 \\[2mm] \dfrac{\partial u}{\partial x} + 1 = 1 \end{cases} \quad \Longrightarrow \quad \begin{cases} \dfrac{\partial v}{\partial x}(0, 1) = 0 \\[2mm] \dfrac{\partial u}{\partial x} = 0. \end{cases}$$

Similarly, we differentiate with respect to y in the implicit equations to deduce

$$\begin{cases} x\dfrac{\partial u}{\partial y} + v + y\dfrac{\partial v}{\partial y} = -\left(u + y\dfrac{\partial u}{\partial y}\right)\sin(yu) \\[2mm] u + y\dfrac{\partial u}{\partial y} + x\dfrac{\partial v}{\partial y} = x\dfrac{\partial v}{\partial y}\cos(xv). \end{cases}$$

Taking again $(x, y) = (0, 1)$ one finds

$$\frac{\partial u}{\partial y}(0, 1) = 0 \quad \text{and} \quad \frac{\partial v}{\partial y}(0, 1) = -1.$$

The Taylor linear approximations at $(0, 1)$ are $L_u(x, y) = 0$ and $L_v(x, y) = 2 - y$.

12.9 Solutions Chapter 9

1. Follow line by line the proof of Theorem 9.1 in which one replaces x by y (and y by x).

2. Let $f : \mathbb{R}^2 \to \mathbb{R}$, $f(x,y) = x^2 y + y^3 - \cos(xy) + 1$.

   (i) Since $\dfrac{\partial f}{\partial y}(1,0) = x^2 + 3y^2 + x\sin(xy)\Big|_{(x,y)=(1,0)} = 1 \neq 0$, one may apply Theorem 9.1 to conclude that the equation $f(x,y) = 0$ can be solved near $(x,y) = (1,0)$ as a continuously differentiable function $y = y(x)$.

   (ii) Since $\dfrac{\partial f}{\partial x}(1,0) = 2xy + y\sin(xy)\Big|_{(x,y)=(1,0)} = 0$, the hypotheses of Theorem 9.1 are not fulfilled to reach the desired conclusion.

3. Let $f : (0,\infty) \times (0,\infty) \to \mathbb{R}$, $f(x,y) = x^y + y^x$. To apply Theorem 9.1 we impose $\dfrac{\partial f}{\partial y}(b,1) \neq 0$. This yields $b(1 + \ln b) \neq 0$. Since $b > 0$ we need to impose $b \neq 1/e$.

4. Let $g : \mathbb{R}^2 \to \mathbb{R}$ be given by $g(x,y) = f(f(x,y), xy)$. Then, g is continuously differentiable and $g(1,2) = 1$. To apply Theorem 9.1 we need to impose $\frac{\partial g}{\partial y}(1,2) \neq 0$. Observe that by Chain Rule one has

$$\frac{\partial g}{\partial y}(x,y) = \frac{\partial f}{\partial x}(f(x,y), xy)\frac{\partial f}{\partial y}(x,y) + x\frac{\partial f}{\partial y}(f(x,y), xy).$$

Hence,

$$\frac{\partial g}{\partial y}(1,2) \neq 0 \implies \frac{\partial f}{\partial x}(1,2)\frac{\partial f}{\partial y}(1,2) + \frac{\partial f}{\partial y}(1,2) \neq 0.$$

This yields

$$\frac{\partial f}{\partial y}(1,2)\left(\frac{\partial f}{\partial x}(1,2) + 1\right) \neq 0.$$

The conditions we need to impose are

$$\frac{\partial f}{\partial y}(1,2) \neq 0 \quad \text{and} \quad \frac{\partial f}{\partial x}(1,2) \neq -1.$$

5. From (9.3) one has

$$\phi'(x) = -\frac{\dfrac{\partial f}{\partial x}(x,\phi(x))}{\dfrac{\partial f}{\partial y}(x,\phi(x))} \quad \text{for all } x \in [a - \delta, a + \delta].$$

Hence, ϕ' is the quotient of two differentiable functions and thus ϕ is twice differentiable. The conclusion follows by differentiating the above equality and using again (9.3).

6. Let $f : \mathbb{R}^2 \to \mathbb{R}$ be given by $f(x, y, z) = e^{x+2y} + e^{y+2z} + e^{z+2x}$. Then, f is continuously differentiable and $\dfrac{\partial f}{\partial z}(0,0,0) = 3 \neq 0$. This shows that the hypotheses of Theorem 9.4 are fulfilled, so we can solve the equation $f(x, y, z) = 3$ in a neighbourhood of $(0,0,0)$ to obtain a continuously differentiable function $z = z(x, y)$.

7. Let $f : \mathbb{R}^2 \to \mathbb{R}$, $f(x, y) = xy + ay^2 z + bz^3$. First we impose $f(3, -2, 1) = 2$ which yields $4a + b = 8$. In order to use Theorem 9.4 we impose $\dfrac{\partial f}{\partial z}(3, -2) \neq 0$, so $4a + 3b \neq 0$. Using $b = 8 - 4a$ one finds $a \neq 3$. Hence: $a, b \in \mathbb{R}$, $a \neq 3$ and $b = 8 - 4a$.

8. Let $f : \mathbb{R}^2 \to \mathbb{R}$, $f(x, y) = xy^2 + yz + \sin(x - y + z)$.

   (i) Since $\dfrac{\partial f}{\partial z}\left(0, 0, \dfrac{\pi}{2}\right) = 0$, the hypotheses of Theorem 9.1 are not fulfilled. Thus, one may not apply Theorem 9.4 to conclude that the equation $f(x, y, z) = 0$ can be solved near $(x, y, z) = (0, 0, \pi/2)$ as a continuously differentiable function $z = z(x, y)$.

   (ii) Since $\dfrac{\partial f}{\partial y}\left(0, 0, \dfrac{\pi}{2}\right) = \dfrac{\pi}{2} \neq 0$, the hypotheses of Theorem 9.4 are fulfilled to conclude that the equation $f(x, y, z) = 0$ can be solved near $(x, y, z) = \left(0, 0, \dfrac{\pi}{2}\right)$ as a continuously differentiable function $y = y(x, z)$.

   (iii) Since $\dfrac{\partial f}{\partial x}\left(0, 0, \dfrac{\pi}{2}\right) = 0$, the hypotheses of Theorem 9.4 are not fulfilled to reach the desired conclusion.

9. Let $g : \mathbb{R}^3 \to \mathbb{R}$ be defined as $g(x, y, z) = f(xz) + zf(y - z) - yf(y - x)$. Since f is continuously differentiable, so is g. Also,

$$\frac{\partial g}{\partial z}(1, 2, 1) = xf'(xz) + f(y-z) - zf'(y-z)\Big|_{(x,y,z)=(1,2,1)} = f(1) \neq 0.$$

Hence, g fulfils the hypotheses of Theorem 9.4 from which the conclusion follows.

10. Let $g(x, y, z) = f(xy, yz, zx)$. Then $g(1, 1, 1) = 1$ and to apply Theorem 9.4 we need to impose $\dfrac{\partial g}{\partial z}(1, 1, 1) \neq 0$. By Chain Rule we find

$$\frac{\partial g}{\partial z}(1, 1, 1) = \frac{\partial f}{\partial y}(1, 1, 1) + \frac{\partial f}{\partial z}(1, 1, 1).$$

Hence, to apply Theorem 9.4 we impose

$$\frac{\partial f}{\partial y}(1, 1, 1) + \frac{\partial f}{\partial z}(1, 1, 1) \neq 0.$$

11. Consider the function $F : \mathbb{R}^3 \to \mathbb{R}^2$ defined by

$$F(x, y, z) = (x + y + z, 2^x + 3^y + 4^z).$$

Then, F is continuously differentiable on $\mathbb{R}^3$ and

$$J_{F(\cdot,\cdot,z)}(x, y) = \begin{pmatrix} 1 & 1 \\ 2^x \ln 2 & 3^y \ln 3 \end{pmatrix}.$$

For $(x, y, z) = (2, 1, 0)$ one has $\det J_{F(\cdot,\cdot,0)}(2, 1) = \ln \frac{27}{16} > 0$, so $J_{F(\cdot,\cdot,0)}(2, 1)$ is invertible. We thus satisfy the hypotheses in Theorem 9.6 for $a = 0$ and $\mathbf{b} = (2, 1)$. Thus, there exist $\delta > 0$, an open set $U \subset \mathbb{R}^2$ that contains $(2, 1)$ and a continuously differentiable function $\phi : [-\delta, \delta] \to \overline{U}$ such that:

(i) $\phi(0) = (2, 1)$;
(ii) $F(x, y, z) = (3, 8)$ for all $(x, y, z) \in \overline{U} \times [-\delta, \delta]$;
(iii) whenever $(x, y, z) \in \overline{U} \times [-\delta, \delta]$ and $F(x, y, z) = (3, 8)$ one has $(x, y) = \phi(z)$.

The last statement shows that we are entitled to write $(x, y) = \phi(z)$, so x and y are both functions of z. By (9.11) one has

$$J_{(x,y)}(z) = -\left[J_{F(\cdot,z)}(x, y) \right]^{-1} J_{F(x,y,\cdot)}(z).$$

Taking $(x, y, z) = (2, 1, 0)$ we find

$$J_{(x,y)}(0) = -\left[J_{F(\cdot,\cdot,0)}(2, 1) \right]^{-1} J_{F(2,1,\cdot)}(0)$$

$$= -\begin{pmatrix} 1 & 1 \\ \ln 16 & \ln 27 \end{pmatrix}^{-1} \begin{pmatrix} 1 \\ \ln 4 \end{pmatrix}$$

$$= -\frac{1}{\ln(27/16)} \begin{pmatrix} \ln 27 & -1 \\ -\ln 16 & 1 \end{pmatrix} \begin{pmatrix} 1 \\ \ln 4 \end{pmatrix}$$

$$= -\frac{1}{\ln(27/16)} \begin{pmatrix} \ln \dfrac{27}{4} \\ -\ln 4 \end{pmatrix}$$

Hence

$$x'(0) = -\frac{\ln(27/4)}{\ln(27/16)} \quad \text{and} \quad y'(0) = \frac{\ln 4}{\ln(27/16)}. \tag{12.33}$$

As in Remark 9.8 we could use directly the implicit differentiation in the simultaneous equations as in Section 5.2 of Chapter 5 to compute $x'(0)$ and $y'(0)$. Indeed, by the implicit differentiation with respect to z variable one has

$$\begin{cases} x'(z) + y'(z) + 1 = 0 \\ 2^{x(z)} x'(z) \ln 2 + 3^{y(z)} y'(z) \ln 3 + 4^z \ln 4 = 0. \end{cases}$$

Letting $z = 0$ for which $x(0) = 2$ and $y(0) = 1$ one has to solve the simultaneous equations

$$\begin{cases} x'(0) + y'(0) + 1 = 0 \\ x'(0) \ln 16 + y'(z) \ln 27 = -\ln 4. \end{cases}$$

From here one obtains (12.33).

12. Define $F : \mathbb{R}^4 \to \mathbb{R}^2$ by

$$F(x, y, u, v) = (x^3 + y^3 - u^2 - v^2, x^4 + y^4 - u^3 + v^3).$$

(i) We have $J_{F(1,1,\cdot,\cdot)}(1, -1) = \begin{pmatrix} -2 & 2 \\ -3 & 3 \end{pmatrix}$. Since $J_{F(1,1,\cdot,\cdot)}(1, -1)$ is not invertible, we cannot apply Theorem 9.6 to reach the conclusion.

(ii) We have $J_{F(\cdot,\cdot,1,-1)}(1, 1) = \begin{pmatrix} 3 & 3 \\ 4 & 4 \end{pmatrix}$. Since $J_{F(\cdot,\cdot,1,-1)}(1, 1)$ is not invertible, we cannot apply Theorem 9.6 to reach the conclusion.

13. Let $F : \mathbb{R}^4 \to \mathbb{R}^2$ be given by

$$F(x, y, u, v) = \left(ax^2v + uy^3 - e^{xu-v}, xu^2 + yv^3 - e^{yv} \right).$$

To apply Theorem 9.6 we impose $J_{F(1,0,\cdot,\cdot)}(1, 1)$ invertible. We have

$$J_{F(x,y,\cdot,\cdot)}(u, v) = \begin{pmatrix} y^3 - xe^{xu-v} & ax^2 + e^{xu-v} \\ 2xu & 3yv^2 - ye^{yv} \end{pmatrix}.$$

Hence, $J_{F(1,0,\cdot,\cdot)}(1, 1)$ is invertible if and only if $a \neq -1$.

14. Define $F : \mathbb{R}^5 \to \mathbb{R}^2$ by

$$F(x, y, z, u, v) = (xy^2 + zv - ve^{uz}, yv + u^2vx - ze^{uv}).$$

Then

$$J_{F(a,b,0,\cdot,\cdot)}(1, c) = \begin{pmatrix} 0 & -1 \\ 2ac & a + b \end{pmatrix}.$$

In order to apply the Implicit Function Theorem 9.6 we impose $J_{F(a,b,\cdot)}(1, c)$ is invertible, that is, $ac \neq 0$. Under this condition one can solve the system and deduce the existence of two continuously differentiable function $u = u(x, y, z)$ and $v = v(x, y, z)$ defined in a ball centred at $(x, y, z) = (a, b, 0)$.

15. Let $F : \mathbb{R}^5 \to \mathbb{R}^2$ be defined by

$$F(x, y, u, v, w) = \begin{pmatrix} xu + yv - w\sin(u + v) \\ yv + xw - u\sin(v + w) \\ xu + yw - v\sin(w + u) \end{pmatrix}.$$

Then

$$J_{F(0,0,\cdot,\cdot,\cdot)}(\pi,\pi,2\pi) = \begin{pmatrix} -2\pi & -2\pi & 0 \\ 0 & \pi & \pi \\ \pi & 0 & \pi \end{pmatrix}.$$

Since $\det J_{F(0,0,\cdot,\cdot,\cdot)}(\pi,\pi,2\pi) = -4\pi^3 \neq 0$, one can use the Implicit Function Theorem 9.6 to reach the conclusion.

16. Define $F : \mathbb{R}^3 \to \mathbb{R}^2$ by

$$F(x,y,z) = \big(xf(y) + yf(z), f(xy) + xf(yz)\big).$$

Observe first that F is continuously differentiable and for all $(x,y) \in \mathbb{R}^2$ one has

$$J_{F(\cdot,\cdot,z)}(x,y) = \begin{pmatrix} f(y) & xf'(y) + f(z) \\ yf'(xy) + f(yz) & xf'(xy) + xzf'(yz) \end{pmatrix}.$$

Since $f(1) = c/2 \neq 0$, at $(x,y,z) = 1$, we have

$$\det J_{F(\cdot,\cdot,1)}(1,1) = -f(1)^2 - f'(1)^2 < 0.$$

It follows that F satisfies the conditions in Theorem 9.6 and thus there exists two continuously differentiable functions $x = x(z)$ and $y = y(z)$ defined in a neighbourhood of $z = 1$ that satisfy the original simultaneous equations.

To find $x'(1)$ and $y'(1)$ we can either use Theorem 9.6 or we can differentiate implicitly in the two simultaneous equations. In the latter case we find

$$\begin{cases} f(1)x'(1) + (f(1) + f'(1))y'(1) = -f'(1) \\ (f(1) + f'(1))x'(1) + 2f'(1)y'(1) = -f'(1). \end{cases}$$

Solving the above system we find

$$x'(1) = \frac{f'(1)(f'(1) - f(1))}{f(1)^2 + f'(1)^2} \quad \text{and} \quad y'(1) = -\frac{f'(1)^2}{f(1)^2 + f'(1)^2}.$$

17. Let $F : \mathbb{R}^4 \to \mathbb{R}^2$ be given by

$$F(u,v,x,y) = (f(x) + y - u, x - f(y) - v).$$

Then, F is continuously differentiable (since f is continuously differentiable) and for all $(x,y) \in \mathbb{R}^2$, we have

$$\det J_{F(u,v,\cdot,\cdot)}(x,y) = \det \begin{pmatrix} f'(x) & 1 \\ 1 & -f'(y) \end{pmatrix}$$

$$= -f'(x)f'(y) - 1 < 0.$$

In the above we used the fact that f is monotone, so either $f'(x), f'(y) \geq 0$ (if f is increasing) or $f'(x), f'(y) \geq 0$ (if f is decreasing); thus $f'(x)f'(y) \geq 0$. Hence, $J_{F(0,0,\cdot,\cdot)}$ is invertible at (x_0, y_0) which shows that (x, y) can be uniquely expressed as a differentiable function of u and v variables on some ball $B_r(0,0)$, $r > 0$.

12.10 Solutions Chapter 10

1. (i) $(0,0)$ is a saddle point, $(3,3)$ and $(-3,-3)$ are minimum points.

   (ii) $(0,0)$ is a saddle point.

   (iii) $\left(-\frac{1}{2}, \frac{1}{2}\right)$ is a minimum point.

   (iv) The critical points are found by solving

   $$\nabla f(x,y) = (0,0) \Longrightarrow \begin{cases} \dfrac{6x}{x^2 + y} = y \\ \dfrac{3}{x^2 + y} = x. \end{cases} \tag{12.34}$$

   From the second equation of (12.34) we deduce $x \neq 0$ and then from the first equation of (12.34) we derive $y \neq 0$. We further obtain

   $$x^2 + y = \frac{6x}{y} \quad \text{and} \quad x^2 + y = \frac{3}{x} \Longrightarrow \frac{6x}{y} = \frac{3}{x} \Longrightarrow y = 2x^2.$$

   Using $y = 2x^2$ in the second equation of (12.34) we find $x = 1$ and then $y = 2$. Then, with Corollary 10.8, we have that $(1,2)$ is a saddle point of f.

2. (i) $\nabla f(x,y,z) = (y^3 + 2z, 3xy^2 + 4z^2 - 7, 8yz + 2x)$. The critical points are found by solving

   $$\begin{cases} y^3 + 2z = 0 \\ 3xy^2 + 4z^2 = 7 \\ 8yz + 2x = 0 \end{cases} \Longrightarrow \begin{cases} z = -\dfrac{y^3}{2} \\ 3xy^2 + 4z^2 = 7 \\ x = -4yz = 2y^4. \end{cases} \tag{12.35}$$

   Use the first and third equation of (12.35) into the second one to deduce $y^6 = 1$ so $(x,y,z) = \left(2,1,-\frac{1}{2}\right)$ or $(x,y,z) = \left(2,-1,\frac{1}{2}\right)$. The Hessian matrix at these points is

   $$H_f\left(2,1,-\frac{1}{2}\right) = \begin{pmatrix} 0 & 3 & 2 \\ 3 & 12 & -4 \\ 2 & -4 & 8 \end{pmatrix}, \quad H_f\left(2,-1,\frac{1}{2}\right) = \begin{pmatrix} 0 & 3 & 2 \\ 3 & -12 & 4 \\ 2 & 4 & -8 \end{pmatrix}.$$

If A_2 is the second leading principal submatrix in either of the above Hessian matrices, then $\det(A_2) = -9 < 0$ so $\left(2, 1, -\frac{1}{2}\right)$ and $\left(2, -1, \frac{1}{2}\right)$ are saddle points.

(ii) Looking for critical points of f one has to solve

$$\nabla f(x, y, z) = (0, 0, 0) \implies \begin{cases} y(1 - 2x^2) = 0 \\ x(1 - 2y^2) = 0 \\ 2xyz = 0. \end{cases} \tag{12.36}$$

If $x = 0$, then the first equation in (12.36) yields $y = 0$ and then $z \in \mathbb{R}$ can be any real number. A similar situation occurs if $y = 0$, then $x = 0$ and $z \in \mathbb{R}$.

Assume $xy \neq 0$. Then, (12.36) yields $x, y = \pm\frac{1}{\sqrt{2}}$ and $z = 0$. Thus, the critical points are

$$(0, 0, z), \quad \left(\pm\frac{1}{\sqrt{2}}, \pm\frac{1}{\sqrt{2}}, 0\right).$$

Also,

$$H_f(0, 0, z) = \begin{pmatrix} 0 & e^{-z^2} & 0 \\ e^{-z^2} & 0 & 0 \\ 0 & 0 & 0 \end{pmatrix},$$

so, $(0, 0, z)$ is a saddle point. Further,

$$H_f\left(\frac{1}{\sqrt{2}}, \frac{1}{\sqrt{2}}, 0\right) = H_f\left(-\frac{1}{\sqrt{2}}, -\frac{1}{\sqrt{2}}, 0\right)$$
$$= \begin{pmatrix} -2e^{-1} & 0 & 0 \\ 0 & -2e^{-1} & 0 \\ 0 & 0 & -e^{-1} \end{pmatrix},$$

and thus $\pm\left(\frac{1}{\sqrt{2}}, \frac{1}{\sqrt{2}}, 0\right)$ are local maximum points. Finally,

$$H_f\left(\frac{1}{\sqrt{2}}, -\frac{1}{\sqrt{2}}, 0\right) = H_f\left(-\frac{1}{\sqrt{2}}, \frac{1}{\sqrt{2}}, 0\right)$$
$$= \begin{pmatrix} 2e^{-1} & 0 & 0 \\ 0 & 2e^{-1} & 0 \\ 0 & 0 & e^{-1} \end{pmatrix},$$

and thus $\pm\left(\frac{1}{\sqrt{2}}, -\frac{1}{\sqrt{2}}, 0\right)$ are local minimum points.

(iii) $\nabla f(x, y, z) = (0, 0, 0)$ implies

$$\begin{cases} ye^{xy+z} = yz \\ xe^{xy+z} = xz \\ e^{xy+z} = xy + 1. \end{cases} \tag{12.37}$$

If $x \neq 0$ and $y \neq 0$, then (12.37) yields $e^{xy+z} = z = xy + 1$ and from here one gets $e^{2z-1} = z$. This further implies $z > 0$. Using the classical inequality $e^t \geq 1 + t$ for all $t \in \mathbb{R}$, we deduce

$$e^{2z-1} \geq (2z - 1) + 1 = 2z > z \quad \text{for all } z > 0.$$

Thus, there are no critical points with $xy \neq 0$. If $x = 0$, then the third equation of (12.37) yields $e^z = 1$ so $z = 0$ and then the first equation in (12.37) implies $x = y = 0$. Thus, $(0, 0, 0)$ is the only critical point of f. The Hessian matrix of f at $(0, 0, 0)$ is

$$H_f(0,0,0) = \begin{pmatrix} 0 & 1 & 0 \\ 1 & 0 & 0 \\ 0 & 0 & 1 \end{pmatrix},$$

and the second leading principal submatrix A_2 satisfies $\det(A_2) = -1 < 0$. Thus, $(0, 0, 0)$ is a saddle point of f.

3. Using the fact that the Hessian matrix $H_f(a, b)$ is symmetric, we have

$$\det H_f(a, b) = - \left(\frac{\partial^2 f}{\partial x \partial y}(a, b) \right)^2 \leq 0.$$

Since $\det H_f(a, b)$ is nonzero, from the above equality it must be negative, so (a, b) is a saddle point of f.

4. We impose $f(1, -1) = 2$, $\dfrac{\partial f}{\partial x}(1, -1) = 0$, $\dfrac{\partial f}{\partial y}(1, -1) = 0$ and obtain $a = 1$, $b = 2$, $c = -4$. Using the Second Order Derivative Test we cannot decide whether $(1, -1)$ is a local or a global extreme point. Instead, we remark that for $a = 1$, $b = 2$, $c = -4$ our function f reads

$$\begin{aligned} f(x, y) &= x^4 + 2y^4 - 4xy^2 + 3 \\ &= (x^2 - 1)^2 + 2(y^2 - x)^2 + 2 \\ &\geq 2 = f(1, -1) \quad \text{for all } (x, y) \in \mathbb{R}^2. \end{aligned}$$

Thus, $(1, -1)$ is a global minimum point.

5. Using Theorem 8.10 we identify the coefficients in the quadratic approximation as follows

$$f(1, 0) = -1, \quad \frac{\partial f}{\partial x}(1, 0) = \frac{\partial f}{\partial y}(1, 0) = 0,$$

and then

$$\frac{\partial^2 f}{\partial x^2}(1, 0) = 6, \quad \frac{\partial^2 f}{\partial x \partial y}(1, 0) = \frac{\partial^2 f}{\partial y \partial x}(1, 0) = 4, \quad \frac{\partial^2 f}{\partial y^2}(1, 0) = 4.$$

In particular $\nabla f(1, 0) = (0, 0)$ so $(1, 0)$ is a critical point of f.

The Hessian matrix of f at $(1,0)$ is $H_f(1,0) = \begin{pmatrix} 6 & 4 \\ 4 & 4 \end{pmatrix}$. Since $\det H_f(1,0) = 8 > 0$ and $\frac{\partial^2 f}{\partial x^2}(1,0) = 6 > 0$, we have that $(1,0)$ is a local minimum point of f.

6. Using implicit differentiation we find

$$\begin{cases} \dfrac{\partial z}{\partial x} = -\dfrac{2y}{3z^2 + \cos z} \\[3mm] \dfrac{\partial z}{\partial y} = -\dfrac{2x}{3z^2 + \cos z} \end{cases} \implies \dfrac{\partial z}{\partial x}(0,0) = \dfrac{\partial z}{\partial y}(0,0) = 0.$$

Hence, $(0,0)$ is a critical point of f. We may differentiate further in the above equalities to deduce

$$\begin{cases} \dfrac{\partial^2 z}{\partial x^2} = 2y\dfrac{6z - \sin z}{(3z^2 + \cos z)^2}\dfrac{\partial z}{\partial x} \\[3mm] \dfrac{\partial^2 z}{\partial x \partial y} = -2\dfrac{(3z^2 + \cos z) - y(6z - \sin z)\frac{\partial z}{\partial y}}{(3z^2 + \cos z)^2} \\[3mm] \dfrac{\partial^2 z}{\partial y^2} = 2x\dfrac{6z - \sin z}{(3z^2 + \cos z)^2}\dfrac{\partial z}{\partial y} \end{cases}$$

which yields

$$\frac{\partial^2 z}{\partial x^2}(0,0) = 0, \quad \frac{\partial^2 z}{\partial x \partial y}(0,0) = -2, \quad \frac{\partial^2 z}{\partial y^2}(0,0) = 0.$$

Hence, $(0,0)$ is a saddle point of f.

7. (i) The function $\phi : \mathbb{R} \to \mathbb{R}$, $\phi(z) = e^z + z$ is bijective. Thus, for any $(x,y) \in \mathbb{R}^2$ there exists a unique $z = z(x,y)$ such that $\phi(z) = \sin y - x^2$. This means that the original implicit equation has a unique solution $z = z(x,y) : \mathbb{R}^2 \to \mathbb{R}$.

(ii) Letting $(x,y) = \left(0, \dfrac{\pi}{2}\right)$ in the implicit equation we find $e^z + z = 1$ so, by the uniqueness of z one has $z\left(0, \dfrac{\pi}{2}\right) = 0$. By implicit differentiation we find

$$\frac{\partial z}{\partial x} = -\frac{2x}{e^z + 1} \quad \text{and} \quad \frac{\partial z}{\partial y} = \frac{\cos y}{e^z + 1}. \tag{12.38}$$

From the above we clearly see that $\left(0, \frac{\pi}{2}\right)$ is a critical point of z.

We differentiate once more in (12.38) to find

$$
\begin{cases}
\dfrac{\partial^2 z}{\partial x^2} = -\dfrac{2(e^z + 1) - 2xe^z \dfrac{\partial z}{\partial x}}{(e^z + 1)^2} \\[3em]
\dfrac{\partial^2 z}{\partial x \partial y} = \dfrac{2xe^z \dfrac{\partial z}{\partial y}}{(e^z + 1)^2} \\[3em]
\dfrac{\partial^2 z}{\partial y^2} = -\dfrac{(e^z + 1)\sin y + e^z(\cos y)\dfrac{\partial z}{\partial y}}{(e^z + 1)^2}
\end{cases}
\implies
\begin{cases}
\dfrac{\partial^2 z}{\partial x^2}\left(0, \dfrac{\pi}{2}\right) = -1 \\[1.5em]
\dfrac{\partial^2 z}{\partial x \partial y}\left(0, \dfrac{\pi}{2}\right) = 0 \\[1.5em]
\dfrac{\partial^2 z}{\partial y^2}\left(0, \dfrac{\pi}{2}\right) = -\dfrac{1}{2}.
\end{cases}
$$

Hence, $\left(0, \dfrac{\pi}{2}\right)$ is a maximum point of f.

8. (i) For $(x, y) = (0, 0)$ in the implicit equations we find

$$
\begin{cases}
2e^{u(0,0)} - e^{v(0,0)} = 1 \\
3e^{v(0,0)} - 2e^{u(0,0)} = 1
\end{cases}
\implies e^{u(0,0)} = e^{v(0,0)} = 1,
$$

so $u(0,0) = v(0,0) = 0$.

(ii) Differentiate with respect to x variable in both implicit equations to obtain

$$
\begin{cases}
2e^{u+y}\dfrac{\partial u}{\partial x} - e^{v-x}\left(\dfrac{\partial v}{\partial x} - 1\right) = 1 \\[1.5em]
3e^{v+y}\dfrac{\partial v}{\partial x} - 2e^{u-x}\left(\dfrac{\partial u}{\partial x} - 1\right) = 2.
\end{cases} \tag{12.39}
$$

Letting $(x, y) = (0, 0)$ and using $u(0,0) = v(0,0) = 0$ one finds

$$
\dfrac{\partial u}{\partial x}(0,0) = 0 \quad \text{and} \quad \dfrac{\partial v}{\partial x}(0,0) = 0.
$$

Now, differentiate implicitly with respect to y variable in the original implicit equations to deduce

$$
\begin{cases}
2e^{u+y}\left(\dfrac{\partial u}{\partial y} + 1\right) - e^{v-x}\dfrac{\partial v}{\partial y} = 2 \\[1.5em]
3e^{v+y}\left(\dfrac{\partial v}{\partial y} + 1\right) - 2e^{u-x}\dfrac{\partial u}{\partial y} = 3.
\end{cases} \tag{12.40}
$$

Similar to the above, for $(x, y) = (0, 0)$ one has

$$
\dfrac{\partial u}{\partial y}(0,0) = 0 \quad \text{and} \quad \dfrac{\partial v}{\partial y}(0,0) = 0.
$$

Thus, $\nabla u(0,0) = \nabla v(0,0) = (0,0)$ so $(0,0)$ is a critical point for both u and v.

In order to classify $(0,0)$ we need to find the Hessian matrix of both u and v at this point. Hence, we differentiable with respect to x the equations (12.39). We find

$$
\begin{cases}
2e^{u+y}\left(\dfrac{\partial u}{\partial x}\right)^2 + 2e^{u+y}\dfrac{\partial^2 u}{\partial x^2} - e^{v-x}\left(\dfrac{\partial v}{\partial x}-1\right)^2 - e^{v-x}\dfrac{\partial^2 v}{\partial x^2} = 0 \\[2mm]
3e^{v+y}\left(\dfrac{\partial v}{\partial x}\right)^2 + 3e^{v+y}\dfrac{\partial^2 v}{\partial x^2} - 2e^{u-x}\left(\dfrac{\partial u}{\partial x}-1\right)^2 - 2e^{u-x}\dfrac{\partial^2 u}{\partial x^2} = 0.
\end{cases}
$$

Take $(x,y)=(0,0)$ in the above equalities and find

$$
\begin{cases}
2\dfrac{\partial^2 u}{\partial x^2}(0,0) - \dfrac{\partial^2 v}{\partial x^2}(0,0) = 1 \\[2mm]
3\dfrac{\partial^2 v}{\partial x^2}(0,0) - 2\dfrac{\partial^2 u}{\partial x^2}(0,0) = 2.
\end{cases}
\quad\Longrightarrow\quad
\begin{cases}
\dfrac{\partial^2 u}{\partial x^2}(0,0) = \dfrac{5}{4} \\[2mm]
\dfrac{\partial^2 v}{\partial x^2}(0,0) = \dfrac{3}{2}.
\end{cases}
$$

We next differentiate with respect to y the equations in (12.39) and let $(x,y)=(0,0)$ to obtain

$$
\frac{\partial^2 u}{\partial x \partial y}(0,0) = \frac{\partial^2 v}{\partial x \partial y}(0,0) = 0.
$$

Finally, we differentiate with respect to y variable in (12.40) and let $(x,y)=(0,0)$ as above to find

$$
\frac{\partial^2 u}{\partial y^2}(0,0) = -\frac{9}{4} \quad \text{and} \quad \frac{\partial^2 v}{\partial y^2}(0,0) = -\frac{5}{2}.
$$

Now, the Hessian matrix at $(0,0)$ of u and v is

$$
H_u(0,0) = \begin{pmatrix} \frac{5}{4} & 0 \\ 0 & -\frac{9}{4} \end{pmatrix} \quad \text{and} \quad H_v(0,0) = \begin{pmatrix} \frac{3}{2} & 0 \\ 0 & -\frac{5}{2} \end{pmatrix}.
$$

It follows that $(0,0)$ is a saddled point of both functions u and v.

9. $(0,0)$ is a degenerate critical point of all three functions.

(i) Observe that for $t>0$ small, we have

$$
f(t,t) = \sin(t^3) > 0 = f(0,0)
$$
$$
f(t,-t) = -\sin(t^3) < 0 = f(0,0).
$$

This shows that $(0,0)$ is a saddle point of f.

(ii) Observe that for all $t>0$, we have

$$
f(0,t) = t^2 > 0 = f(0,0) \quad \text{and} \quad f(t,-t) = -t^4 < 0 = f(0,0).
$$

Thus, $(0,0)$ is a saddle point of f.

(iii) Note that $f(x,y) \geq 0 = f(0,0)$ for all $(x,y) \in \mathbb{R}^2$, so $(0,0)$ is a global minimum of f.

10. We have

$$\nabla f(x,y,z) = (0,0,0) \implies \begin{cases} 2xy + bz = 0 \\ 2ayz + x^2 = 0 \\ ay^2 + bx = 0 \end{cases} \implies \begin{cases} z = -\dfrac{2}{b}xy \\ x^2 = -2ayz \\ x = -\dfrac{a}{b}y^2. \end{cases}$$

Use the first and the third equation in the second equation to deduce $x = y = z = 0$. The Hessian matrix at $(0,0,0)$ is

$$H_f(0,0,0) = \begin{pmatrix} 0 & 0 & b \\ 0 & 0 & 0 \\ b & 0 & 0 \end{pmatrix}$$

so that $(0,0,0)$ is a degenerate critical point of f. Since $b \neq 0$ we observe that

$$f(t,t,t) = \big(b + (1+a)t\big)t^2$$
$$f(-t,t,t) = \big(-b + (1+a)t\big)t^2.$$

Thus, for t close to zero, $f(t,t,t)$ and $f(-t,t,t)$ are nonzero and have opposite sign. This shows that in any ball centred at $(0,0,0)$, the function f takes both positive and negative values. Thus, $(0,0,0)$ is a saddle point of f.

11. (i) The critical points are solutions of $\nabla f(x,y) = (0,0)$, that is,

$$\begin{cases} -\sin(x+y) + \cos(x-y) = 0 \\ -\sin(x+y) - \cos(x-y) = 0 \end{cases} \implies \sin(x+y) = \cos(x-y) = 0.$$

This yields

$$x + y = k\pi \quad \text{and} \quad x - y = \frac{(2\ell+1)\pi}{2} \quad \text{for some integers } k, \ell.$$

At these critical points the Hessian matrix is $H_f(x,y) = \begin{pmatrix} A & B \\ B & A \end{pmatrix}$ where

$$\begin{cases} A = -\cos(x+y) - \sin(x-y) = (-1)^{k+1} - (-1)^\ell \\ B = -\cos(x+y) + \sin(x-y) = (-1)^{k+1} + (-1)^\ell. \end{cases}$$

Hence

$$\det H_f(x,y) = A^2 - B^2 = (A+B)(A-B) = 4(-1)^{k+\ell+2} \neq 0.$$

(ii) For $(x,y) = \left(\frac{5\pi}{4}, \frac{3\pi}{4}\right)$, we have $k = 2$, $\ell = 0$ and then $A = -2$, $B = 0$ in the above calculations. We apply Corollary 10.8(ii) to reach the conclusion.

12. (i) $\nabla f(x, y, z) = (0, 0, 0)$ implies

$$\begin{cases} 2x \sin z = 0 \\ 4y \sin z = 0 \\ \sin z + (x^2 + 2y^2 + z) \cos z = 0. \end{cases}$$

(i1) If $\sin z = 0$, then $x^2 + 2y^2 = n\pi$ and $z = -n\pi$, where $n \geq 0$ is an integer.

(i2) If $\sin z \neq 0$, then $x = y = 0$ and $\sin z + z \cos z = 0$, so that $\tan z = -z$. In this case, the critical points are $(x, y, z) = (0, 0, z)$ where z satisfies $\tan z = -z$. Note that there are infinitely many solutions of this trigonometric equation (see Figure 12.5).

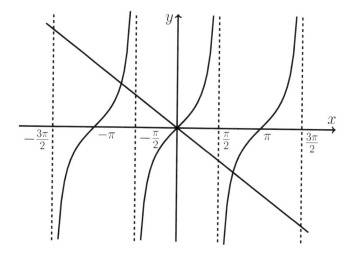

Figure 12.5
The infinitely many solutions of $\tan z = -z$.

We have thus obtained two infinite sets of critical points.

Let us consider the first class of critical points $(x, y, -n\pi)$ with $x^2 + 2y^2 = n\pi$. The Hessian matrix at these points reads

$$H_f(x, y, -n\pi) = \begin{pmatrix} 0 & 0 & 2(-1)^n x \\ 0 & 0 & 4(-1)^n y \\ 2(-1)^n x & 4(-1)^n y & 0 \end{pmatrix}$$

and $\det H_f(x, y, -n\pi) = 0$. Thus, all these critical points are degenerate.

(ii) Observe from Figure 12.5 that the equation $\tan z = -z$ has infinitely many solutions that lie in the third quadrant. Precisely, the equation $\tan z = -z$ has a solution in each of the intervals

$$\left(-\pi, -\frac{\pi}{2}\right), \ \left(-3\pi, -\frac{5\pi}{2}\right), \ \left(-5\pi, -\frac{9\pi}{2}\right), \ \dots$$

For each solution z in the above intervals, we have $\sin z < 0$ and $\cos z < 0$. The Hessian matrix at $(0, 0, z)$ reads

$$H_f(0,0,z) = \begin{pmatrix} 2\sin z & 0 & 0 \\ 0 & 4\sin z & 0 \\ 0 & 0 & (2+z^2)\cos z \end{pmatrix}.$$

Then, using the fact that $\sin z < 0$, $\cos z < 0$, the three leading principal minors satisfy $\det(A_1) < 0$, $\det(A_2) > 0$ and $\det(A_3) < 0$, so $(0, 0, z)$ is a local maximum point of f.

(iii) $(0, 0, 0)$ is a degenerate critical point of f. We observe that along the curve $(x, y, z) = (t, t, t^2)$, $t > 0$, we have

$$f(x, y, z) = 4t^2 \sin(t^2) > 0 \quad \text{for } t > 0 \text{ small.}$$

Also, along the curve $(x, y, z) = (t, t, -t^2)$, $t > 0$, we have

$$f(x, y, z) = -2t^2 \sin(t^2) < 0 \quad \text{for } t > 0 \text{ small.}$$

Thus, $(0, 0, 0)$ is a saddle point of f.

13. (i) By direct computations, the Hessian matrix of g at $(0, 0, 0)$ is

$$H_g(0,0,0) = \begin{pmatrix} 0 & f(0,0,0) & 0 \\ f(0,0,0) & 0 & 0 \\ 0 & 0 & 0 \end{pmatrix}.$$

Since $\det(A_2) = -f(0, 0, 0)^2 < 0$, we can use Corollary 10.7(iii) to deduce that $(0, 0, 0)$ is a saddle point.

(ii) Let $f(x, y, z) = xy$. Then $f(0, 0, 0) = 0$ and $g(x, y, z) = x^2 y^2$. One can check that $(0, 0, 0)$ is a degenerate critical point of g. On the other hand,

$$g(x, y, z) \geq 0 = g(0,0,0) \quad \text{for all } (x, y, z) \in \mathbb{R}^3,$$

so $(0, 0, 0)$ is a global minimum point of g.

14. The critical points are found by solving

$$\nabla f(x, y, z) = (0,0,0) \implies \begin{cases} \sin(x - y) = 0 \\ \cos(y + z) = 0. \end{cases}$$

Hence,

$$\begin{cases} x - y = k\pi \\ y + z = \ell\pi + \dfrac{\pi}{2}, \end{cases} \tag{12.41}$$

where k, ℓ are integers. The Hessian matrix is

$$\begin{pmatrix} \cos(x-y) & -\cos(x-y) & 0 \\ -\cos(x-y) & -\sin(y+z)+\cos(x-y) & -\sin(y+z) \\ 0 & -\sin(y+z) & -\sin(y+z) \end{pmatrix}.$$

(i) If k is even and ℓ is odd, then the Hessian matrix at these critical points is

$$H_f(x,y,z) = \begin{pmatrix} 1 & -1 & 0 \\ -1 & 2 & 1 \\ 0 & 1 & 1 \end{pmatrix} \quad \text{and} \quad \det(H_f(x,y,z)) = 0.$$

This shows that (x,y,z) is a degenerate critical point. Since $-1 \leq \sin t, \cos t \leq 1$ we see that $-2 \leq f \leq 2$. Thus, the critical points (x,y,z) that satisfy (12.41) with k even and ℓ odd one has $f(x,y,z) = -2$. Hence, all these critical points are global minima.

If k is odd and ℓ is even, similar to above all these critical points are degenerate. Since $f(x,y,z) = 2$, they yield global maxima of f.

(ii) If k and ℓ are both odd or both even, then the Hessian matrix is

$$H_f(x,y,z) = \pm\begin{pmatrix} 1 & -1 & 0 \\ -1 & 0 & -1 \\ 0 & -1 & -1 \end{pmatrix} \quad \text{and} \quad \det(H_f(x,y,z)) = 0.$$

We observe that the second leading principal submatrix in the above Hessian is

$$A_2 = \pm\begin{pmatrix} 1 & -1 \\ -1 & 0 \end{pmatrix} \quad \text{and} \quad \det(A_2) = -1 < 0,$$

so all critical points in this case are saddle points of f.

15. (i) Differentiate with respect to x and y in the implicit equation to deduce

$$\begin{cases} x\dfrac{\partial z}{\partial x} + z + 2y\dfrac{\partial z}{\partial x}e^{yz} = 0 \\[2mm] x\dfrac{\partial z}{\partial y} + 2(y-1) + 2\left(z + y\dfrac{\partial z}{\partial y}\right)e^{yz} = 0. \end{cases} \tag{12.42}$$

We let $(x,y) = (1,1)$ in the above and use $z(1,1) = 0$ to find

$$\frac{\partial z}{\partial x}(1,1) = \frac{\partial z}{\partial y}(1,1) = 0.$$

We differentiate further with respect to x and y in the first equation of (12.42) and with respect to y in the second equation of (12.42). We find

$$\begin{cases} 2\dfrac{\partial z}{\partial x} + x\dfrac{\partial^2 z}{\partial x^2} + 2\left(y\dfrac{\partial z}{\partial x}\right)^2 e^{yz} + 2y\dfrac{\partial^2 z}{\partial x^2}e^{yz} = 0 \\[2mm] x\dfrac{\partial^2 z}{\partial x\partial y} + \dfrac{\partial z}{\partial y} + 2y\dfrac{\partial z}{\partial x}\left(z + y\dfrac{\partial z}{\partial y}\right)e^{yz} + 2\left(\dfrac{\partial z}{\partial x} + y\dfrac{\partial^2 z}{\partial x\partial y}\right)e^{yz} = 0 \\[2mm] x\dfrac{\partial^2 z}{\partial y^2} + 2 + 2\left(z + y\dfrac{\partial z}{\partial y}\right)^2 e^{yz} + 2\left(2\dfrac{\partial z}{\partial y} + y\dfrac{\partial^2 z}{\partial y^2}\right)e^{yz} = 0. \end{cases}$$

Letting $(x, y) = (1, 1)$ in the above equalities we derive the Hessian matrix of z at $(1, 1)$ as $H_z(1, 1) = \begin{pmatrix} 0 & 0 \\ 0 & -\frac{2}{3} \end{pmatrix}$. Thus, $(1, 1)$ is a degenerate critical point of z.

(ii) We claim that $z(x, y) \leq 0$ for all $(x, y) \in B_r(1, 1)$. Indeed, if there exists $(x, y) \in B_r(1, 1)$ with $z(x, y) > 0$, then, since $r \in (0, 1)$ one has $x > 0$ and $y > 0$. This implies

$$xz(x, y) > 0 \quad \text{and} \quad yz(x, y) > 0$$

so that

$$xz(x, y) + (y - 1)^2 + 2e^{yz(x,y)} > 2e^0 = 2, \quad \text{contradiction.}$$

Hence, $z(x, y) \leq 0 = z(1, 1)$ for all $(x, y) \in B_r(1, 1)$ which shows that $(1, 1)$ is a global maximum of z.

12.11   Solutions Chapter 11

1. Let $g(x, y) = x^2 + y^2$. Since $\nabla g(x, y) \neq (0, 0)$ on the set $\{g(x, y) = 5\}$, we apply the Lagrange Multipliers Method and solve

$$\begin{cases} \nabla f(x, y) = \lambda g(x, y) \\ g(x, y) = 20 \end{cases} \implies \begin{cases} 3x^2 + 3y^2 - 3y = 2\lambda x \\ 6xy - 3x = 2\lambda y \\ x^2 + y^2 = 5. \end{cases} \qquad (12.43)$$

We use the third equation in the first one of (12.43) to re-write the above as

$$\begin{cases} 15 - 3y = 2\lambda x \\ 6xy - 3x = 2\lambda y \\ x^2 + y^2 = 5. \end{cases} \qquad (12.44)$$

If $x = 0$, then the first equation of (12.44) yields $y = 5$ which contradicts the last equation of (12.44). If $y = 0$, then the second equation of (12.44) implies $x = 0$ and again this contradicts the last equation of (12.44). Hence, $x \neq 0$ and $y \neq 0$. From the first two equations of (12.44) one deduces

$$2\lambda = \frac{15 - 3y}{x} = \frac{6xy - 3x}{y} \implies \frac{5 - y}{x} = \frac{2xy - x}{y}.$$

This last equality yields $5y - y^2 = 2x^2y - x^2$ and then, using $x^2 = 5 - y^2$ we find the cubic equation $2y^3 - 2y^2 - 5y + 5 = 0$ with the real solutions $y = 1$ and $y^2 = \frac{5}{2}$. We obtain

$$(x, y) = (\pm 2, 1), \left(\pm \frac{\sqrt{10}}{2}, \pm \frac{\sqrt{10}}{2} \right).$$

The global maximum is $\dfrac{5(2\sqrt{10} + 3)}{2}$ and the global minimum is $-\dfrac{5(2\sqrt{5} + 3)}{2}$.

2. Let $g(x, y) = x^2 + 2y^2$. We want to find the extrema of $f(x, y) = (x - 2)^2 + y^2$ subject to the constraint $g(x, y) = 88$. Observe first that $\nabla g(x, y) \neq (0, 0)$ on the set $\{g(x, y) = 88\}$, so by the Lagrange Multipliers Method we solve

$$\begin{cases} \nabla f(x, y) = \lambda g(x, y) \\ g(x, y) = 88 \end{cases} \implies \begin{cases} x - 2 = \lambda x \\ y = 2\lambda y \\ x^2 + 2y^2 = 88. \end{cases} \quad (12.45)$$

From the second equation of (12.45) we find $y = 0$ or $\lambda = \frac{1}{2}$.

(i) If $y = 0$, then the third equation of (12.45) implies $x = \pm 2\sqrt{22}$.

(ii) If $\lambda = \frac{1}{2}$, then the first and third equation of (12.45) yield $x = 4$ and $y = \pm 6$.

The largest distance from $(2, 0)$ to the ellipse is $2\sqrt{22} + 2$ and the shortest distance is $2\sqrt{10}$.

3. Let $g(x, y) = x^2 + y^2$. Since $\nabla g(x, y) \neq (0, 0)$ on the constraint set $\{g(x, y) = 5\}$, it follows that all extreme points of f subject to $g(x, y) = 5$ are found by solving

$$\begin{cases} \nabla f(x, y) = \lambda g(x, y) \\ g(x, y) = 5 \end{cases} \implies \begin{cases} ay = 2\lambda x \\ ax + 6y = 2\lambda y \\ x^2 + y^2 = 5. \end{cases} \quad (12.46)$$

Since $(2, -1)$ is one extreme point, it must satisfy the above conditions, so $a = -4\lambda$ and $2a - 6 = -2\lambda$. This yields $\lambda = -1$ and

$a = 4$. Next, by solving (12.46) with $a = 4$ we derive that $(2, -1)$ is a global minimum of f.

4. Let $g(x, y) = x^2 + y^2$ and note that $\nabla g(x, y) \neq (0, 0)$ on the constraint $\{g(x, y) = 25\}$. If $(3, 4)$ was an extreme point of f subject to the constraint $\{g(x, y) = 25\}$, then,

$$\nabla f(3, 4) = \lambda \nabla g(3, 4) = 2\lambda(3, 4), \qquad (12.47)$$

for some real parameter λ. On the other hand f increases the most rapidly at $(3, 4)$ in the direction $(-1, 2)$ so $\nabla f(3, 4) = a(-1, 2)$ for some real number $a \neq 0$. This last equality together with (12.47) yields $a(-1, 2) = 2\lambda(3, 4)$, which is impossible.

5. Let $g(x, y) = \sin x + \sin y$ and it is easy to check that $\nabla g(x, y) \neq (0, 0)$ on the constraint $g(x, y) = 1$. By the Lagrange Multipliers Method we solve

$$\begin{cases} \nabla f(x, y) = \lambda g(x, y) \\ g(x, y) = 1 \end{cases} \implies \begin{cases} -\sin(x - y) = \lambda \cos x \\ \sin(x - y) = \lambda \cos y \\ \sin x + \sin y = 1. \end{cases} \qquad (12.48)$$

From the first two equations in (12.48) we find $\lambda \cos x = -\lambda \cos y$, so $\lambda = 0$ or $\cos x = -\cos y$.

(i) If $\lambda = 0$, then (12.48) yields $\sin(x - y) = 0$ so $x - y = k\pi$ for some integer k. This further implies $x = k\pi + y$ and then $\sin x = (-1)^k \sin y$. Thus, the constraint $\sin x + \sin y = 1$ may hold only if k is even, so $k = 2\ell$ for some integer ℓ and then $x = 2\ell\pi + y$. Hence $f(x, y) = \cos(x - y) = 1$ and this is a global maximum of f.

(ii) If $\cos x = -\cos y$, then $\cos x = \cos(\pi + y)$ which yields $x = (2k + 1)\pi \pm y$, where k is an integer.

(ii1) If $x = (2k + 1)\pi + y$, then $\sin x = -\sin y$ and this contradicts the constraint $\sin x + \sin y = 1$.

(ii2) If $x = (2k + 1)\pi - y$, then $\sin x = \sin y$ and the constraint yields $\sin x = \sin y = \frac{1}{2}$. Then

$$f(x, y) = \cos(x - y) = \cos\left((2k + 1)\pi - 2y\right)$$
$$= -\cos(2y) = 2\sin^2 y - 1 = -\frac{1}{2}.$$

The global minimum of f is $-\frac{1}{2}$ and the global maximum value of f is 1.

6. Let $g(x, y, z) = x^2 + y^2 + z^2$ and observe that $\nabla g(x, y, z) \neq (0, 0, 0)$ on the constraint $\{g(x, y, z) = 5\}$. Hence, by Lagrange Multipliers

Method we solve

$$\begin{cases} \nabla f(x,y,z) = \lambda g(x,y,z) \\ g(x,y,z) = 5 \end{cases} \implies \begin{cases} 2x = 2\lambda x \\ -6y + 4z = \lambda y \\ 8y = 2\lambda z \\ x^2 + y^2 + z^2 = 5. \end{cases} \quad (12.49)$$

From the first equation of (12.49) we derive $x = 0$ or $\lambda = 1$.

(i) If $x = 0$, then we exploit the third equation of (12.49) which yields $y = \frac{\lambda}{4}z$ which used in the second equation of (12.49) implies $z(\lambda^2 + 6\lambda - 16) = 0$, so $z = 0$ or $\lambda = 2$ or $\lambda = -8$. We find $(x,y,z) = (0,1,2)$, $(0,-1,-2)$, $(0,-2,1)$, $(0,2,-1)$.

(ii) If $\lambda = 1$, then the third and the second equation of (12.49) yield $y = z = 0$ and then $(x,y,z) = (\pm\sqrt{5},0,0)$.

The minimum is $f(0,-2,1) = f(0,2,-1) = -40$ and the maximum is $f(0,1,2) = f(0,-1,-2) = 10$.

7. Let $g(x,y,z) = x^2 + y^2 + 3z^2$ and observe first that $\nabla g(x,y,z) \neq (0,0,0)$ on the constraint $g(x,y,z) = 20$. Thus, we solve

$$\begin{cases} \nabla f(x,y,z) = \lambda g(x,y,z) \\ g(x,y,z) = 20 \end{cases} \implies \begin{cases} y + 3z = 2\lambda x \\ x + 3z = 2\lambda y \\ x + y = 2\lambda z \\ x^2 + y^2 + 3z^2 = 20. \end{cases} \quad (12.50)$$

We add the first two equations and use the third equation of (12.50) to obtain

$$(x + y) + 6z = 2\lambda(x + y) \implies 2\lambda z + 6z = 4\lambda^2 z.$$

Hence, $z = 0$ or $2\lambda^2 - \lambda - 3 = 0$ which yields $\lambda = -1$ or $\lambda = \frac{3}{2}$.

(i) If $z = 0$, then the last two equations of (12.50) yield $x + y = 0$ and $x^2 + y^2 = 20$ so $(x,y,z) = \pm(\sqrt{10}, -\sqrt{10}, 0)$.

(ii) If $\lambda = -1$, then (12.50) yields $x + y = -2z$ and $y + 3z = -2x$. Subtracting these equations we find $x = -z$ and then $y = -z$. We obtain $(x,y,z) = \pm(2,2,-2)$.

(iii) If $\lambda = \frac{3}{2}$, then (12.50) yields $x + y = 3z$ and $y + 3z = 3x$. By addition we find

$$x = y = \frac{3z}{2} \quad \text{and then} \quad (x,y,z) = \pm\left(\sqrt{6}, \sqrt{6}, \frac{2\sqrt{6}}{3}\right).$$

The global minimum of f is

$$f(2,2,-2) = f(-2,-2,2) = -20$$

and the global maximum value is

$$f\left(\sqrt{6}, \sqrt{6}, \frac{2\sqrt{6}}{3}\right) = f\left(-\sqrt{6}, -\sqrt{6}, -\frac{2\sqrt{6}}{3}\right) = 30.$$

8. (i) If $\mathbf{a} = \mathbf{0}$, then both sides of the inequality are zero.

(ii) Dividing the whole inequality with $\|\mathbf{a}\| \neq 0$ we arrive at

$$(b_1 x_1 + b_2 x_2 + \cdots + b_m x_m)^2 \leq x_1^2 + x_2^2 + \cdots + x_m^2$$

for all $(x_1, x_2, \ldots, x_m) \in \mathbb{R}^m$. This reads

$$(\mathbf{b} \bullet \mathbf{x})^2 \leq \|\mathbf{x}\|^2 \quad \text{for all } \mathbf{x} = (x_1, x_2, \ldots, x_m) \in \mathbb{R}^m. \quad (12.51)$$

(iii) If $\mathbf{x} = \mathbf{0}$, then (12.51) holds. Also, (12.51) is homogeneous in the sense that if it holds for $\mathbf{x} \in \mathbb{R}^m$, then it also holds for $\lambda \mathbf{x}$ for any $\lambda > 0$. Thus, if it holds for all $\mathbf{x} \in \mathbb{R}^m$ with $\|\mathbf{x}\| = 1$ it holds for all $\mathbf{x} \in \mathbb{R}^m$. Let us consider $f : \mathbb{R}^n \to \mathbb{R}$, $f(\mathbf{x}) = (\mathbf{b} \bullet \mathbf{x})^2$ subject to $\|\mathbf{x}\| = 1$. The conditions in the Lagrange Multipliers Method are fulfilled and if $\mathbf{x} = (x_1, x_2, \ldots, x_m)$ is an extreme point of f subject to $\|\mathbf{x}\| = 1$ we solve

$$\begin{cases} b_i(\mathbf{b} \bullet \mathbf{x}) = \lambda x_i & \text{for all } 1 \leq i \leq m, \\ x_1^2 + x_2^2 + \cdots + x_m^2 = 1. \end{cases} \quad (12.52)$$

We multiply the first equation of (12.52) by x_i and sum over $1 \leq i \leq m$ to obtain $(\mathbf{b} \bullet \mathbf{x})^2 = \lambda \|\mathbf{x}\|^2 = \lambda$.

If $\lambda = 0$, then $\mathbf{b} \bullet \mathbf{x} = 0$ which shows that $f(\mathbf{x}) = \mathbf{0}$. This yields a global minimum of f, since $f \geq 0$ on $\mathbb{R}^m$.

If $\lambda \neq 0$, then (12.52) and $\lambda = (\mathbf{b} \bullet \mathbf{x})^2$ imply

$$x_i = \frac{b_i}{\lambda}(\mathbf{b} \bullet \mathbf{x}) = \frac{b_i}{\mathbf{b} \bullet \mathbf{x}}, \quad \text{so} \quad \mathbf{x} = \frac{\mathbf{b}}{\mathbf{b} \bullet \mathbf{x}}.$$

Then $\|\mathbf{b}\| = \|\mathbf{x}\| = 1$ implies $\mathbf{b} \bullet \mathbf{x} = \pm 1$ and thus

$$f(\mathbf{x}) = (\mathbf{b} \bullet \mathbf{x})^2 = 1.$$

This shows that $f(\mathbf{x}) \leq 1$ for all $\mathbf{x} \in \mathbb{R}^m$ with $\|\mathbf{x}\| = 1$. This establishes (12.51) whence our original inequality follows.

9. Let $D = \{(x, y, z) \in \mathbb{R}^3 : x, y, z \geq 0\}$ and

$$f : D \to \mathbb{R}, \quad f(x, y, z) = \sin(2x) + \sin(2y) + \sin(2z).$$

We want to find the extrema of f subject to $x + y + z = \pi$. By Lagrange Multipliers Method (whose hypotheses are easy to verify) one solves

$$\cos(2x) = \cos(2y) = \cos(2z) = \frac{\lambda}{2} \quad \text{and} \quad x + y + z = \pi.$$

Since $0 \le y, z \le \pi$ it follows that $y, z \in \{x, \pi - x\}$. As in Example 11.7, three main situations along with their permutations may occur:

$$\begin{cases} x = y \\ z = \pi - x \end{cases} \qquad \text{or} \qquad z = y = \pi - x \qquad \text{or} \qquad x = y = z.$$

We find $(x, y, z) = (\pi, 0, 0)$, $(0, \pi, 0)$, $(0, 0, \pi)$, $\left(\frac{\pi}{3}, \frac{\pi}{3}, \frac{\pi}{3}\right)$ and then $f(\pi, 0, 0) = f(0, \pi, 0) = f(0, 0, \pi) = 0$ is the minimum value of f while $f\left(\frac{\pi}{3}, \frac{\pi}{3}, \frac{\pi}{3}\right) = 3\sqrt{3}/2$ is the maximum value. Finally, we note that the upper bound in the original inequality is achieved if and only if the triangle ABC is equilateral.

10. Let $D = \{(x, y, z, t) \in \mathbb{R}^4 : x, y, z, t \ge 0\}$ and

$$f : D \to \mathbb{R}, \quad f(x, y, z, t) = \cos x + \cos y + \cos z + \cos t.$$

We want to find the extrema of f under the constraint $x + y + z + t = 2\pi$. Using the Lagrange Multipliers Method we solve

$$\sin x = \sin y = \sin z = \sin t = -\frac{\lambda}{2} \quad \text{and} \quad x + y + z + t = 2\pi.$$

From the above we find $y, z, t \in \{x, \pi - x\}$. The values $(2\pi, 0, 0)$, $(0, 2\pi, 0, 0)$, $(0, 0, 2\pi, 0)$ and $(0, 0, 0, 2\pi)$ yield a global maximum for f. The minimum value is achieved at $(x, x, \pi - x, \pi - x)$, $0 \le x \le \pi$ and all possible permutations. In such a case $ABCD$ is either a trapeze or a cyclic quadrilateral.

11. Let $g_1(x, y, z) = x^2 + y^2 + z^2$ and $g_2(x, y, z) = y(x + z)$. One can check that the Jacobian matrix of $G = (g_1, g_2)$ has rank 2 at all points (x, y, z) which satisfy the two constraints. Thus, by the Lagrange Multipliers Method we solve

$$\begin{cases} 1 = 2\lambda_1 x + \lambda_2 y \\ 1 = 2\lambda_1 y + \lambda_2(x + z) \\ 1 = 2\lambda_1 z + \lambda_2 y \\ x^2 + y^2 + z^2 = 3 \\ y(x + z) = 2. \end{cases} \qquad (12.53)$$

From the first and third equation of (12.53) we find $\lambda_1 x = \lambda_1 z$ so $\lambda_1 = 0$ or $x = z$.

(i) If $\lambda_1 = 0$, then we solve

$$\begin{cases} 1 = \lambda_2 y \\ 1 = \lambda_2(x + z) \\ x^2 + y^2 + z^2 = 3 \\ y(x + z) = 2 \end{cases} \implies \begin{cases} y = x + z \\ x^2 + y^2 + z^2 = 3 \\ y(x + z) = 2 \end{cases} \implies \begin{cases} y = x + z \\ x^2 + y^2 + z^2 = 3 \\ y^2 = 2. \end{cases}$$

We find
$$(x, y, z) = \pm\left(\frac{1}{\sqrt{2}}, \sqrt{2}, \frac{1}{\sqrt{2}}\right).$$

(ii) If $x = z$, then the last two equations of (12.53) yield
$$\begin{cases} 2x^2 + y^2 = 3 \\ xy = 1 \end{cases} \implies \begin{cases} y = \dfrac{1}{x} \\ 2x^2 + \dfrac{1}{x^2} = 3. \end{cases}$$

We find
$$(x, y, z) = \pm(1, 1, 1), \pm\left(\frac{1}{\sqrt{2}}, \sqrt{2}, \frac{1}{\sqrt{2}}\right).$$

The global maximum of f is $f(1, 1, 1) = 3$ and the global minimum of f is $f(-1, -1, -1) = -3$.

12. (i) Geometrically, the set E is the intersection between the cylinder $x^2 + y^2 = 10$ and the plane $x + y - z = 3$, so, an ellipse. Algebraically, observe first that the sets $\{g_1(x, y, z) = 10\}$ and $\{g_2(x, y, z) = 3\}$ are closed. Hence, E is closed. Further, $x^2 + y^2 = 10$ implies $x^2, y^2 \leq 10$, so $|x| \leq \sqrt{10}$ and $|y| \leq \sqrt{10}$. This means that x and y take values in the compact interval $[-\sqrt{10}, \sqrt{10}]$. From the second constraint, we have $z = x + y - 3$ so z takes values in the bounded interval $[-2\sqrt{10} - 3, 2\sqrt{10} + 3]$. Thus, E is bounded and being also closed, it is a compact set in $\mathbb{R}^3$.

(ii) We first make sure that the hypotheses in Theorem 11.8 are fulfilled. Then, we solve
$$\begin{cases} 2x = 2\lambda_1 x + \lambda_2 \\ -2y + 2z = \lambda_2 \\ 2y = 2\lambda_1 z - \lambda_2 \\ x^2 + z^2 = 10 \\ x + y - z = 3. \end{cases} \tag{12.54}$$

We add the second and the third equation of (12.54) to obtain $2z = 2\lambda_1 z$, so $z = 0$ or $\lambda_1 = 1$.

(ii1) If $z = 0$, then the last two equations in (12.54) yield
$$(x, y, z) = (\sqrt{10}, 3 - \sqrt{10}, 0), (-\sqrt{10}, 3 + \sqrt{10}, 0).$$

(ii2) If $\lambda_1 = 1$, then from the first two equations of (12.54) one has $\lambda_2 = 0$ and $y = z$. We obtain $(x, y, z) = (3, 1, 1), (3, -1, -1)$.

The global maximum of f is $f(3, 1, 1) = f(3, -1, -1) = 10$ and the global minimum of f is $f(-\sqrt{10}, 3 + \sqrt{10}, 0) = -6\sqrt{10} - 9$.

13. We find the extrema of $f(x, y, z) = x^2 + (y - 1)^2 + (z + 2)^2$ subject to the constraints $g_1(x, y, z) = 10$ and $g_2(x, y, z) = -2$ where $g_1(x, y, z) = x^2 + y^2$ and $g_2(x, y, z) = x + z$. We first check that the hypotheses in Theorem 11.8 are fulfilled. Thus, we solve

$$\begin{cases} 2x = 2\lambda_1 x + \lambda_2 \\ 2y - 2 = 2\lambda_1 y \\ 2z + 4 = \lambda_2 \\ x^2 + y^2 = 10 \\ x + z = -2. \end{cases} \qquad (12.55)$$

We use the third equation of (12.55) in the first equation to obtain $x(1 - \lambda_1) = z + 2$. Note also that the second equation of (12.55) yields $y(1 - \lambda_1) = 1$. Hence, dividing these two equations (note that we are entitled to divide them as $y(1 - \lambda_1) = 1$ implies $y \neq 0$ and $1 - \lambda_1 \neq 0$) we arrive at $x = y(z + 2)$. We solve now

$$\begin{cases} x^2 + y^2 = 10 \\ x + z = -2 \\ x = y(z + 2). \end{cases} \qquad (12.56)$$

The second equation of (12.56) yields $z + 2 = -x$ which we use it in the last equation of (12.56) to obtain $x = -xy$ so $x = 0$ or $y = -1$.

(i) If $x = 0$, then $(x, y, z) = (0, \pm\sqrt{10}, -2)$.
(ii) If $y = -1$, then $(x, y, z) = (3, -1, -5), (-3, -1, 1)$.

We find that $(0, \sqrt{10}, -2)$ is the closest point on the ellipse to $(0, 1, -2)$ while the points $(3, -1, -5), (-3, -1, 1)$ are the farthest on the ellipse to $(0, 1, -2)$.

14. We find the extrema of $f(x, y, z) = x^2 + y^2 + z^2$ subject to the constraints $g_1(x, y, z) = 0$ and $g_2(x, y, z) = 6$ where $g_1(x, y, z) = x^2 + y^2 - z$ and $g_2(x, y, z) = x + y + 2z$. It is straightforward to check that the requirements in Theorem 11.8 are satisfied. Then, we solve

$$\begin{cases} 2x = 2\lambda_1 x + \lambda_2 \\ 2y = 2\lambda_1 y + \lambda_2 \\ 2z = -\lambda_1 + 2\lambda_2 \\ x^2 + y^2 = z \\ x + y + 2z = 6. \end{cases} \qquad (12.57)$$

The first two equations of (12.57) imply

$$2(1 - \lambda_1)x = \lambda_2 = 2(1 - \lambda_1)y, \quad \text{so} \quad \lambda_1 = 1 \text{ or } x = y.$$

(i) If $\lambda_1 = 1$, then the first and the third equation of (12.57) yield $\lambda_2 = 0$ and $z = -\dfrac{1}{2}$. Using $z = -\dfrac{1}{2}$ in the fourth equation of (12.57) we reach a contradiction. Hence, there are no solutions in this case.

(ii) If $x = y$ we consider the last two equations of (12.57) to derive $z = 2x^2$ and $x + z = 3$. We find $(x, y, z) = (1, 1, 2), \left(-\frac{3}{2}, -\frac{3}{2}, \frac{9}{2}\right)$.

Thus, $(1, 1, 2)$ is closest point on the ellipse to the origin while the point $\left(-\frac{3}{2}, -\frac{3}{2}, \frac{9}{2}\right)$ is the farthest point on the ellipse from the origin.

15. (i) Observe that for all $y \in \mathbb{R}$, the point $(-3, y, -y)$ lies on both the hyperboloid and on the plane. This shows that $\mathcal{C}$ is unbounded.

(ii) One can check that all points on the constraint except $(-3, -3, 3)$ satisfy the conditions in Theorem 11.8. Thus, if $(x, y, z) \neq (-3, -3, 3)$, by Lagrange Multipliers Method we solve

$$\begin{cases} 2x = 2\lambda_1 x - \lambda_2 \\ 2y = -2\lambda_1 y + \lambda_2 \\ 2z = 2\lambda_1 z + \lambda_2 \\ x^2 - y^2 + z^2 = 9 \\ -x + y + z = 3. \end{cases} \qquad (12.58)$$

The first and the third equation of (12.58) yield

$$(1 - \lambda_1)x = -(1 - \lambda_1)z, \quad \text{so} \quad \lambda_1 = 1 \text{ or } x = -z.$$

(ii1) If $\lambda_1 = 1$, then the first two equations in (12.58) imply $\lambda_2 = 0$ and $y = 0$. Now, the last two equations of (12.58) yield $(x, y, z) = (0, 0, 3), (-3, 0, 0)$.

(ii2) If $x = -z$, then the last two equations of (12.58) yield

$$\begin{cases} 2x^2 - y^2 = 9 \\ y - 2x = 3 \end{cases} \implies (x, y, z) = (-3, -3, 3).$$

The closest points on $\mathcal{C}$ to the origin are $(0, 0, 3), (-3, 0, 0)$.

(iii) No! Since $\mathcal{C}$ is unbounded there are point on $\mathcal{C}$ arbitrarily distant from the origin.

16. The only critical point of f in D° is $(0, 0)$. Letting $g(x, y) = x^2 + 2y^2$ we see first that $\nabla g(x, y) \neq (0, 0)$ on the constraint $\{g(x, y) = 24\}$ and thus, by the Lagrange Multipliers Method we solve

$$\begin{cases} 3x^2 - 2y^2 = 2\lambda x \\ -4xy = 4\lambda y \\ x^2 + 2y^2 = 24. \end{cases} \qquad (12.59)$$

The second equation yields $-xy = \lambda y$ so $y = 0$ or $x = -\lambda$.

If $y = 0$ then the last equation implies $x^2 = 24$, so $(x, y) = (\pm\sqrt{24}, 0)$.

If $x = -\lambda$, then the first equation yields $5x^2 = 2y^2$. Using this equality in the last equation above we deduce $x^2 = 4$, so $(x, y) = (\pm 2, \pm\sqrt{10})$.

The global maximum is $f(2\sqrt{6}, 0) = 48\sqrt{6}$ and the global minimum is $f(-2\sqrt{6}, 0) = -48\sqrt{6}$.

17. There are no critical points in the interior of the domain. To find the extrema on the boundary, we check that the hypotheses of Lagrange Multipliers Theorem 11.1 are fulfilled and solve

$$\begin{cases} 1 = 2\lambda x \\ 3y^2 = 6\lambda y \\ x^2 + 3y^2 = 4. \end{cases} \tag{12.60}$$

If $y = 0$, then from the last equation of (12.60) we find $x^2 = 4$ and thus $(x, y) = (2, 0), (-2, 0)$.

If $y \neq 0$, then the first two equations of (12.60) imply $y = 2\lambda = \frac{1}{x}$. Thus, using $y = \frac{1}{x}$ in the third equation of (12.60) one finds

$$x^2 + \frac{3}{x^2} = 4 \implies x^4 - 4x^2 + 3 = 0 \implies x^2 = 1 \text{ or } x^2 = 3.$$

We find $(x, y) = (1, 1), (-1, -1), \left(\sqrt{3}, \frac{1}{\sqrt{3}}\right), \left(-\sqrt{3}, -\frac{1}{\sqrt{3}}\right)$. The global maximum is $f(1, 1) = f(2, 0) = 2$ and the global minimum is $f(-1, -1) = f(-2, 0) = -2$.

18. To find the critical points in the interior of the disc we solve $\nabla f(x, y) = (0, 0)$ which yields

$$\begin{cases} y(1 + x) = 0 \\ x(1 + y) = 0 \end{cases} \implies (x, y) = (0, 0), (-1, -1). \tag{12.61}$$

To find the extrema of f on the circle $x^2 + y^2 = 8$, we check that the hypotheses of Lagrange Multipliers Method are fulfilled and solve

$$\begin{cases} y(1 + x)e^{x+y} = 2\lambda x \\ x(1 + y)e^{x+y} = 2\lambda y \\ x^2 + y^2 = 8. \end{cases} \tag{12.62}$$

If $\lambda = 0$, then we end up with the same equations as in (12.61) when we found the critical points of f in the interior of the disc. Hence, we may assume $\lambda \neq 0$.

If $x = 0$, then the first equation of (12.62) yields $y = 0$ but then the last condition in (12.62) cannot hold. Thus, $x \neq 0$ and similarly $y \neq 0$. We may now divide the first two equations in (12.62) to deduce

$$\frac{y(1+x)}{x(1+y)} = \frac{x}{y} \implies (y-x)(x+y+xy) = 0.$$

(i) If $x = y$, then $x^2 + y^2 = 8$ yields $(x, y) = \pm(2, 2)$.

(ii) If $x + y + xy = 0$, denote $S = x + y$ and $P + xy$. Then $S = -P$ and

$$S^2 = (x+y)^2 = (x^2 + y^2) + 2P = 8 + 2P.$$

Hence $P^2 - 2P - 8 = 0$ with solutions $P = -S = -2$ and $P = -S = 4$. There is no need to further determine the values of x and y since $f(x,y) = Pe^S$.

Comparing the values of f at the above points we find the global maximum $f(2,2) = 4e^4$ and the global minimum $-2e^2$.

19. There are no critical points inside of the domain. The hypotheses of Lagrange Multipliers Method are fulfilled and thus we solve

$$\begin{cases} y = 2\lambda x \\ x = 2\lambda y \\ 2 = 2\lambda z \\ x^2 + y^2 + z^2 = 12. \end{cases} \tag{12.63}$$

From the first two equations of (12.63) we find

$$y = 2\lambda x = 2\lambda(2\lambda y) = 4\lambda^2 y \implies y = 0 \text{ or } \lambda = \pm\frac{1}{2}.$$

We discuss separately the above three cases and find $(x, y, z) = (0, 0, \pm 2\sqrt{3})$, $(2, 2, 2)$, $(-2, -2, 2)$, $(2, -2, -2)$, $(-2, 2, -2)$. The global maximum of f is $f(2, 2, 2) = f(-2, -2, 2) = 8$ and the global minimum is $f(2, -2, -2) = f(-2, 2, -2) = -8$.

20. The only critical point of f in the interior of the domain is $(0,0,0)$.

To find the extrema on the boundary, we first check that the hypotheses of Lagrange Multipliers Method are fulfilled and we are led to solve

$$\begin{cases} 2x = 2\lambda x \\ -2y + z^2 = 2\lambda y \\ 2yz = 2\lambda z \\ x^2 + y^2 + z^2 = 5. \end{cases} \tag{12.64}$$

The first equation in (12.64) yields $(1 - \lambda)x = 0$ so $x = 0$ or $\lambda = 1$.

(i) If $x = 0$, from the third equation in (12.64) we deduce $z = 0$ or $y = \lambda$.

If $z = 0$, we find $(x, y, z) = (0, \pm\sqrt{5}, 0)$.

If $y = \lambda$, then the second equation of (12.64) yields $z^2 = 2y^2 + 2y$ which we use in the last equation of (12.64) to deduce $3y^2 + 2y - 5 = 0$. Hence

$$(x, y, z) = (0, 1, \pm 2), \ \left(0, -\frac{5}{3}, \pm\frac{2\sqrt{5}}{3}\right).$$

(ii) If $\lambda = 1$, then (12.64) reads

$$\begin{cases} z^2 = 4y \\ yz = z \\ x^2 + y^2 + z^2 = 5. \end{cases} \tag{12.65}$$

From the second equation of (12.65) one gets $y = 1$ or $z = 0$. We find $(x, y, z) = (0, 1, \pm 2), (\pm\sqrt{5}, 0, 0)$.

The global maximum is $f(\pm\sqrt{5}, 0, 0) = 5$ and the global minimum is $f\left(0, -\frac{5}{3}, \pm\frac{2\sqrt{5}}{3}\right) = -\frac{175}{27}$.

21. The constraint function is $g(x, y) = 6x + 8y$ and the budget constraint is $g(x, y) = 168$. Note that $\nabla g(x, y) = (6, 8) \neq (0, 0)$. Thus, by Lagrange Multipliers Method we solve

$$\begin{cases} \dfrac{y^3}{2\sqrt{x}} = 6\lambda \\ 3y^2\sqrt{x} = 8\lambda \\ 6x + 8y = 168. \end{cases}$$

From the first two equations we deduce

$$\lambda = \frac{y^3}{12\sqrt{x}} = \frac{3y^2\sqrt{x}}{8}$$

Thus, $x = \frac{2}{9}y$. From $6x + 8y = 168$ we deduce $(x, y) = (4, 18)$.

22. (i) The constraints are $g(x, y) = 80$ and $x, y > 0$, where $g(x, y) = 4x + 8y$. By Lagrange Multipliers Method we solve

$$\begin{cases} (y + xy)e^{x+2y} = 4\lambda \\ (x + 2xy)e^{x+2y} = 8\lambda \\ 4x + 8y = 80. \end{cases}$$

Since $x, y > 0$, we divide the first two above equations to obtain $x = 2y$ and then from $4x + 8y = 80$ we find $(x, y) = (10, 5)$. Thus, the maximum utility is $U(10, 5) = 50e^{20}$ and $\lambda = \frac{55}{4}e^{20}$.

(ii) From (11.33) we find

$$\text{New } U \simeq \text{Old } U + \lambda\big(\text{New } g - \text{Old } g\big)$$
$$\simeq 50e^{20} + \frac{55}{4}e^{20}(100 - 80)$$
$$\simeq 325e^{20}.$$

23. The constraints are $g(x, y) = t$ and $x, y > 0$, where $g(x, y) = rx + sy$. By the Lagrange Multipliers Method we solve

$$\begin{cases} C\alpha x^{\alpha-1}y^{\beta} = \lambda r \\ C\beta x^{\alpha}y^{\beta-1} = \lambda s \\ rx + sy = t. \end{cases}$$

As in Example 11.15 we divide the above first two equations and obtain

$$\frac{y}{x} = \frac{\beta}{\alpha} \cdot \frac{r}{s}.$$

Then, using $rx + sy = t$ we find

$$x = \frac{t}{r} \cdot \frac{\alpha}{\alpha + \beta} \quad \text{and} \quad y = \frac{t}{s} \cdot \frac{\beta}{\alpha + \beta}.$$

Using the above values of x and y in the expression of the production function q we find the desired result.

A

Useful Facts in Linear Algebra

This chapter presents an account on the most important results in Linear Algebra which are needed in the textbook. The focus lies on determinants and their properties, inverse matrices as well as on the features of positive and negative definite matrices.

A.1 Basics

A matrix A is a $m \times n$ array consisting of m rows and n columns. The element found at the intersection of row i and column j of A is denoted a_{ij} and we write

$$A = (a_{ij})_{1 \le i \le m, 1 \le j \le n}.$$

We shall call the numbers a_{ij} the *entries* of the matrix A. In line with this terminology, vectors $\mathbf{u} = (u_1, u_2, \ldots, u_m) \in \mathbb{R}^m$ may be regarded as $1 \times m$ matrices. If the number of rows and columns in a matrix A are equal, we call A a *square matrix*. In this case, the sequence $a_{11}, a_{22}, \ldots, a_{mm}$ is called the main diagonal of A. Two particular matrices are: the *zero matrix* whose entries are all zero and the *identity matrix* I_m whose entries are 1 on the main diagonal and zero elsewhere.

If $A = (a_{ij})_{1 \le i \le m, 1 \le j \le n}$ and $B = (b_{ij})_{1 \le i \le m, 1 \le j \le n}$, then we can define the *sum* $A + B$ and the *scalar multiplication* λA, $\lambda \in \mathbb{R}$ by

$$A + B = (a_{ij} + b_{ij})_{1 \le i \le m, 1 \le j \le n} \quad \text{and} \quad \lambda A = (\lambda a_{ij})_{1 \le i \le m, 1 \le j \le n}.$$

The *multiplication* between a $m \times n$ matrix $A = (a_{ij})_{1 \le i \le m, 1 \le j \le n}$ and a $p \times q$ matrix $B = (b_{ij})_{1 \le i \le p, 1 \le j \le q}$ is possible only if the number of columns of A equals the number of rows of B, that is, $n = p$. In this case the product AB is a $m \times q$ matrix whose entries are given by

$$(AB)_{ij} = \sum_{k=1}^{n} a_{ik} b_{kj}.$$

DOI: 10.1201/9781003449652-A

For instance,

$$\begin{pmatrix} 2 & -3 & 1 \\ 1 & 0 & -2 \end{pmatrix} \begin{pmatrix} 3 & 1 & -1 \\ 0 & -2 & 1 \\ 1 & 1 & -1 \end{pmatrix} = \begin{pmatrix} 7 & 9 & -6 \\ 1 & -1 & 1 \end{pmatrix}.$$

If $A = (a_{ij})_{1\le i \le m, 1 \le j \le n}$ is a $m \times n$ matrix with real entries, then its *transpose* A^T is the $n \times m$ matrix obtained from A by writing its rows as columns, that is, $A^T = (a_{ji})_{1 \le j \le n, 1 \le i \le m}$. For instance, if

$$A = \begin{pmatrix} 2 & -1 & -2 \\ 0 & 3 & 1 \end{pmatrix} \quad \text{then} \quad A^T = \begin{pmatrix} 2 & 0 \\ -1 & 3 \\ -2 & 1 \end{pmatrix}.$$

The matrix A will be called *symmetric* if $A = A^T$ (and implicitly, this means that A is a square matrix). The following properties arise naturally:

Theorem A.1 (Properties of the transpose matrix)
  Let $m, n, p \ge 1$ be positive integers.

  (i) *If A, B are $m \times n$ matrices and $\lambda \in \mathbb{R}$, then*

  $$(A \pm B)^T = A^T \pm B^T, \quad (\lambda A)^T = \lambda A^T, \quad (A^T)^T = A;$$

  (ii) *If A is an $m \times n$ matrix and B is an $n \times p$ matrix, then*

  $$(AB)^T = B^T A^T.$$

A.2   Determinants

The determinant is a quantity attached only to square matrices. For a 2×2 matrix this is computed by the formula

$$\begin{vmatrix} a_{11} & a_{12} \\ a_{21} & a_{22} \end{vmatrix} = a_{11}a_{22} - a_{12}a_{21}.$$

For a 3×3 matrix, the computation of its determinant requires a bit of attention. One possible way to do it is to apply the so-called *Rule of Sarrus* which is schematically presented in Figure A.1.
Technically, we proceed as follows. Let $A = (a_{ij})_{1 \le i,j \le 3}$ be a 3×3 matrix.

Step 1: Copy the first two columns of the determinant to the right.

Step 2: Compute the product of the elements highlighted on the two sets of diagonals:

- From top left corner to bottom right corner (continuous arrow).

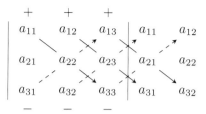

Figure A.1
The Rule of Sarrus for a 3 × 3 matrix.

• From bottom left corner to top right corner (discontinuous arrows).

Step 3: Compute the sum of the products on continuous arrows from which one subtracts the products on discontinuous arrows.

Finally, write the expression of the determinant as

$$\det(A) = a_{11}a_{22}a_{33} + a_{12}a_{23}a_{31} + a_{13}a_{21}a_{32} - a_{31}a_{22}a_{13} - a_{32}a_{23}a_{11} - a_{33}a_{21}a_{12}.$$

Example A.2 *Compute the determinant of* $A = \begin{pmatrix} 2 & -1 & 3 \\ 3 & 1 & 0 \\ 1 & 2 & -4 \end{pmatrix}$.

Solution. With the Rule of Sarrus, we have

$$\det(A) = 2 \cdot 1 \cdot (-4) + (-1) \cdot 0 \cdot 1 + 3 \cdot 3 \cdot 2 - 1 \cdot 1 \cdot 3 - 2 \cdot 0 \cdot 2 - (-4) \cdot 3 \cdot (-1) = -5.$$

We next remind the definition of a determinant of order n.

Definition A.3 *If* $A = (a_{ij})_{1 \leq i,j \leq m}$, *then the determinant of* A *is given by*

$$\det(A) = \sum_{\sigma} \text{sgn}(\sigma) a_{1\sigma(1)} a_{2\sigma(2)} \cdots a_{m\sigma(m)}, \tag{A.1}$$

where the above sum is taken over all bijective functions $\sigma : \{1, 2, \ldots, m\} \to \{1, 2, \ldots, m\}$ *and* $\text{sgn}(\sigma)$ *denotes the sign (or signature) of* σ.

There are exactly $m!$ bijective functions so that the above definition of the determinant is the sum of $m!$ products of the entries of A. We rarely use the above expression to numerically compute a determinant. However, with the help of (A.1) one deduces the following:

Theorem A.4 (Properties of determinants)
Let A, B *be two* $m \times m$ *matrices. Then:*

(i) $\det(A) = \det(A^T)$ *and* $\det(AB) = \det(A) \cdot \det(B)$;

(ii) *If we add a multiple of a row of A to another row of A, then* $\det(A)$ *does not change;*

(iii) *If we swap two rows of A, then the determinant of the new matrix equals* $-\det(A)$. *In particular, if A has two equal rows, then its determinant is zero;*

(iv) *If we multiply one row with a constant $c \in \mathbb{R}$, then the determinant of the new matrix equals* $c \det(A)$.

Since $\det(A) = \det(A^T)$, the properties (ii)–(iv) in Theorem A.4 also apply to columns (instead of rows).

A more convenient formula to compute the determinant of a matrix is to use the *cofactors method*. Precisely, if $A = (a_{ij})_{1 \leq i,j \leq m}$ is a $m \times m$ matrix, we denote by A_{ij} the $(m-1) \times (m-1)$ matrix obtained from A by removing the ith row and the jth column of A.

Theorem A.5 (The cofactors method)
    If $A = (a_{ij})_{1 \leq i,j \leq m}$, then

$$\det(A) = \sum_{j=1}^{m} (-1)^{i+j} a_{ij} \det(A_{ij}) \quad \text{for all } 1 \leq i \leq m. \tag{A.2}$$

The quantity $(-1)^{i+j}\det(A_{ij})$ is called the *cofactor* of A corresponding to a_{ij} and (A.2) says that $\det(A)$ is a linear combination of m such cofactors. We thus reduce the computation of a $m \times m$ determinant to the computation of m determinants of order $m-1$. For instance, if $m = 3$, we have the following alternative to the Rule of Sarrus for the computation of a 3×3 determinant:

$$\det(A) = (-1)^{1+1} a_{11}\det(A_{11}) + (-1)^{1+2} a_{12}\det(A_{12}) + (-1)^{1+3} a_{13}\det(A_{13}).$$

We should point out that the formula (A.2) holds if the summation is done over columns instead of rows (this is again a consequence of the fact that a matrix and its transpose have the same determinant). Thus, one has

$$\det(A) = \sum_{i=1}^{m} (-1)^{i+j} a_{ij}\det(A_{ij}) \quad \text{for all } 1 \leq j \leq m.$$

Also, if we replace row k with row i, $i \neq k$ in the matrix A, then by property (iii) in Theorem A.4, we have that the determinant of the new matrix is zero, so

$$\sum_{j=1}^{m} (-1)^{k+j} a_{ij}\det(A_{kj}) = 0. \tag{A.3}$$

Thus, combining (A.2) and (A.3), we have

$$\sum_{j=1}^{m} (-1)^{k+j} a_{ij}\det(A_{kj}) = \delta_{ik}\det(A) \quad \text{for all } 1 \leq i, k \leq m. \tag{A.4}$$

For computational reasons, we select the row or column of A with the largest number of zero entries. Following the above Example A.2, to compute the determinant of

$$A = \begin{pmatrix} 2 & -1 & 3 \\ 3 & 1 & 0 \\ 1 & 2 & -4 \end{pmatrix},$$

we select row 2 (or column 3) as we spot there one zero entry. Thus, we have to compute only two cofactors of order 2:

$$\det(A) = (-1)^{2+1}a_{21}\det(A_{21}) + (-1)^{2+2}a_{22}\det(A_{22}) + (-1)^{2+3}a_{23}\det(A_{23})$$

$$= -3 \cdot \begin{vmatrix} -1 & 3 \\ 2 & -4 \end{vmatrix} + 1 \cdot \begin{vmatrix} 2 & 3 \\ 1 & -4 \end{vmatrix} - 0 \cdot \begin{vmatrix} 2 & -1 \\ 1 & 2 \end{vmatrix}$$

$$= -3 \cdot (4 - 6) + 1 \cdot (-8 - 3) = -5.$$

A.3   Inverse Matrices

Definition A.6 *A $m \times m$ matrix A is called invertible if there exists another $m \times m$ matrix B such that*

$$AB = BA = I_m.$$

We call the matrix B the inverse of A and denote $B = A^{-1}$.

A matrix which is invertible it is also called *nonsingular* as opposed to noninvertible matrices which are called *singular*.

The properties of inverse matrices are stated below.

Theorem A.7 (Properties of inverse matrices)
    Let A, B be two $m \times m$ matrices.

    (i) *A is invertible if and only if $\det(A) \neq 0$;*

    (ii) *If A is invertible, then its transpose A^T is invertible and*

$$(A^T)^{-1} = (A^{-1})^T;$$

    (iii) *If A and B are invertible, then AB is invertible and*

$$(AB)^{-1} = B^{-1}A^{-1}.$$

Proof (i) If A is invertible, then there exists a $m \times m$ matrix A^{-1} such that $AA^{-1} = I_m$. Taking the determinants in this last equality and using Theorem A.4(i) one finds

$$\det(A) \cdot \det(A^{-1}) = \det(I_m) = 1, \quad \text{so} \quad \det(A) \neq 0.$$

Conversely assume that $\det(A) \neq 0$. Let us introduce the *cofactor matrix* C of A by $C = (C_{ij})_{1 \leq i,j \leq m}$ where $C_{ij} = (-1)^{i+j}\det(A_{ij})$. Then, we further define the *adjoint matrix of A* by

$$\mathrm{adj}(A) = C^T. \tag{A.5}$$

We claim that

$$A^{-1} = \frac{1}{\det(A)}\mathrm{adj}(A). \tag{A.6}$$

Indeed, by (A.4) and (A.5), for $1 \leq i, k \leq m$ one has

$$(AA^{-1})_{ik} = \frac{1}{\det(A)}(AC^T)_{ik}$$

$$= \frac{1}{\det(A)}\sum_{j=1}^{m} a_{ij}C_{kj}$$

$$= \frac{1}{\det(A)}\sum_{j=1}^{m}(-1)^{k+j}a_{ij}\det(A_{kj})$$

$$= \delta_{ik},$$

and similarly $(A^{-1}A)_{ik} = \delta_{ik}$. Thus $AA^{-1} = A^{-1}A = I_m$ which shows that A is invertible.

(ii) Using the properties of the transpose matrix in Theorem A.1, we have

$$A^T(A^{-1})^T = (A^{-1}A)^T = I_m^T = I_m$$

$$(A^{-1})^T A^T = (AA^{-1})^T = I_m^T = I_m.$$

Thus, A^T is invertible and $(A^T)^{-1} = (A^{-1})^T$.

(iii) As in part (ii) above one checks that

$$(AB)(B^{-1}A^{-1}) = A(BB^{-1})A^{-1} = AI_mA^{-1} = AA^{-1} = I_m$$

and

$$(B^{-1}A^{-1})(AB) = B^{-1}(A^{-1}A)B = B^{-1}I_mB = B^{-1}B = I_m,$$

so AB is invertible and $(AB)^{-1} = B^{-1}A^{-1}$. $\qquad\square$

The proof of Theorem A.7(i) provides one way of constructing the inverse of a matrix as given by (A.6).

If $A = \begin{pmatrix} a & b \\ c & d \end{pmatrix}$ is such that $\det(A) = ad - bc \neq 0$, then its inverse is given by

$$A^{-1} = \frac{1}{ad - bc}\begin{pmatrix} d & -b \\ -c & a \end{pmatrix}.$$

For 3×3 matrices, we illustrate the construction of the inverse by taking the matrix from Example A.2.

Example A.8 *Find the inverse of* $A = \begin{pmatrix} 2 & -1 & 3 \\ 3 & 1 & 0 \\ 1 & 2 & -4 \end{pmatrix}$.

Solution. We first compute the cofactor matrix C of A.

$$C_{11} = (-1)^{1+1}\det(A_{11}) = \begin{vmatrix} 1 & 0 \\ 2 & -4 \end{vmatrix} = -4$$

$$C_{12} = (-1)^{1+2}\det(A_{12}) = -\begin{vmatrix} 3 & 0 \\ 1 & -4 \end{vmatrix} = 12$$

$$C_{13} = (-1)^{1+3}\det(A_{13}) = \begin{vmatrix} 3 & 1 \\ 1 & 2 \end{vmatrix} = 5$$

$$C_{21} = (-1)^{2+1}\det(A_{21}) = -\begin{vmatrix} -1 & 3 \\ 2 & -4 \end{vmatrix} = 2$$

and similarly $C_{22} = -11$, $C_{23} = -5$, $C_{31} = -3$, $C_{32} = 9$ and $C_{33} = 5$. Thus,

$$A^{-1} = \frac{1}{\det(A)}C^T = \frac{1}{5}\begin{pmatrix} -4 & 2 & -3 \\ 12 & -11 & 9 \\ 5 & -5 & 5 \end{pmatrix}.$$

A.4 The Rank of a Matrix

If A is a $m \times n$ matrix and $1 \le p \le m$, $1 \le q \le n$ are positive integers, the matrix whose entries are located at the intersection of p rows and q columns of A is called a $p \times q$ *submatrix* of A.

Definition A.9 *Let A be a $m \times n$ matrix. The rank of A, denoted $rank(A)$, is the largest number p so that A has a $p \times p$ nonsingular submatrix.*

There are several ways to introduce the rank of a matrix; here we discuss it using the properties of determinants and inverse matrices which were presented above. Let us first notice that if A is a $m \times n$ matrix, then $rank(A) \le \min\{m, n\}$. For instance, the matrix

$$A = \begin{pmatrix} 2 & 1 & -3 \\ -2 & -1 & 0 \end{pmatrix}$$

has rank 2 since its 2×2 submatrix $\begin{pmatrix} 1 & -3 \\ -1 & 0 \end{pmatrix}$ is nonsingular. Similarly, the matrix

$$A = \begin{pmatrix} 3 & 0 & 3 & 0 \\ 1 & 2 & 3 & -2 \\ 1 & -1 & 0 & 1 \end{pmatrix}$$

has rank 2 since all its 3×3 submatrices are singular while its 2×2 submatrix $\begin{pmatrix} 3 & 0 \\ 1 & 2 \end{pmatrix}$ is nonsingular. Next, we state below without proof the properties of the rank.

Theorem A.10 (Properties of the rank of a matrix)
   Let A be a $m \times n$ matrix.

   (i) *$rank(A) = 0$ if and only if A is the zero matrix;*

   (ii) *$rank(I_m) = m$;*

   (iii) *$rank(A) = rank(A^T)$;*

   (iv) *$rank(A)$ is the number of rows (respectively, columns) of A which are linearly independent;*

   (v) *If we add a multiple of a row (respectively, column) of A to another row (respectively, column) of A, then $rank(A)$ does not change;*

   (vi) *If we swap two rows (respectively, columns) of A, then $rank(A)$ does not change;*

   (vii) *If we multiply one row (respectively, column) with a nonzero constant, then the rank of the new matrix equals $rank(A)$.*

A.5   Positive and Negative Definite Matrices

Definition A.11 *A symmetric matrix $A = (a_{ij})_{1 \leq i,j \leq m}$ is called positive definite if*
$$\mathbf{u} A \mathbf{u}^T > 0 \quad \text{for all row vectors } \mathbf{u} \in \mathbb{R}^m \setminus \{\mathbf{0}\}. \tag{A.7}$$
Reverting the inequality in (A.7) we obtain the definition of a negative matrix.

Example A.12 *The matrix $A = \begin{pmatrix} 2 & -1 & 1 \\ -1 & 3 & 0 \\ 1 & 0 & 1 \end{pmatrix}$ is positive definite.*

Solution   Indeed, if $\mathbf{u} = (u_1, u_2, u_3) \in \mathbb{R}^3 \setminus \{\mathbf{0}\}$, then

$$\mathbf{u} A \mathbf{u}^T = (u_1, u_2, u_3) \begin{pmatrix} 2 & -1 & 1 \\ -1 & 3 & 0 \\ 1 & 0 & 1 \end{pmatrix} \begin{pmatrix} u_1 \\ u_2 \\ u_3 \end{pmatrix}$$
$$= 2u_1^2 + 3u_2^2 + u_3^2 - 2u_1 u_2 + 2u_1 u_3.$$

By completion of squares, we have

$$\mathbf{u} A \mathbf{u}^T = (u_1 - u_2)^2 + (u_1 + u_3)^2 + 2u_2^2 \geq 0.$$

If the equality with zero holds, then $u_1 - u_2 = 0$, $u_1 + u_3 = 0$ and $u_2 = 0$ which implies $u_1 = u_2 = u_3 = 0$. This is however not possible since $\mathbf{u} \neq (0,0,0)$. Hence $\mathbf{u}A\mathbf{u}^T > 0$ for all $\mathbf{u} \in \mathbb{R}^3 \setminus \{\mathbf{0}\}$, which shows that A is positive definite.

In practice, we have a good number of methods to check the positive definiteness of matrices which are shorter than the above approach; these methods will be described below. Firstly, for $1 \leq k \leq m$ we denote by A_k the $k \times k$ submatrix formed with the entries located on the first k rows and the first k columns of A, that is,

$$A_k = \begin{pmatrix} a_{11} & a_{12} & \cdots & a_{1k} \\ a_{21} & a_{22} & \cdots & a_{2k} \\ \cdots & \cdots & \cdots & \cdots \\ a_{k1} & a_{k2} & \cdots & a_{kk} \end{pmatrix}.$$

We call A_1, A_2, ..., A_m the *leading principal submatrices of A*. For instance, in case of the matrix A in Example A.12, we have

$$A_1 = (2), \quad A_2 = \begin{pmatrix} 2 & -1 \\ -1 & 3 \end{pmatrix}, \quad A_3 = \begin{pmatrix} 2 & -1 & 1 \\ -1 & 3 & 0 \\ 1 & 0 & 1 \end{pmatrix}.$$

Definition A.13 *A complex number $\lambda \in \mathbb{C}$ is called an eigenvalue of the $m \times m$ matrix A if there exists a vector $\mathbf{u} \in \mathbb{C}^m \setminus \{\mathbf{0}\}$ such that $A\mathbf{u}^T = \lambda \mathbf{u}^T$.*

One can show that λ is an eigenvalue of A if and only if $\det(\lambda I_m - A) = 0$. The quantity $P_A(\lambda) = \det(\lambda I_m - A)$ is called the *characteristic polynomial* of A. It is a monic polynomial, that is, the coefficient of the leading term is 1. If $\lambda_1, \lambda_2, \ldots, \lambda_m$ denote the eigenvalues of A, then these are the roots of $P_A(\lambda)$ so

$$P_A(\lambda) = (\lambda - \lambda_1)(\lambda - \lambda_2) \cdots (\lambda - \lambda_m).$$

In particular,

$$P_A(0) = \det(-A) = (-1)^m \det(A) = (-1)^m \lambda_1 \lambda_2 \cdots \lambda_m,$$

so

$$\det(A) = \lambda_1 \lambda_2 \cdots \lambda_m. \tag{A.8}$$

Proposition A.14 (Properties of positive definite matrices)
Let A be a positive definite matrix. Then:

(i) *All eigenvalues of A are real and positive.*

In particular $\det(A) > 0$;

(ii) *There exists $c > 0$ such that*

$$\mathbf{u}A\mathbf{u}^T \geq c\|\mathbf{u}\|^2 \quad \text{for all } \mathbf{u} \in \mathbb{R}^m \setminus \{\mathbf{0}\}.$$

(iii) *All leading principal submatrices A_k, $1 \leq k \leq m$, are positive definite.*

Proof (i) Let $\lambda \in \mathbb{C}$ be an eigenvalue of A and $\mathbf{u} \in \mathbb{C}^m \backslash \{\mathbf{0}\}$ be a corresponding eigenvector. Then $A\mathbf{u}^T = \lambda \mathbf{u}^T$, and then

$$\lambda \|\mathbf{u}\|^2 = \lambda \bar{\mathbf{u}} \mathbf{u}^T = \bar{\mathbf{u}}(\lambda \mathbf{u}^T) = \bar{\mathbf{u}}(A\mathbf{u}^T) = \bar{\mathbf{u}}(\mathbf{u}A)^T = (\mathbf{u}A\bar{\mathbf{u}}^T)^T$$
$$= \left[\mathbf{u}\overline{(A\mathbf{u}^T)}\right]^T = \left[\mathbf{u}\overline{(\lambda \mathbf{u}^T)}\right]^T = \bar{\lambda}\mathbf{u}\bar{\mathbf{u}}^T = \bar{\lambda}\|\mathbf{u}\|^2.$$

Since $\mathbf{u} \neq \mathbf{0}$, we have $\|\mathbf{u}\| \neq 0$, so $\lambda = \bar{\lambda}$ which yields $\lambda \in \mathbb{R}$. Further, from $A\mathbf{u}^T = \lambda \mathbf{u}^T$, by taking the complex conjugate we obtain $A\bar{\mathbf{u}}^T = \lambda \bar{\mathbf{u}}^T$. Thus, replacing $\mathbf{u}$ by $\frac{1}{2}(\mathbf{u} + \bar{\mathbf{u}})$ we may assume $\mathbf{u} \in \mathbb{R}^m \backslash \{\mathbf{0}\}$. We also have

$$\mathbf{u}A\mathbf{u}^T = \lambda \mathbf{u}\mathbf{u}^T = \lambda \|\mathbf{u}\|^2.$$

Since $\mathbf{u}A\mathbf{u}^T > 0$ and $\|\mathbf{u}\| > 0$ it follows that

$$\lambda = \frac{\mathbf{u}A\mathbf{u}^T}{\|\mathbf{u}\|^2} > 0.$$

By (A.8) it now follows that $\det(A) > 0$.

(ii) We use the Extreme Value Theorem 10.2 . Let $S = \{\mathbf{v} \in \mathbb{R}^m : \|\mathbf{v}\| = 1\}$ and $g : S \to \mathbb{R}$, $g(\mathbf{v}) = \mathbf{v}A\mathbf{v}^T$. Then, g is continuous on the compact set S. It follows that g achieves its minimum value at a certain $\mathbf{w} \in S$ so, letting $c = g(\mathbf{w}) > 0$ one has

$$g(\mathbf{v}) \geq c \quad \text{for all } v \in S.$$

Now, if $\mathbf{u} \in \mathbb{R}^m \backslash \{\mathbf{0}\}$, then $\mathbf{v} = \frac{\mathbf{u}}{\|\mathbf{u}\|} \in S$ and we have

$$g\left(\frac{\mathbf{u}}{\|\mathbf{u}\|}\right) \geq c \Longrightarrow \mathbf{u}A\mathbf{u}^T \geq c\|\mathbf{u}\|^2.$$

(iii) Let $1 \leq k \leq m$ and write

$$A = \begin{pmatrix} A_k & B \\ C & D \end{pmatrix}$$

for some matrices B, C and D. Let also $\mathbf{v}$ be a nonzero vector in $\mathbb{R}^k$ and $\mathbf{u} = (\mathbf{v}, \mathbf{0}) \in \mathbb{R}^m$. Then

$$\mathbf{u}A\mathbf{u}^T = \mathbf{v}A_k\mathbf{v}^T > 0, \quad \text{so } A_k \text{ is positive definite.}$$

$\square$

Theorem A.15 *Let A be a $m \times m$ symmetric matrix. The following statements are equivalent:*

(i) *A is positive definite;*

(ii) $\det(A_k) > 0$ *for all* $1 \le k \le m$;

(iii) *There exists a unique upper triangular matrix U with positive entries on the main diagonal such that $A = U^T U$.*

Proof We shall prove the sequence of implications

$$(\text{iii}) \Longrightarrow (\text{i}) \Longrightarrow (\text{ii}) \Longrightarrow (\text{iii})$$

(iii) $\Longrightarrow$ (i) Let $\mathbf{u} \in \mathbb{R}^m \setminus \{\mathbf{0}\}$. Then

$$\mathbf{u} A \mathbf{u}^T = \mathbf{u}(U^T U)\mathbf{u}^T = (U\mathbf{u}^T)^T (U\mathbf{u}^T) = \|U\mathbf{u}^T\|^2 > 0.$$

(i) $\Longrightarrow$ (ii) By Proposition A.14(iii) one has that A_k is positive definite for all $1 \le k \le m$ and by part (i) of Proposition A.14 its determinant is positive.

(ii) $\Longrightarrow$ (iii) We proceed by induction over $m \ge 1$.

If $m = 1$, then $A = (a)$ and by Proposition A.14(ii) one has $a > 0$. In this case we take $U = (\sqrt{a})$. Assume now that the statement holds for $(m-1) \times (m-1)$ matrices, in particular for A_{m-1}. Hence, there exists an $(m-1) \times (m-1)$ upper triangular matrix V such that $A_{m-1} = V^T V$. Since A is symmetric, we have

$$A = \begin{pmatrix} A_{m-1} & \mathbf{w}^T \\ \mathbf{w} & a \end{pmatrix}$$

where $\mathbf{w} \in \mathbb{R}^{m-1}$ and $a \in \mathbb{R}$. If we let $\mathbf{u} = \mathbf{w}V^{-1}$ and $b = a - \|\mathbf{u}\|^2$, then

$$A = \begin{pmatrix} V^T V & \mathbf{w}^T \\ \mathbf{w} & a \end{pmatrix} = \begin{pmatrix} V^T & \mathbf{0}^T \\ \mathbf{u} & 1 \end{pmatrix} \begin{pmatrix} V & \mathbf{u}^T \\ \mathbf{0} & b \end{pmatrix}. \tag{A.9}$$

By taking the determinant in the above equality we find $\det(A) = b(\det(V))^2 > 0$, so $b > 0$. Now, letting

$$U = \begin{pmatrix} V & \mathbf{u}^T \\ \mathbf{0} & \sqrt{b} \end{pmatrix}$$

it follows that $A = U^T U$ where U is an upper triangular matrix with positive entries on the main diagonal. To check the uniqueness of U with the above property, assume

$$A = U^T U = V^T V,$$

where U, V are upper triangular matrices with positive entries on the main diagonal. Let

$$D = VU^{-1} = (V^T)^{-1}U^T.$$

Since V, U^{-1} are upper triangular matrices, so is their product D. Similarly, $(V^T)^{-1}$ and U^T are lower triangular, so D is also lower triangular. Thus, D is a diagonal matrix. From the above equality we deduce $V = DU$ and $U^T = V^T D$, so $U = DV$. Thus, $V = DU = D^2 V$ and this implies, since V

is invertible, that $D^2 = I_m$. Using the fact that D has positive entries on the main diagonal, one gets $D = I_m$ and $U = V$. $\qquad\qquad\qquad\qquad\qquad\square$

Using the fact that a symmetric matrix A is negative definite if and only if $-A$ is positive definite, Theorem A.15 yields:

Corollary A.16 *Let A be a $m \times m$ symmetric matrix.*

Then, A is negative definite if and only if $(-1)^k \det(A_k) > 0$ for all $1 \leq k \leq m$.

Bibliography

[1] J.B. Conway, A First Course in Analysis, Cambridge Mathematical Textbooks, Cambridge University Press, 1st Edition, 2018.

[2] S. Dineen, Functions of Two Variables, 2nd Edition, Chapman & Hall/CRC, 2000.

[3] P.M. Fitzpatrick, Advanced Calculus, Pure and Applied Undergraduate Texts No. 5, 2nd Edition, Amer. Math. Soc. 2006.

[4] F.J. Garcia-Pacheco, Abstract Calculus: A Categorical Approach, 1st Edition, Monographs and Research Notes in Mathematics, Chapman & Hall/CRC, 2021.

[5] W. Rudin, Principles of Mathematical Analysis, McGraw-Hill, 3rd Edition, 1976.

[6] J.L. Taylor, Foundations of Analysis, Pure and Applied Undergraduate Texts No. 18, Amer. Math. Soc. 2012.

Index

leading principal submatrices,
299
multiplication, 291
negative definite, 157, 298
nonsingular, 295
positive definite, 157, 298, 299
rank of, 181, 297
rows and columns, 291
singular, 295
square, 291
submatrix of, 297
symmetric, 116, 292
transpose, 292
Mean Value Theorem
for one variable, 64, 76
for several variables, 94
Multi-index, 126

Objective function, 171

Parametric equation of the line, 8, 12
Parametrized curve, 38
Partial derivative, 62
Point in $\mathbb{R}^m$
boundary, 17
exterior, 17
interior, 17
isolated, 17
regular, 142
Polar coordinates, 45, 86, 118
Production function, 192
Proof by contradiction, 43

Quadric surface, 35

Range of a function, 22, 30
Rate of change
in the x direction, 66
in the y direction, 66
maximum or minimum, 103
of a function, 97
of the volume, 86, 87
Regular point, 63, 142
Rule of Sarrus, 292

Saddle point, 156, 157

Scalar, 3, 4, 8
Sequence
bounded, 14
convergent, 14
subsequence of, 14
Set
boundary, 17
bounded, 20
closed, 17
compact, 20
complement of, 17
open, 17
Sided limits, 43, 63, 70, 99
Spherical coordinates, 27, 89
Squeeze Theorem, 49
Standard unit vectors, 3, 62
Stereographic projection, 24
Submatrix, 297
Subsequence, 14
Surface, 27
cylindrical, 35
quadric, 35
traces, 28, 97

Tangent
plane, 66
line to a curve, 39
vector to a curve, 39
Taylor
linear approximation, 131, 190
polynomial, 130
quadratic approximation, 133
remainder, 130
theorem, 128
Theorem
Critical Point, 155, 172, 182, 186
Extreme Value, 154, 186, 300
Implicit Function, 138, 139, 144
Mean Value, 94, 240
Rolle, 95
Taylor, 128
Triangle inequality, 7
Two-path test, 43

Unit sphere, 35

Printed and bound by CPI Group (UK) Ltd, Croydon, CR0 4YY

17/10/2024

01775682-0013